The Minerals, Metals & Materials Series

Steven Barela • Aeriel Leonard •
Petra Maier • Neale R. Neelameggham •
Victoria M. Miller
Editors

Magnesium Technology 2023

TMS

Springer

Editors
Steven Barela
Terves Inc./Magnesium-USA
Euclid, OH, USA

Aeriel Leonard
The Ohio State University
Columbus, OH, USA

Petra Maier
University of Applied Sciences Stralsund
Stralsund, Germany

Neale R. Neelameggham
IND LLC
South Jordan, UT, USA

Victoria M. Miller
University of Florida
Gainesville, FL, USA

ISSN 2367-1181 ISSN 2367-1696 (electronic)
The Minerals, Metals & Materials Series
ISBN 978-3-031-22644-1 ISBN 978-3-031-22645-8 (eBook)
https://doi.org/10.1007/978-3-031-22645-8

This Springer imprint is published by the registered company Springer Nature Switzerland AG
The registered company address is: Gewerbestrasse 11, 6330 Cham, Switzerland

Preface

Magnesium (Mg) and its alloys continue to be extensively investigated for applications in the automotive and aerospace industries as well as biomedical. Although the largest use of Mg is desulfurization of steel, followed by alloying of aluminum, its primary attributes as the lightest structural metal include high strength-to-weight ratio, good vibration damping, electromagnetic and RF shielding, low toxicity, and controllable corrosion rates for dissolvable applications. It is because of these attributes that Mg will play a crucial role in reducing the carbon footprint and forwarding the development of sustainable technology, specifically in the replacement of conventional materials such as steels and aluminum alloys. Compared to carbon fiber composites, Mg is more cost effective and can be readily recycled and offers metallic ductile behavior.

Collaborative research and development by researchers, scientists, engineers, industry, government agencies/laboratories, and academic institutions are actively working to develop roadmaps for next-generation products and address these challenges through innovative alloy designs and methods. The TMS Magnesium Committee has been actively involved in providing a platform for these institutions to disseminate the latest information, developments, and cutting-edge research and development, and to present the latest research and development trends related to magnesium and its alloys through the Magnesium Technology Symposium held each year at the TMS Annual Meeting & Exhibition.

The twenty-fourth volume in the series, *Magnesium Technology 2023*, is the proceedings of the Magnesium Technology Symposium held during the 152nd TMS Annual Meeting & Exhibition in San Diego, California, March 19–23, 2022. The volume captures papers and extended abstracts from 12 different countries. The papers have been categorized based on topics pertaining to alloy design, fundamentals of plastic deformation, primary production, recycling and ecological issues, characterization, joining, machining, forming, degradation and biomedical applications, corrosion and surface protection, and computational materials engineering.

The symposium began with keynote sessions that featured several distinguished invited speakers from industry, government organizations, and academia, who provided their perspectives on the state of the art, goals, and opportunities in magnesium alloy research and development. Andrew Sherman from Terves Inc./Magnesium-USA spoke about defending patents/intellectual property. Aaron Palumbo of Big Blue Technologies addressed reductant consideration in thermal pathways to primary magnesium metal production. Carolyn Woldring, co-founder of Magsorbeo Biomedical, discussed engineered bioabsorption for medical implant applications. Mr. Michael Ren, Sunlightmetal Consulting Inc., spoke about primary Mg production, specifically development of compound-vertical-retort technology for magnesium production and its application. Norbert Hort with Helmholtz-Zentrum Hereon, Geesthacht, Germany spoke about the concept of metastable vs. stable. Mohammadreza Yaghoobi with the University of Michigan covered recent advances in PRISMS-Plasticity software for simulation of deformation in Mg alloys. Mariyappan Arul Kumar from Los Alamos National Laboratory gave a presentation about the combined effect of alloying and grain size on the deformation behavior of magnesium alloys.

In conclusion, the 2022–2023 Magnesium Committee would like to thank and express its deep appreciation to all authors who contributed to the success of the symposium; our panel of distinguished keynote speakers for sharing the newest developments and valuable thoughts on the future of magnesium technology; all the reviewers for their best efforts in reviewing the manuscripts; and the session chairs, judges, TMS staff members, and other volunteers for their excellent support, which allowed us to develop a successful, high-quality symposium and proceedings volume.

Steven Barela
Aeriel Leonard
Petra Maier
Neale R. Neelameggham
Victoria M. Miller

Contents

About the Editors

Steven Barela is from Pueblo, Colorado, home to the Colorado Fuel & Iron Steel Mill (now Evraz). Driven by the need to resolve welding problems encountered when fabricating race cars, he attended the University of S. Colorado (now CSU-Pueblo) earning an A.A.S. in Metallurgical Engineering Technology. This led to a position at Rocky Flats/DOE nuclear assembly facility in the Non-Nuclear Joining R&D group as an Intern Engineer while simultaneously earning a B.S. in Metallurgical & Materials Engineering at the Colorado School of Mines (CSM) where he specialized in joining and was involved in the CSM Joining Research Center. Mr. Barela then went on to work at Martin Marietta Astronautics Group in the Advanced Manufacturing Technology Group which oversaw procedures and the production of the Titan family of launch vehicles, specifically all joining operations (braze, TIG, MIG, Variable Polarity Plasma) on aluminum, stainless steel, titanium, and metal matrix composites materials. He also participated in the development of Weldalite, a program to produce the Al-Li external tank for the NASA space shuttle program. In the late 1990s, Mr. Barela transitioned his career from welding metallurgy technology to marketing, product development, and sales of welding, forging, and fabricated products as a Technical Sales Engineer at Timminco Extruded Magnesium Products. He then worked for 10 years for Solikamsk Magnesium Works (Russia) by running the U.S. subsidiary Magnesium.com, Inc. During his tenure, Mr. Barela championed forged Magnesium (Mg) wheel projects, generating, and overseeing sales of various Mg products worldwide. Currently, Mr. Barela is with Terves Inc./Magnesium-USA overseeing marketing, product development, and technical sales for extruded and forged products. Over the 20+ years as a member of the TMS Magnesium Committee, Mr. Barela has brought practical industrial end-use knowledge, experience, and insight to the proceedings.

Aeriel Leonard is an Assistant Professor in the Materials Science and Engineering Department at The Ohio State University. She was awarded the Department of Energy Early Career Award in 2022 and the Office of Naval Research Young Investigator Award in 2021. She earned her B.A. in Metallurgical and Materials Engineering from the University of Alabama in 2012. In 2013, she began her Ph.D. journey at the University of Michigan in Materials Science and Engineering where she earned her Ph.D. in 2018. Dr. Leonard's Ph.D. work investigated real-time microstructural and deformation evolution in magnesium alloys using advanced characterization techniques such as high energy diffraction microscopy and electron back scatter diffraction. During her time at the University of Michigan, she led and worked on many teams aimed at increasing the number of underrepresented minorities in engineering including developing and implementing a leadership camp for female engineering students in Monrovia, Liberia. Dr. Leonard was awarded an NRC Postdoctoral Fellowship at the US Naval Research Laboratory in Washington, DC where she worked for two years. During this time, she used advanced characterization techniques such as x-ray computed tomography and high energy diffraction microscopy to understand damage and texture evolution during in-situ loading in additive manufactured materials. She also runs a lifestyle blog titled AerielViews aimed at young graduate and professional students.

Petra Maier received her doctoral degree from Loughborough University, United Kingdom, in 2002 in Materials Science, in the field of grain boundary segregation in steel. After completing her Ph.D., she worked at the University of Applied Sciences Wildau, Germany, as a postdoctoral fellow under the supervision of Prof. Asta Richter with a focus on mechanical properties by nanoindentation. From 2004 to 2006 she worked as a research associate in the MagIC under the supervision of Dr. Norbert Hort at the Helmholtz-Zentrum Geesthacht, Germany, being a part of the Institute of Materials Research lead by Prof. Karl Ulrich Kainer. Her focus was on magnesium recycling and high-temperature alloy development.

From 2006 to 2008, Dr. Maier was a research associate at the Technical University Berlin, Germany, in the Institute of Material Sciences and Technologies, Department of Materials Engineering of Prof. Claudia Fleck. There, her research specialties included corrosion fatigue on magnesium.

Since 2008, Dr. Maier is a professor of Materials and Production Engineering in the School of Mechanical Engineering at the University of Applied Sciences Stralsund in Germany. Since October 2022, she is a Lise Meitner professor at Lund University in the Department of Mechanical Engineering Sciences. She enjoys working in the field of Mg-based biodegradable implants. Her research is focused on corrosion under stress, the influence of the corrosion morphology on mechanical properties and how crack initiation and propagation is influenced by the microstructure. She is currently a past-chair of the TMS Magnesium Committee.

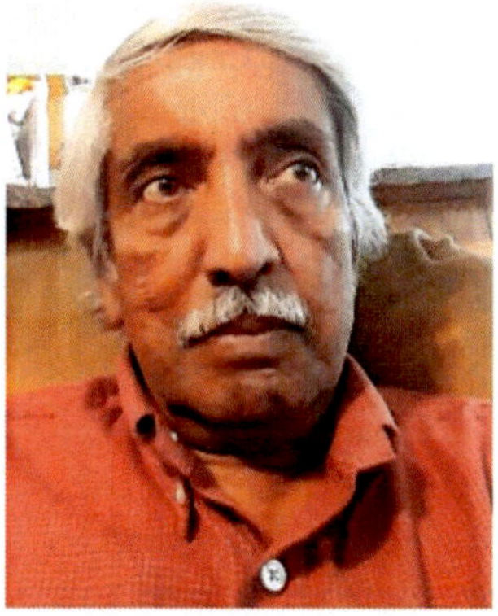

Neale R. Neelameggham IND LLC, is involved in international technology and management licensing for metals and chemicals, thiometallurgy, energy technologies, Agricoal, lithium-ion battery, energy efficient low cost OrangeH2, Netzero sooner with Maroon gas and Pink hydrogen, rare earth oxides, etc. He has more than 38 years of expertise in magnesium production and was involved in the process development of its startup company NL Magnesium to the present US Magnesium LLC, UT until 2011, during which he was instrumental in process development from the solar ponds to magnesium metal foundry. His expertise includes competitive magnesium processes worldwide and related trade cases. In 2016, Dr. Neelameggham and Brian Davis authored the ICE-JNME award-winning paper "Twenty-First Century Global Anthropogenic Warming Convective Model." He is working on Agricoal® to greening arid soils, and at present energy efficient Orange hydrogen, and methane abatement. He authored the ebook *The Return of Manmade CO_2 to Earth: Ecochemistry*. Dr. Neelameggham holds 16 patents and applications and has published several technical papers. He has served in the Magnesium Committee of the TMS Light Metals Division (LMD) since its inception in 2000, chaired in 2005, and since 2007 has been a permanent advisor for the Magnesium Technology Symposium. He has been a member of the Reactive Metals Committee, Recycling Committee, Titanium Committee, and Program Committee for LMD and LMD council. Dr. Neelameggham was the Inaugural Chair, when in 2008, LMD and the TMS Extraction and Processing Division (EPD) created the Energy Committee and has been a Co-Editor of the Energy Technology Symposium through the present. He received the LMD Distinguished Service Award in 2010. As Chair of the Hydrometallurgy and Electrometallurgy Committee, he initiated the Rare Metal Technology Symposium in 2014 and has been a co-organizer to the present. He organized the 2018 TMS Symposium on Stored Renewable Energy in Coal.

Victoria M. Miller is an Assistant Professor in the Department of Materials Science and Engineering at the University of Florida, a position she started in September 2019. She was previously an assistant professor at North Carolina State University from 2017 to 2019. Originally from Michigan, she received her B.S.E. in Materials Science and Engineering from the University of Michigan in 2011 and completed her Ph.D. in Materials at the University of California Santa Barbara in 2016.

After graduate school, she worked for a year at UES, Inc. onsite in the Materials and Manufacturing Directorate of the Air Force Research Laboratory in Dayton, Ohio. She also previously worked at Ford Motor Company, Toyota Engineering and Manufacturing, and Lockheed Martin Aeronautics.

Her primary research interest is microstructural evolution during thermomechanical processing of metals and alloys, particularly for those with low symmetry crystal structures. She has been researching Mg alloys since the age of 16.

Professionally, Dr. Miller has served on many committees within TMS, is Associate Editor for *JOM*, and is a Key Reader for *Metallurgical and Materials Transactions A*. She was a recipient of the 2017 TMS Young Leaders Professional Development Award, the 2020 ASM Bronze Medal Award, and the 2022 TMS-JIMM Young Leaders International Scholar Award.

Session Chairs

Magnesium Technology

Keynote Session

Steven Barela, Terves Inc./Magnesium-USA
Petra Maier, University of Applied Sciences Stralsund
Aeriel Leonard, The Ohio State University

Microstructure Evolution

Benjamin Anthony, University of Florida
Kiran Solanki, Arizona State University

Corrosion and Coatings

Petra Maier, University of Applied Sciences Stralsund
Joshua Caris, Terves Inc.

Deformation and Advanced Processing

Domonkos Tolnai, Helmholtz-Zentrum Hereon
Vineet Joshi, Pacific Northwest National Laboratory

Primary Production and Recycling/Alloy Development

Aaron Palumbo, Big Blue Technologies

Reviewer Pool

Benjamin Anthony, University of Florida, USA
Christopher D. Barrett, Mississippi State University, USA
Benjamin Begley, University of Florida, USA
Tracy Berman, University of Michigan, USA
Jan Bohlen, Magnesium Innovation Centre, Germany
Joshua Caris, Terves Inc., USA
Hajo Dieringa, Helmholtz-Zentrum Hereon, Germany
Jarek Drehlich, Michigan Technological University, USA
Adam Griebel, Fort Wayne Metals, USA
Daniel Höche, TU Darmstadt, Germany
Norbert Hort, Helmholtz-Zentrum Hereon, Germany
Steven Johnson, Central Connecticut State University, USA
Brian Jordon, Baylor University, USA
Vineet V. Joshi, Pacific Northwest National Laboratory, USA
Alan Luo, The Ohio State University, USA
Petra Maier, University of Applied Sciences Stralsund, Germany
Michele Manuel, University of Florida, USA
Suveen Mathaudhu, Colorado School of Mines, USA
Victoria Miller, University of Florida, USA
Neale Neelameggham, IND LLC, USA
Aaron Palumbo, Big Blue Technologies, USA
Regine Willumeit-Römer, Helmholtz Center Hereon, Germany

Part I
Keynote Session

Sustainable Domestic Manufacturing and Protecting IP in a Post AIA World

Andrew Sherman

Abstract

Ten years ago, the American Invents Act was passed by Congress. Despite lofty intentions, the AIA significantly degrades the value of IP and patents. Terves has recently gone through an enforcement action against a large importer of foreign magnesium products. The current system is rigged against small company inventors, with non-technical administrative judges invalidating 84% of patents in favor of infringers, versus <50% by multiple skilled patent examiners during reexaminations. Enforceable IP is essential for American competitiveness to complete with subsidized, unregulated, and lower cost offshore locations. Terves is a member of the US Inventor, representing 60,000 inventors focused on restoring individual patent rights post AIA. Terves' experience enforcing IP rights in today's climate as well as US inventors pending bill to restore patent rights will be discussed along with potential strategies and actions that inventors can take to mitigate AIA limitations.

Keywords

Magnesium • Intellectual property • US inventors

Background

Middle market materials manufacturers face stiff competition for overseas competitors that have unfair advantages in labor, regulatory, energy, and capital costs as a result of operating out of non-free market economies. In order to recreate a domestic manufacturing industry for critical materials, including magnesium, public policy and investment is required to offset these substantial disadvantages. Recent developments include investment and production tax credits, proposed rebates/credits for domestic sourced materials, enhanced buy-American and domestic content requirements, and public investment (loans, grants) to reduce and spread capital risk. Intellectual property, including patents, trade secrets, and trademarks (branding) can also establish a barrier to offshoring. However, the American Invents Act (AIA) has changed the patent landscape, making enforcement more difficult and expensive.

Introduction to PMT Group and Terves LLC

PMT group is a group of critical materials manufacturers, including Terves LLC, which is North America's only vertically integrated producer of wrought magnesium. Parent Company Powdermet Inc. was formed in 1996, when it acquired the powder metallurgy assets of refractory materials producer Ultramet Inc. Powdermet and its affiliates still provide powder metal and cermet feedstocks and toll powder production services for highly engineered materials. In 2013, Terves LLC was formed to commercialize "engineered response" materials which provide tailored responses to the environment, including changing dimensions (expanding), disintegrating, generating heat, releasing chemicals, or producing a signal. In 2016, to meet demand and reduce costs for domestically produced Tervalloy™ dissolvable magnesium alloys, Terves built a permanent mold foundry, added a 4000 ton extrusion facility and CNC machine shop, representing a significant investment in critical materials and magnesium production. Today, Terves has established capacity for over 1000 tons/annum of critical materials (Mg) production, and has over 40 patents on numerous magnesium alloys and engineered response materials and their applications. In addition to extruded products and powder metal and cermet feedstocks, Terves and Powdermet also provide metal matrix composite materials, forging

A. Sherman (✉)
Terves LLC, Euclid, OH, USA
e-mail: asherman@tervesinc.com

S. Barela et al. (eds.), *Magnesium Technology 2023*, The Minerals, Metals & Materials Series,
https://doi.org/10.1007/978-3-031-22645-8_1

stock, fully machined parts to customer specifications, and full assemblies built to OEM specifications.

As part of our efforts to diversify markets and products, in 2021, Terves commissioned Wagstaff to design and construct North America's only magnesium vertical direct chill (VDC) casting system to be built in the last 35 years. With a capacity of 3000–5000 tons/annum of magnesium billets, Terves can support additional markets in recreation/sporting goods, defense, aerospace, biomedical, energy storage, and transportation lightweighting markets. To support these markets, Terves has developed and licensed a number of new heat treatable Magnesium alloys, including new magnesium alloys specifically engineered for high plasticity (energy absorption), high strength, lower cost (lower alloy content and higher extrusion, forming, and rolling speeds), and with reduced or eliminated high value rare earth contents. When the new VDC comes on-line (spring, 2023), the current permanent mold foundry will be converted to the metallothermal production of rare earth metals, including Nd, Gd, and Dy. Parent company Powdermet Inc. also broke ground in August, 2022 on a new 30,000 sq ft facility to produce MnBi (licensed from the critical materials institute) rare earth free magnets, as well as NdFeB rare earth magnetic materials via strip casting using purified, recycled, or internally reduced rare earth metals.

In 2014, Terves introduced its patent pending Tervalloy™ dissolvable magnesium alloys. This development transformed the oil and gas completions industry, reducing water use and emissions during completion operations by up to 92% by eliminating the need to drill out plugs while flushing the produced debris from the 3–5 mile well-string in order to reestablish communication with the geologic formation. Plugs made from Tervalloy simply dissolve/disintegrate upon exposure to fluid, salinity, and temperature, returning the magnesium to the seawater from which it was originally extracted (Mg is 2% of seawater) in a controlled manner. By 2017, when our first patents issued, the majority of the market had been offshored to low cost Chinese suppliers who use the pigeon process (high CO_2) and low cost labor, utilities, and regulations to out-compete western manufacturers. Magnesium has been designated as a critical material for lightweighting (including its use in aluminum alloys, titanium production, and steel processing) primarily due to unfair Chinese trade practices (https://doi.org/10.3133/ofr20181021).

Patents are one strategy that act as a sword to attack infringers, including unfair foreign competitors. Terves has been working to enforce its patents, and after tracking a large importer of infringing magnesium alloys, we filed for patent enforcement/infringement in 2018. *Terves LLC v. Ecometal Inc.*, Case No. 1:19-cv-1611-DCN (N.D. Ohio 2019). After more than three years, Terves prevailed at trial, and after surviving two IPR's and an ex-party reexamination, the patent validity, infringement, and lost profit damages were established in district court, and a permanent injunction was issued against the infringing parties. Terves continues to enforce its patents to sustain North American manufacturing and is increasing its investment and commitment to critical materials production in the USA with a planned $27 M investment in expansion and additional production at our Euclid, Ohio facilities.

Patents, IP, and Enforcement Post AIA

The right to a patent was granted to authors and inventors in the US constitution. It was the first time in history that the common "man or woman" were allowed to own their inventions. Prior to the US constitution, grants were by the state, generally to landed and gentry class, and the common person had no right to ownership at the national level. To support sustainable domestic manufacture of critical materials, PMT group has over 100 patents issued or pending and has invested heavily in IP creation and enforcement as a core strategy to create barriers to (mainly) foreign competition and to try and prevent rapid commoditization of products to allow faster payback and acceptable returns on investment in the capital intensive and long product life cycle materials industry. After 26 years, (including several prior freedom to operate, patent infringement, and theft of trade secret litigations), the current enforcement experience has highlighted the strengths and weaknesses, and expenses, of IP strategy. In particular, the patent trial and appeals board (PTAB), created under the AIA, and set up by big tech, has overturned 85% of all patents brought before it, using administrative court judges with no requirement for any patent or technology expertise. The PTAB was designed with that in mind by big tech. In fact, Googles' main patent strategist, Michell Lee, was appointed head of the patent and trademark office by president Obama and instituted the rules and procedures designed to greatly weaken patent holders rights and remedies, and to shield large corporations from patent infringement damages. Ostensibly, the PTAB was supposed to be a faster and cheaper method of resolving patent validity as compared to district court proceedings. In reality, it was designed in favor of large corporations to allow them to utilize other peoples intellectual "property" without compensation.

We are part of US Inventor, an organization of over 60,000 inventors working to restore patent rights that were largely stripped by the PTAB under the American Invention Act (AIA). According to the US Inventor website "The America Invents Act of 2011 (AIA) created an easier way to invalidate (revoke) an issued patent. The PTAB is an administrative court with no jury and much less due process than a real court. Rather than a lifetime-appointed judge, a

PTAB trial typically has three attorneys who are called Administrative Patent Judges (APJs). Since inception, 84% of the patents that go through a PTAB process get fully or partially invalidated (partially usually means the parts of the patent that matter).

When you attempt to stop a large corporation from infringing your patent, they will try to use the PTAB to invalidate your patent. If you win one PTAB attack, you can still be pulled into additional ones by the same or other infringers. According to the AIPLA (American Institute of Patent Law Association), a reasonable PTAB defense costs $400,000 to $800,000. Historically, the typical inventor would hire an attorney on a contingency basis to fight an infringer (where the inventor doesn't pay much up front and the attorney gets a percentage of the verdict award or settlement amount). Today, it is extremely rare for an attorney to take any PTAB case on contingency."

Furthermore, the AIA largely removed the ability to stop infringers, we believe against the constitution (Supreme Court ruled that patents are no longer a property right, to allow the PTAB to exist). Per US Inventor " The U.S. Supreme Court decided that it was in the "public interest" for a proven infringer to continue infringing because it could serve the market better than a startup (Ebay, 2006). As a result, even if you win your case, you will have to pass a "public interest" test before an injunction can be issued to stop the infringer. A startup vs an entrenched corporation will typically fail this test, so you can't stop the infringer. You end up with a court-ordered royalty that you cannot negotiate, and the infringer keeps your invention and the market. This is often an impossible barrier for what would have been, until recently, the next great American disruptive startup."

US Inventor is working to restore US Inventor rights, including the ability of inventor-operators to opt out of the IPR in favor of district court (IPRs are almost exclusively filed in response to a claim of infringement), and the restore the right no not be forced to license to a corporation that has been found to infringe.

Terves' experience in litigating was that it was long (over three years), mainly due to the ability of the infringer to delay and stall litigation and particularly discovery. We were unable to serve or get discovery in china, and have abandoned our Chinese patents as worthless due to their extremely poor legal system (no experts, no discovery, mainly works on behalf of government interests, not private party interests). Cost of litigating, including defending two IPR's and one reexamination, discovery, motions, and trial approached $2M. We did receive a permanent injunction and lost profits, proving to the jury that no reasonable alternative to Tervalloy is available in the market, a major win.

One of Terves strategies to offset the power of the PTAB has been to obtain a large number of patents and claims, to increase costs of litigation against our IP portfolio. Terves currently has more than five patents issued and over 350 claims issued on our dissolvable magnesium materials and their application. However, this comes at significant cost to both prosecute (obtain), defend (against claims of invalidity), and enforce (attack infringers). This is in addition to substantial investment in vertical integration, inventory, and productivity to reduce costs, speed delivery, and meet all demand for wrought magnesium in North America. We also invest and collaborate to develop new alloys, heat treatments, and production processes to enhance performance, customize products for individual clients, and meet customer demands in order to better compete with foreign producers and importers.

In addition to our district court enforcement actions, we have been researching ITC 337 exclusion processes. These have been used by some startups (Aspen Aerogels, for example) to block import of infringing product. The advantage of 337 proceedings is they cannot be stalled by legal maneuverings, and reach conclusion in 18 months or less under statutory requirements, leading to an exclusion order blocking imports of infringing products into the USA. Disadvantages are that the costs are high and cannot be recovered, nor can damages be recovered, and the exclusion order needs to be enforced by an overworked customs and border patrol. However, violators of the exclusion order are subject to punishing fines and penalties that can reach $100,000/day, and shipments and product can be confiscated. Those that have successfully pursued 337 have indicated that they largely rely on the risk aversion of large public users to enforce and comply, in addition to working with customs and border patrol to identify and seize inbound infringing products. This was successful in the case of Aspen Aerogels, while it was unsuccessful in the case of crucible materials, who won a general exclusion order for NdFeB magnets but did not prevent offshoring (no magnets are currently produced in the USA, despite the fact that they were invented and at one time solely produced in the USA). Many of the ITC procedures and precedents were set in the crucible materials action, and ITC proceedings are increasingly being used to enforce and enhance patent rights for composition of matter and other utility patents.

Summary

In summary, patents, trademarks, and trade secrets are one set of tools to support domestic manufacturing. Innovation, particularly process improvements to reduce costs and labor content as well as creating new markets/demand, are more powerful in the short run, but patents provide for longer term sustainability, but only if they can be successfully enforced and defended. Novel composition of matter patents are the easiest to defend and enforce (as in the Tervalloy case, we enforced composition of matter claims). Branding, including trademarks, is an incredibly powerful and relatively easily enforced tool to create a preference and loyalty but is more difficult to achieve in the materials and manufacturing sector serving OEM's, as opposed to the consumer sector. Copy-writes, including material performance specifications, quality documents, and validated property datasets, can also create preferences and increase cost of competition in critical applications. Finally, government intervention and public policy efforts such as domestic content rules, tariffs, tax incentives, and targeted investments are required to offset the actions of foreign governments and non-free market economies to create fair trade, as opposed to free trade.

Reductant Considerations in Thermal Pathways to Primary Magnesium Metal Production: A Case for Aluminothermic Reduction

Aaron W. Palumbo and Boris A. Chubukov

Abstract

Thermal routes to produce primary magnesium metal are generally demarcated with lower capital costs. However, the range of operational costs are high depending on reduction reactor efficiencies and configuration, labor intensity, and the cost of reductant. Choosing a reductant material can be a challenge in terms of logistics, supply security, cost, and technical performance. Presented is an assessment of possible reductant materials and the interplay between heat of reaction, reactor and condenser design, byproducts, and the behavior and role of impurities. On one end of the reductant spectrum is the reaction of carbon and magnesia, generating exclusively a byproduct gas, along with the product magnesium metal, and virtually no solid residue. Conversely, the reaction between calcined dolomite and ferrosilicon generates virtually no byproduct gas and nearly 6× more mass of calcium silicates than magnesium metal. These chemistries share engineering challenges that center on the method of recovery products.

Keywords

Magnesium • Pyrometallurgy • Process technology • Primary production • Aluminothermic reduction • Techno-economics

Extended Abstract

Thermal production routes to produce primary magnesium metal are generally demarcated with lower capital costs over electrolytic processes. However, the range of operational costs are high depending on reduction reactor efficiencies and configuration, labor intensity, and the cost of reductant. Choosing a reductant material can be a challenge in terms of logistics, supply security, cost, and technical performance. For silicothermic reduction chemistry, three different reactor configurations have been commercialized: Pidgeon, Bolzano, and Magnetherm. While the theoretical energy required to drive the reaction are all about the same, depending on reactor temperature, the reactor configurations and heating methods give rise to large variation in applied energy consumption, as reported in Table 1.

In addition to ferrosilicon, other reductants have been explored including carbon, aluminum, carbides, and metal alloys containing combinations of calcium, silicon, and/or aluminum. On one end of the spectrum is the reaction of carbon and magnesia, which generates the byproduct gas carbon monoxide along with the product magnesium metal and virtually no solid residue. Conversely, the reaction between calcined dolomite and ferrosilicon generates virtually no byproduct gas and nearly six times more mass of calcium silicate than magnesium metal. This solid byproduct is practically valueless in industrial markets.

Use of aluminum metal, when used with calcined dolomite, produces a calcium aluminate solid byproduct, a value-add output (Reaction 1). Similarly, when reacted with magnesia, the byproduct is magnesium aluminate spinel, a high-value refractory (Reaction 2). Laboratory experiments confirm the formation of spinel and high conversion of MgO at relatively low reduction temperatures in the range of 1273–1473 K. Total gross energy consumption for an MgO-to-ingot process was calculated to be 8.8 kWh/kgMg. For all unit operations, electrified equipment is used or assumed. Since there no direct emissions from the process, all indirect emissions are related to the source of power. In the USA, the national average of the electrical grid is 0.418 $kgCO_{2eq}$/kWh, which results in estimated process emissions of 3.7 $kgCO_{2eq}$/kgMg with no direct greenhouse gas emissions.

A. W. Palumbo (✉) · B. A. Chubukov
Big Blue Technologies, 11901 W 48th Ave, Wheat Ridge, CO 80033, USA
e-mail: palumbo@bigbluetec.com

S. Barela et al. (eds.), *Magnesium Technology 2023*, The Minerals, Metals & Materials Series,
https://doi.org/10.1007/978-3-031-22645-8_2

Table 1 Comparison of theoretical and apparent energy consumption for various reduction chemistries to produce magnesium metal

Reduction chemistry	Nominal operating temp (K)	Theoretical minimum energy (kWh/kgMg)	Reported energy consumption (kWh/kgMg)	GHG impact ($kgCO_2/kgMg$)	Source
Silicothermic					
Pidgeon	1473	4.26	13.3–14.8	21.8–25.3	[1, 2]
Bolzano	1473	4.26	7.0–7.3	10.1	[3]
Magnetherm	1873	4.77	~16	NR	[4]
Carbothermic	2073	8.39	10–12	0.7–12.6	[5]
Electrolysis	973	7.32	12–19	6–19	[3, 6]
Other			No commercial operations		[7]
Aluminothermic	1473	3.12			
CaC_2	1473	4.17			
SiC	1773	6.20			
Ca–Si [1:3] Alloy	1473	3.88			
Al–Si [1:1] Alloy	1473	3.88			

GHG impact for the overall cradle-to-gate process is also reported for commercialized processes

R1: $3MgO + 3CaO + 2Al = 3Mg_{(g)} + Ca_3Al_2O_6$ $\Delta H(298 \rightarrow 1473\,K) = 950\,kJ$
R2: $4MgO + 2Al = 3Mg_{(g)} + MgAl_2O_4$ $\Delta H(298 \rightarrow 1473\,K) = 818\,kJ$

Using laboratory and pilot data, a discounted cash flow techno-economic model revealed a base-case capital intensity of $3120 per ton-capacity of magnesium metal produced for the MgO-to-ingot process boundary. Mining, transportation, and calcination were not included. The value-add spinel serves to offset the relatively high cost of aluminum scrap, whose normalized cost is on par with ferrosilicon, $550–1300/tMg. However, without the solid byproduct revenue stream, a commercial plant on the order of 50,000 tpy results in a 24.9% internal rate of return and a $163.7 million net present value using a 12% discount rate and a sell price of $4.00/kgMg. Sensitivity analyses, Fig. 1, revealed that the sell price of magnesium was the most significant factor affecting economic metrics along with the prices of aluminum and MgO.

The use of aluminum as a reductant material was shown to be effective with the benefit of a value-add byproduct. The volatility of pricing over the past several years implies some degree of uncertainty in economic forecasting which increases the incentive to implement a production process with more than one revenue stream.

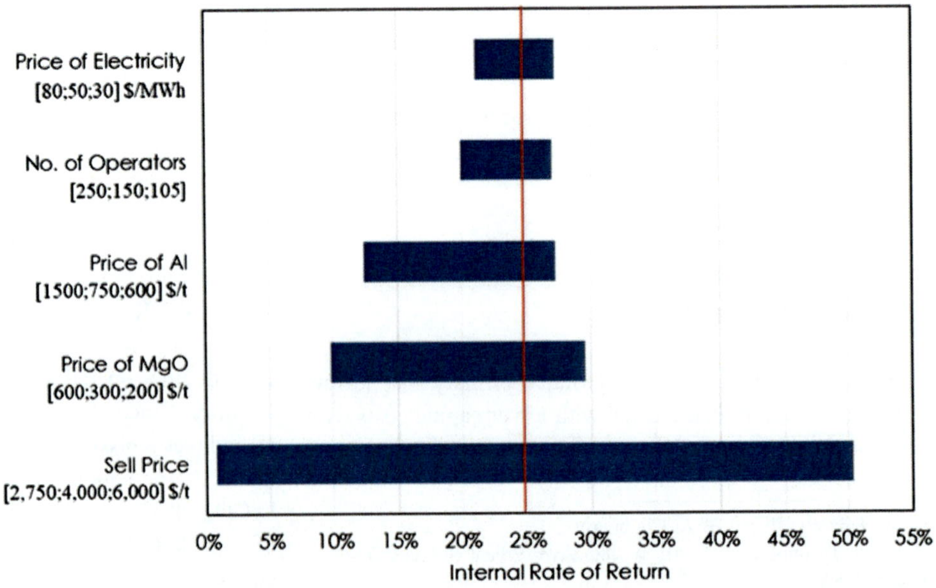

Fig. 1 Sensitivity analysis for selected factors on IRR

References

1. Ehrenberger, S., H. Dieringa, and H.E. Friedrich, *Life Cycle Assessment of Magnesium Components in Vehicle Construction.* 2020, German Aerospace Centre, DLR.
2. Ehrenberger, S., and M. Brost, *Life Cycle Assessment of a New Pidgeon Process - ESAN.* 2015, German Aerospace Center, DLR
3. Friedrich, H.E. and B.L. Mordike, *Magnesium Technology: Metallurgy, Design Data, Applications.* 2006, Germany: Springer
4. Sever, J.C. and M. Ballain, *Evolution of the Magnetherm Magnesium Reduction Process*, in *Magnesium Technology 2013.* 2013, John Wiley & Sons, Inc. p. 69–74.
5. Palumbo, A. *Techno-economics and Life Cycle Analysis of a Carbothermal Reduction Process, 78th Annual World Magnesium Virtual Conference, August 26, 2021.*
6. Thayer, R.L. and R. Neelameggham, *Improving the Electrolytic Process for Magnesium Production.* JOM, 2001.
7. Bale, C.W., et al., *FactSage Thermochemical Software and Databases, 2010–2016.* 2016, Calphad.

Metastable–Stable

Norbert Hort

Abstract

Most metallic materials are often cast. This is a highly dynamic process where composition and solidification conditions can become a challenge due to the appearance of porosity, shrinkholes, segregations, and the formation of intermetallic phases. Heat treatments are applied to get rid of segregations, stable, and metastable intermetallic phases for homogenization purposes and to obtain a material that is homogeneous with an adjustable property profile. However, often it is observed that the dissolution especially of metastable phases really takes long times (hundreds or even thousands of hours) and high temperature which is accepted in the academic world but not acceptable for real industrial applications due to time and costs. Especially for biodegradable metals, metastable phases perhaps could be regarded as "stable" when the time of absorption is in range of a few weeks only and could be used to adjust properties.

Keywords

Stable/metastable phase • Cast/wrought processing • Heat treatments • Surface modification • Phase diagrams

Extended Abstract

To develop new metallic materials (alloys), it is useful to know the requirements of the application (unless it is just basic academic issues). This can be a desired strength, a forming ratio, surface roughness, corrosion rate, etc. Further, it also makes sense to know what process route is required to meet the application requirements. If the starting material is produced via a casting route, if it is subsequently heat treated and formed, and if the surface is modified, then the new material should meet the process requirements in addition to the material requirements. All in all, this is a rather complex process that can definitely be mapped experimentally. However, for some time now, there have also been approaches to first create a "digital twin" with suitable prediction tools before actually starting the experiments.

Melting, casting, and solidification are usually very dynamic processes. There are many influencing factors such as the melting and boiling point of the matrix or alloying elements, vapour pressures, density, whether phases are already formed in the melt or during solidification, etc. The melting and solidification process is also very dynamic. Also of importance are addition and removal, the reaction of the elements involved with tools. Furthermore, the geometry of the materials to be cast also has an influence on the microstructure that forms. Segregation also occurs, as does the formation of intermetallic phases, mainly at the end of solidification (usually at eutectic temperatures).

After the Bronze Age came the Iron Age. Around 1400 BC, iron was successfully produced in Asia Minor. The superiority of this material compared to bronzes quickly became apparent. However, until the nineteenth century, it was costly to produce iron in large quantities. Iron could be produced in a racing furnace and forged steel could be produced through subsequent forging processes. In the Middle Ages, the first blast furnaces were developed that were actually capable of smelting iron. However, one problem was to control the carbon content. Therefore, it was first cast iron that was used in larger quantities. To obtain malleable iron, the pig iron from the blast furnace was melted again in a fresh fire. The properties of the iron could be specifically influenced by tempering, carburising, quenching, and annealing. It was not until the second half of the eighteenth century that usable methods were available for producing cast steel, which, through forging processes, became the steel that was already similar to today's steel.

N. Hort (✉)
Helmholtz-Zentrum Hereon, Geesthacht, Germany
e-mail: norbert.hort@hereon.de

S. Barela et al. (eds.), *Magnesium Technology 2023*, The Minerals, Metals & Materials Series,
https://doi.org/10.1007/978-3-031-22645-8_3

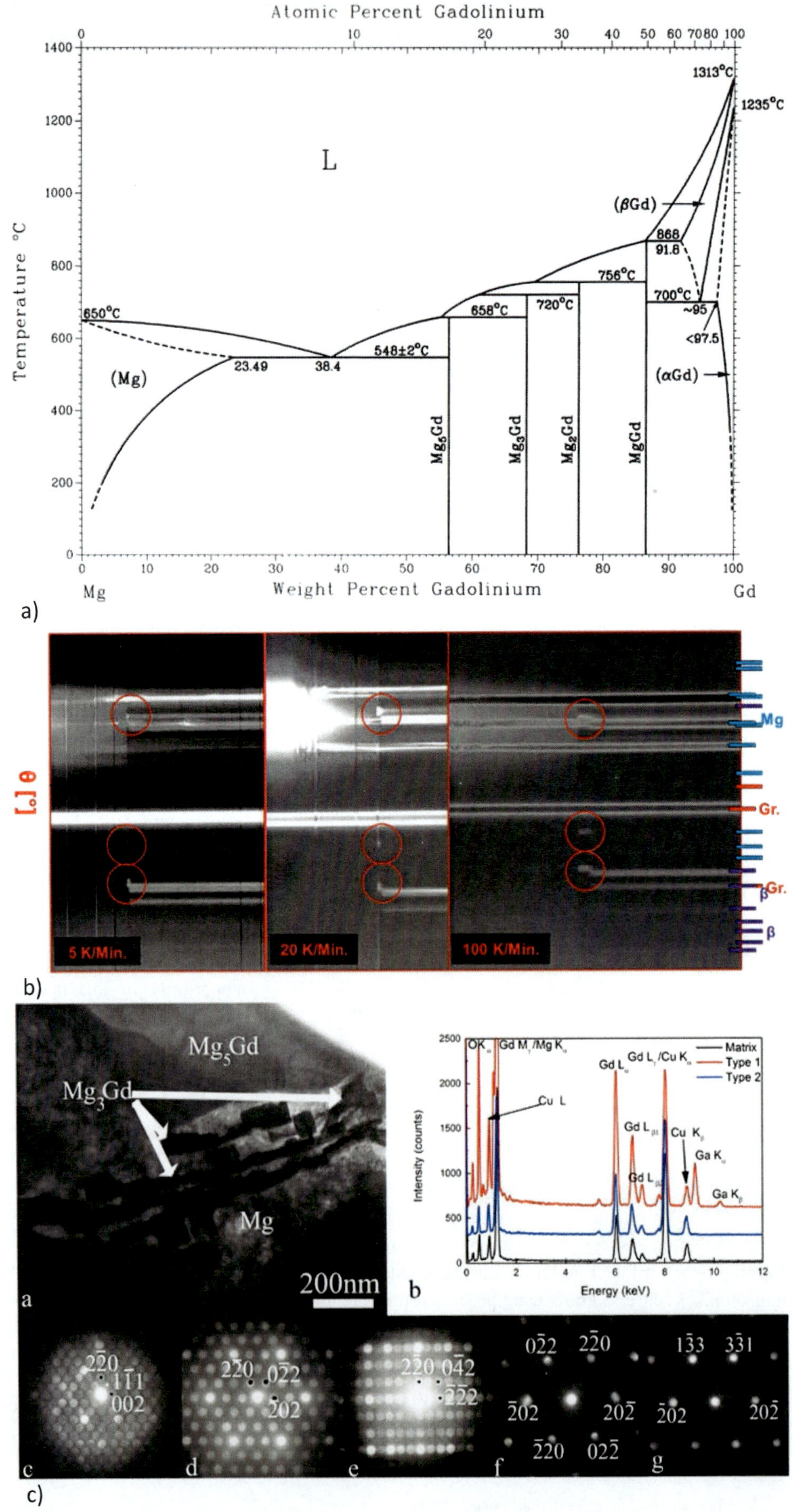

Fig. 1 **a** Mg–Gd phase diagram [2], **b** appearance of Mg_3Gd (in red circles) during in situ synchrotron radiation diffraction investigations (Gr represents the graphite crucible, β = Mg_5Gd), **c** TEM and EDX results showing the presence of Mg_3Gd [1]

In any case, this was one of the essential processes for industrialisation.

What does this have to do with stable and metastable? If you look at an iron-carbon state diagram, you usually see two phase diagrams: the stable one (cast iron: Fe + graphite) and the metastable one (Fe + Fe_3C). Cast iron has largely lost its importance. Cast steel is also still used. The basis of modern industry, however, is steel, which is usually produced by continuous casting, heat treated, and shaped into the desired form and processed by forming processes. Essential, however, is the fact that we consider and apply the metastable Fe–C system. However, Fe_3C, which is actually not in thermodynamic equilibrium, is so stable under the right process conditions that it can actually no longer be called metastable. Even the C dissolved in Fe can be kept in Fe by suitable processes and alloying elements and often does not transform into Fe_3C or even graphite. Thus, the metastable Fe–C system (steel) can, indeed must, be considered a stable system over a long period of time.

Virtually, all magnesium alloys are firstly produced by melting, casting, and solidification. The selection of alloying elements was often empirical and, as already mentioned, segregation, intermetallic phases, shrinkage, etc. were observed. Very often, however, it was not the intermetallic phases that were expected based on looking at state diagrams. Something else had formed, and the reason for this is that metastable phases had often formed first under real conditions. These can indeed be eliminated by suitable heat treatments. However, it was often observed that metastable phases did not transform directly into stable phases. An example is shown in Fig. 1 regarding the Mg–Gd system [1]. With respect to cooling rates, a Mg_3Gd intermetallic forms first rather than the stable Mg_5Gd intermetallic. This is not in agreement with the published Mg–Gd phase diagram that shows Mg_5Gd as the stable β phase [2]. Moreover, a large number of metastable phases often first formed from the saturated solution in complex precipitation sequences before a stable state was reached. Again, the Mg–Gd system can be used as an example [3]. Diffusion is the cause as well as time and temperature.

The literature frequently reports heat treatment times ranging from a few hours to days often weeks at selected temperatures. While the former is still acceptable for industrial processes, the latter is really only of academic interest. For a creep-resistant magnesium alloy, however, exposure under load at high temperatures is an application that ultimately involves heat treatment over a long period of time. Therefore, information is needed here as to which phases transform when and at what temperature. But what if the application temperature is 37 °C and the material is to dissolve within a few months (1 month, 31 days = 744 h)? This is the requirement for a degradable magnesium implant. Here, too, heat treatments are often carried out to obtain a stable homogeneous material. But what if the metastable phases we eliminate in this way would have a positive influence on the application profile? Why would we want to avoid something like that? Why don't we use this specifically to adapt to the required property profile?

References

1. G. Szakács, C. L. Mendis, D. Tolnai, A. Stark, N. Schell, K. U. Kainer, N. Hort, In situ synchrotron radiation diffraction during solidification of Mg15Gd: effect of on various cooling rates. In: M. V. Manuel, M. Alderman, A. Singh, N. R. Neelameggham, Magnesium Technology 2015, John Wiley & Sons, Hoboken, New Jersey, USA, 2015, 79–84
2. ASM Handbook, Vol. 3 – Alloy Phase Diagrams, ASM International, 2016
3. J. Čížek, I. Procházka, B. Smola, I. Stulíková, V. Očenášek, Influence of deformation on precipitation process in Mg–15 wt.% Gd alloy, Journal of Alloys and Compounds 430 (2007) 92–96

Engineered Bioabsorption for Implant Applications

Jacob Edick, Carolyn Woldring, Joshua Caris, Nicholas Farkas, Anuvi Gupta, and Andrew Sherman

Abstract

Bioabsorbable magnesium alloys have often been studied in the context of absorbing too quickly, with the goal of obtaining a slower absorption rate for adequate use as a temporary implant material. While this is an important challenge to overcome, it oversimplifies the needs of surgeons and patients while not accounting for the requirements of various applications and the technological complexities of absorption. Through our presented work, we are building magnesium alloy capabilities to engineer the bioabsorption profile for optimizing implant performance based on requirements such as implant size, time to healing, and anatomical differences.

Keywords

Biomaterials • Magnesium alloy • Biodegradation

Extended Abstract

A large challenge in bioabsorbable material technology is designing implants that can meet the needs of implant indications. Often, bioabsorbable metals are limited by absorption profiles that cause complications (e.g., alloys that degrade too quickly to maintain mechanical integrity through healing or alloys that degrade too slowly to be considered a temporary implant). Micro-alloying, along with material processing, is a technique that has been applied to magnesium alloys to produce slower degrading alloys to accommodate strength through healing. These alloys were engineered to have a reduced absorption rate; however, they lack characteristics necessary for oxide breakdown and thus absorption rate remains slow after healing [1, 2].

Increasing the alloying content of Zn and Ca in a Mg–Zn–Ca alloy system, along with optimizing processing parameters, allows for manipulating the formation and morphology of secondary phases (Mg_2Ca and ternary IM1). Differing only in Zn content, a study of two Mg–Zn–Ca alloys demonstrated ternary IM1 phases act as nano-cathodes and increase the absorption rate and the acceleration of the absorption rate of an alloy with increased Zn content versus an alloy containing only Mg_2Ca secondary phase [2]. Balancing the formation of these secondary phases by altering processing to achieve an optimized absorption profile needs further study.

Presented is a summary of the work to date on optimizing the formation of Mg_2Ca phase and IM1 phase IMPs and their effects on the material's absorption profile. In addition, results will characterize the unique stages of absorption to model an alloy capable of maintaining mechanical integrity through the useful life of an implant and then increasing in absorption rate to fully absorb in a timely manner. This work was designed to develop a magnesium alloy with an absorption profile optimized based on the biomedical inputs of an application. By engineering the absorption profile, implant life and function will be able to meet surgeon needs for a target anatomy.

References

1. Holweg P, Berger L, Cihova M, Donohue N, Clement B, Schwarze U, Sommer N, Hohenberger G, van den Beucken J, Seibert F, Leithner A, Löffler J, Weinberg A (2020) A lean magnesium–zinc–calcium alloy ZX00 used for bone fracture stabilization in a large growing-animal model. Acta Biomaterialia 113:646-659. https://doi.org/10.1016/j.actbio.2020.06.013

J. Edick · C. Woldring (✉)
Magsorbeo Biomedical Corp, Detroit, MI 48214, USA
e-mail: carolyn.woldring@magsorbeo.com

J. Caris · N. Farkas · A. Gupta · A. Sherman
Terves LLC, Euclid, OH 44117, USA

S. Barela et al. (eds.), *Magnesium Technology 2023*, The Minerals, Metals & Materials Series,
https://doi.org/10.1007/978-3-031-22645-8_4

2. Cihova M, Martinelli E, Schmutz P, Myrissa A, Schäublin R, Weinberg AM, Uggowitzer PJ, Löffler JF (2019) The role of zinc in the biocorrosion behavior of resorbable Mg–Zn–Ca alloys. Acta Biomaterialia 100:398-414. https://doi.org/10.1016/j.actbio.2019.09.021

Part II

Microstructure Evolution

Static Recrystallization Kinetics and Texture Evolution in Wrought Mg–Zn–Ca Alloys

T. D. Berman and J. E. Allison

Abstract

It has been established that additions of Ca and Zn to magnesium can yield wrought materials with a weak crystallographic texture that is desirable for formability. This study explores how Zn and Ca additions affect the static recrystallization kinetics and mechanisms in magnesium alloys compressed using a Gleeble thermomechanical processing simulator. The effect of these alloying additions on the texture evolution and grain growth kinetics during post-deformation annealing will be discussed.

Keywords

Recrystallization • Mg alloy sheet • Thermomechanical processing

Extended Abstract

It has been established that additions of Ca and Zn to magnesium can yield wrought materials with a weak crystallographic texture that is desirable for formability [1–6]. This study explores how Zn (0, 1, and 3 wt %) and Ca (0 and 0.1 wt %) additions affect the static recrystallization kinetics and mechanisms in magnesium alloys compressed using a Gleeble thermomechanical processing simulator. The effect of these alloying additions on the texture evolution and grain growth kinetics during post-deformation annealing will be explored.

Rectangular prism specimens (20 mm × 15 mm × 10 mm) were machined from 30 mm diameter extrusions such that the prior extrusion direction is parallel to the compression direction. A Gleeble 3500 thermomechanical processing simulator was used to conduct the plane strain compression (PSC) tests and subsequent annealing treatments. The Gleeble utilizes direct resistance heating that is controlled using feedback from a thermocouple spot welded to the specimen surface. After a thermal soak of 600 s at the deformation temperature, the specimens were subjected to a compressive strain of 0.2. Immediately following deformation, the specimen temperature was ramped to the annealing temperate at a rate of 5 °C per second. For tests in which the deformation and annealing temperature are equal, annealing in-situ within the Gleeble allows one to study very short annealing times with the confidence that the specimen temperature is being held consistently throughout the process. Following annealing, the specimens are immediately quenched using forced air. Optical microscopy and electron backscatter diffraction (EBSD) was used to characterize the grain size, grain morphology, and crystallographic texture. Grains with a grain orientation spread (GOS) less than or equal to 1 degree were classified as recrystallized. Grains with a higher misorientation were characterized as being deformed.

Figure 1 illustrates the microstructure evolution following annealing at 350 °C in Mg–3Zn–0.1Ca (wt %) alloy specimens after being deformed at 350 °C using a strain rate of 0.5 s^{-1}. The mean grain size is stable at approximately 20 μm up to 60 s (Fig. 2a). Grain growth occurs when the annealing duration is increased to 600 s. Though the average size is stable at short times, the grain size distribution does shift towards smaller grains as the annealing duration increases to 60 s. Little recrystallization is observed after 10 s, but the extent of recrystallization increases notably after 60 s (Fig. 2b). The majority of the microstructure is recrystallized after annealing for 600 s.

This microstructure evolution suggests that the Mg–3Zn–0.1Ca alloy subjected to these deformation conditions recrystallizes through a continuous process. There was little evidence of nucleation and growth of recrystallized

T. D. Berman (✉) · J. E. Allison
Department of Materials Science and Engineering, University of Michigan, Ann Arbor, MI 48109, USA
e-mail: tradiasa@umich.edu

S. Barela et al. (eds.), *Magnesium Technology 2023*, The Minerals, Metals & Materials Series,
https://doi.org/10.1007/978-3-031-22645-8_5

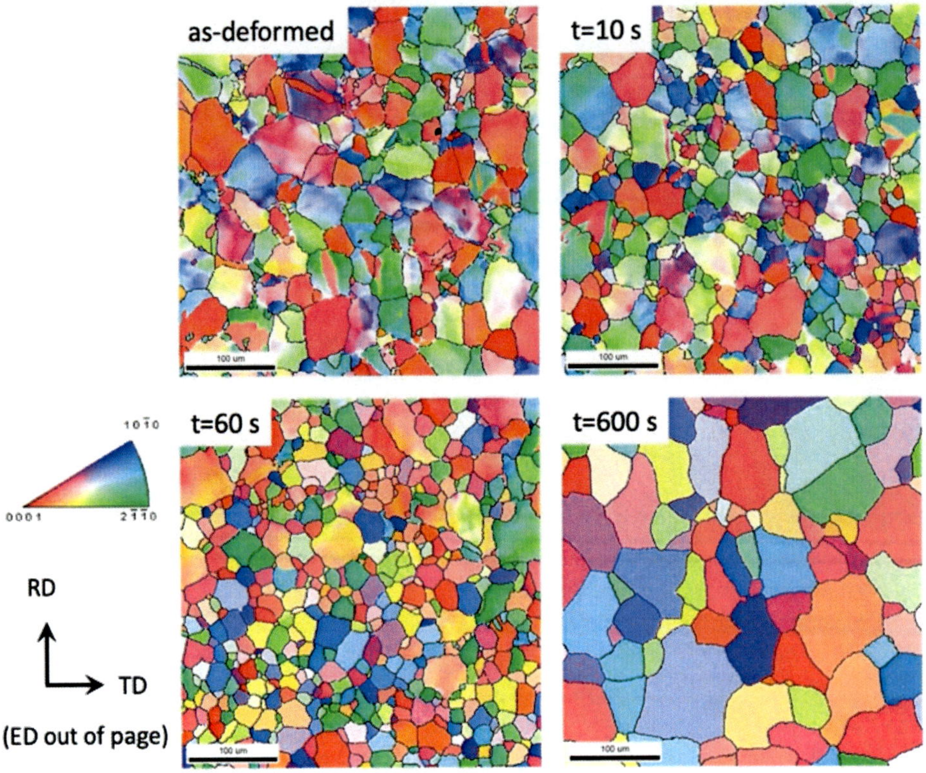

Fig. 1 Inverse pole figure maps of a Mg–3Zn–0.1Ca (wt %) alloy subjected to a compressive pass of 0.2 true strain at 350 °C using a strain rate of 0.5 s^{-1} and subsequently annealed at 350 °C for the time (*t*) indicated. All maps are 400 μm wide (For interpretation of the figure legends, the reader is referred to the web version of this article.)

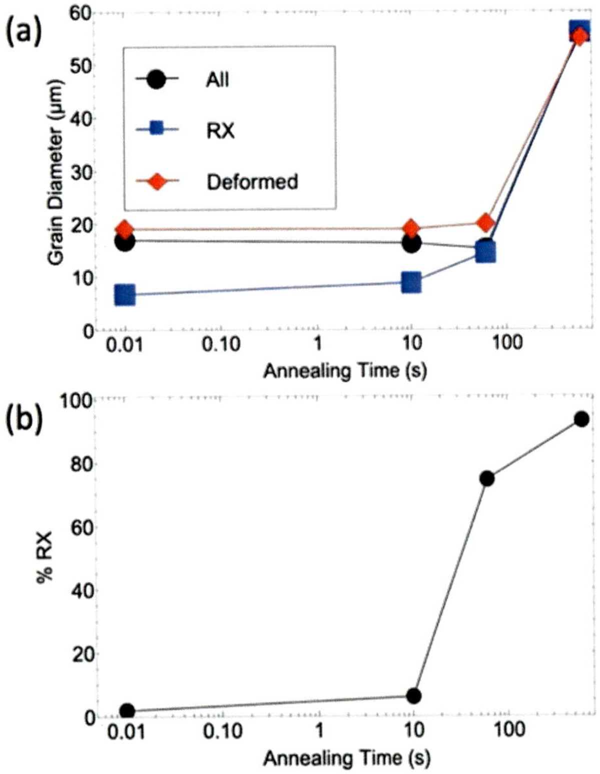

Fig. 2 Evolution of **a** the mean grain size and **b** the area fraction recrystallized as a function of annealing time at 350 °C

(RX) grains. Up to 60 s the deformed grain size is constant. Rather than the deformed grains being slowly consumed by growing recrystallized grains, the deformed grains recover and recrystallize through transformation of the low angle subgrain boundaries into high angle grain boundaries. An example of the subgrain structures prevalent in this alloy can be seen in the sample annealed for 10 s in Fig. 3.

Static recrystallization in this alloy under similar processing conditions results in weak, off-basal textures [1, 7]. The mechanisms by which this texture reduction occurs are still under investigation. One contributing factor, nucleation of randomly oriented nuclei within shear bands and twins [8–10] was not observed in this case. Though continuous recrystallization is not associated with a significant reduction in texture [11], a growth advantage of off-basal grains [7, 12] or the absence of the a growth preference for basal grains (as seen in the AZ31 alloy [13]) [12, 14, 15] can preserve and slightly reduce the already weak as-deformed textures seen in some of the Mg-Zn-Ca ternary alloys. Co-segregation of Zn and Ca to the grain boundaries may alter grain boundary mobilities and contribute to these orientation-dependent growth rates [4, 12]. Victoria-Hernández et al. have observed the retention of off-basal (here transverse direction-split) grains during continuous static recrystallization (extended recovery) in Mg–0.6Zn–0.6Ca–0.1Zr (wt %) [16]. Future work will explore how changes in the Zn and Ca content impact the recrystallization mechanisms and kinetics.

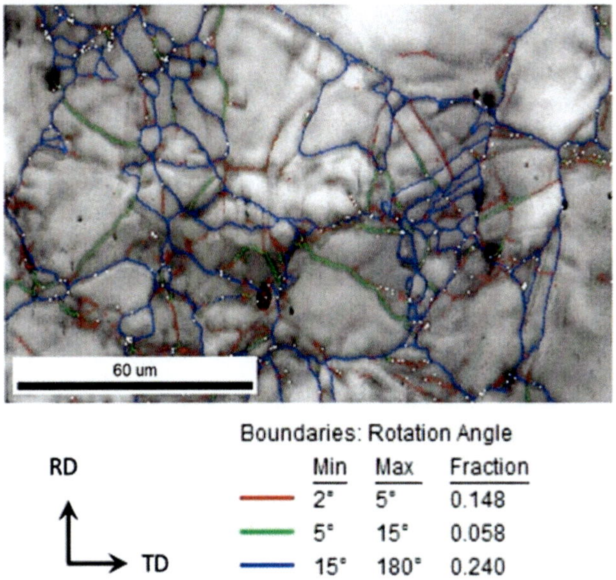

Fig. 3 Grain boundaries overlaid on an EBSD image quality map for the sample annealed for 10 s. (For interpretation of the figure legends, the reader is referred to the web version of this article.)

Acknowledgements This work was supported by the US Department of Energy, Office of Basic Energy Sciences, Division of Materials Sciences and Engineering under Award#DE-SC0008637 as part of the Center for Predictive Integrated Structural Materials Science (PRISMS Center) at University of Michigan. We also acknowledge the financial cost-share support of University of Michigan College of Engineering and Office of the Vice President for Research.

References

1. T.D. Berman, J.E. Allison, Coupling Thermomechanical Processing and Alloy Design to Improve Textures in Mg–Zn–Ca Sheet Alloys, JOM. 73 (2021) 1450–1459. https://doi.org/10.1007/s11837-021-04630-0.
2. D.W. Kim, B.C. Suh, M.S. Shim, J.H. Bae, D.H. Kim, N.J. Kim, Texture evolution in Mg–Zn–Ca alloy sheets, Metall. Mater. Trans. A Phys. Metall. Mater. Sci. 44 (2013) 2950–2961. https://doi.org/10.1007/s11661-013-1674-2.
3. J. Bohlen, J. Wendt, M. Nienaber, K.U. Kainer, L. Stutz, D. Letzig, Calcium and zirconium as texture modifiers during rolling and annealing of magnesium–zinc alloys, Mater. Charact. 101 (2015) 144–152. https://doi.org/10.1016/j.matchar.2015.02.002.
4. Z.R. Zeng, Y.M. Zhu, S.W. Xu, M.Z. Bian, C.H.J. Davies, N. Birbilis, J.F. Nie, Texture evolution during static recrystallization of cold-rolled magnesium alloys, Acta Mater. 105 (2016) 479–494. https://doi.org/10.1016/j.actamat.2015.12.045.
5. H. Ding, X. Shi, Y. Wang, G. Cheng, S. Kamado, Texture weakening and ductility variation of Mg-2Zn alloy with CA or RE addition, Mater. Sci. Eng. A. 645 (2015) 196–204. https://doi.org/10.1016/j.msea.2015.08.025.
6. J. Luo, H. Yan, N. Zheng, R.S. Chen, Effects of zinc and calcium concentration on the microstructure and mechanical properties of hot-rolled Mg–Zn–Ca sheets, Acta Metall. Sin. (English Lett. 29 (2016) 205–216. https://doi.org/10.1007/s40195-016-0378-1.
7. R. Roumina, S. Lee, T.D. Berman, K.S. Shanks, J.E. Allison, A. Bucsek, The dynamics of recrystallized grains during static recrystallization in a hot-compressed Mg–3.2Zn–0.1Ca wt.% alloy using in-situ far field high-energy diffraction microscopy, Acta Mater. 234 (2022) 118039. https://doi.org/10.1016/j.actamat.2022.118039.
8. N. Stanford, M.R. Barnett, The origin of "rare earth" texture development in extruded Mg-based alloys and its effect on tensile ductility, Mater. Sci. Eng. A. 496 (2008) 399–408. https://doi.org/10.1016/j.msea.2008.05.045.
9. D. Guan, W.M. Rainforth, J. Gao, J. Sharp, B. Wynne, L. Ma, Individual effect of recrystallisation nucleation sites on texture weakening in a magnesium alloy: Part 1- double twins, Acta Mater. 135 (2017) 14–24. https://doi.org/10.1016/j.actamat.2017.06.015.
10. D. Guan, W.M. Rainforth, J. Gao, L. Ma, B. Wynne, Individual effect of recrystallisation nucleation sites on texture weakening in a magnesium alloy: Part 2- shear bands, Acta Mater. 145 (2018) 399–412. https://doi.org/10.1016/j.actamat.2017.12.019.
11. F.J. Humphreys, M. Hatherly, Recrystallization and Related Annealing Phenomena, 2nd ed., Elsevier, Netherlands, 2004.
12. R. Pei, Y. Zou, M. Zubair, D. Wei, T. Al-Samman, Synergistic effect of Y and Ca addition on the texture modification in AZ31B magnesium alloy, Acta Mater. 233 (2022) 117990. https://doi.org/10.1016/j.actamat.2022.117990
13. J.J. Bhattacharyya, S.R. Agnew, G. Muralidharan, Texture enhancement during grain growth of magnesium alloy AZ31B, Acta Mater. 86 (2015) 80–94. https://doi.org/10.1016/j.actamat.2014.12.009.
14. D. Guan, X. Liu, J. Gao, L. Ma, B.P. Wynne, W.M. Rainforth, Exploring the mechanism of " Rare Earth " texture evolution in a lean Mg–Zn–Ca alloy, Sci. Rep. 9 (2019) 7152. https://doi.org/10.1038/s41598-019-43415-z.
15. C.D. Barrett, A. Imandoust, H. El Kadiri, The effect of rare earth element segregation on grain boundary energy and mobility in magnesium and ensuing texture weakening, Scr. Mater. 146 (2018) 46–50. https://doi.org/10.1016/j.scriptamat.2017.11.004.
16. J. Victoria-Hernández, S. Yi, D. Klaumünzer, D. Letzig, Recrystallization behavior and its relationship with deformation mechanisms of a hot rolled Mg–Zn–Ca–Zr alloy, Mater. Sci. Eng. A. 761 (2019) 138054. https://doi.org/10.1016/j.msea.2019.138054.

First-Principles Calculations and a Theoretical Model for Predicting Stacking Fault Energies in Common Ternary Magnesium Alloys

Qiwen Qiu and Jun Song

Abstract

Magnesium (Mg) and its alloys are the lightest structural metals with high specific strength. Yet they suffer from low ductility, which limits their wide industrial applications. The stacking fault energy (SFE), which governs core structures and movability of dislocations, is an important property for understanding plastic behavior of Mg alloys. Although the SFEs of Mg alloys have been widely studied, general quantitative models to accurately predict SFEs in Mg alloys are still absent. Moreover, the SFE of common ternary alloys is rarely studied. We carry out high-throughput calculations to show the effects of single solutes and solute pairs on SFEs in ternary Mg systems. With the help of machine learning, a theoretical model for predicting SFE has been developed. The work provides some fundamental mechanistic insights for understanding dislocation behavior in Mg alloys, and useful ICME tools in developing rational alloy design recipes towards Mg alloys with enhanced ductility.

Keywords

Magnesium alloys • Stacking fault energy • First-principles calculations • Dislocation slip

Introduction

The stacking fault energy (SFE) is crucial to deformation and failure of metallic materials. In particular, it plays a major role in prescribing key mechanical parameters, such as dislocation energy, ideal shear strength, Peierls stress, and cross-slip stress, among others [1]. It also governs in part the core structures and mobilities of dislocations and thus plastic behaviors of materials [2]. SFEs have been therefore widely studied in face-centered cubic (fcc), body-centered cubic (bcc), and hexagonal close-packed (hcp) metals [3]. Unlike the cases of fcc and bcc metals, hcp metals possess maximum four independent slip systems which are insufficient for plastic deformation according to von Mises requirement. This intrinsic property leads to the low ductility of hcp metals, thereby becoming a bottleneck that limits their applications. Especially for magnesium (Mg), one of the lightest structural materials with high specific stiffness, it also suffers from poor formability at room temperature [4]. A detailed insight into stacking fault energy is needed to understand the fundamental deformation mechanism. Experimental studies of Zhao et al. [5] indicated that low SFE enabled high density of stacking faults leading to the promotion of dislocation accumulation. Investigation of Yuasa et al. [6] into Mg–Ca–Zn alloys showed that the plastic anisotropy is not only related to the value of SFE in single or multiple slip planes but also controlled by the ratio of SFE between different slip planes and the disembrittlement parameter (ratio of the surface energy to SFE). The addition of rare earth elements, like Y and Gd, has been proved to be an effective approach for improving room-temperature ductility of Mg alloys [7]. However, there appeared clear divergence and debate on the underlying improvement mechanism among scholars. One opinion thought that the I_1 stacking fault acted as the nucleation source for pyramidal (Pyr.) $\langle c + a \rangle$ dislocation which is critical for deformation in c axis. The addition of Y reduced I_1 SFE thus enables the movement of dislocations on Pyr. planes [8]. The opponents considered the enhanced ductility is related to the different effects of rare earth elements on Pyr. I and Pyr. II planes. One reason is that the effects of Y on I_1 SFE can also be achieved by Al at similar concentrations [9]. So, we can see that comprehensive calculation of SFEs among different slip plane and alloying elements is necessary to avoid making one-sided conclusion. In fact, the

Q. Qiu (✉) · J. Song
Department of Mining and Materials Engineering, McGill University, Montréal, QC H3A 0C5, Canada
e-mail: qiwen.qiu@mail.mcgill.ca

S. Barela et al. (eds.), *Magnesium Technology 2023*, The Minerals, Metals & Materials Series,
https://doi.org/10.1007/978-3-031-22645-8_6

effects of common alloying elements have already been studied, and some relationships between atomic radius, electronegativity or chemical effects and SFEs have been found [10]. Unfortunately, these relations are either irregular or qualitative, making it ineligible for predicting SFEs of Mg alloys. Besides, to calculate SFE of each atom on each slip plane is time-consuming. Therefore, the establishment of a quantitative model may help to quickly screen for an alloying element with specific SFE. This is also favorable for understanding the deformation mechanism, developing Mg alloys with high-ductility, and saving calculation resources. Besides, most of the research about SFE are on the effect of single solute. The SFE in ternary alloys is rarely studied. So, the effects of solute pairs on SFE were also considered in this paper.

Method

First-principles DFT calculations using Vienna Ab-initio Simulation Package (VASP) have been performed on a series of binary Mg alloys [12–14]. The workflow to calculate SFE using slab shearing method is shown in Fig. 1. Structure settings were used to determine the configuration size, slip direction, and substituted solutes. VASP settings, including INCAR and KPOINTS settings, can be obtained whether by pymatgen [11] package or manual input. Below for simplicity we refer to the binary Mg alloy as Mg-*X* where *X* stands for the alloying (solute) element. The solute elements in Mg considered are categorized into three groups, being (*i*) main group elements (Li, Na, K, Rb, Ca, and Sr), (*ii*) transition elements (all elements in *3d* and *4d*), and (*iii*) RE elements (La, Nd, Sm, Gd, Dy, Er, and Yb). In calculations, the influence of solutes on unstable SFEs (E_{usf}) and stable SFEs (E_{sf}) of basal, prismatic, and Pyramidal II (abbreviated as Pyr. II below) slips was investigated, with the solute atom introduced by substituting a Mg atom on the slip plane. The generalized stacking fault energy curves (GSFEs) were calculated using the slab shearing method [15, 16] where atoms above the slip plane are displaced by a shift along Burgers vector (***b***) relative to the atoms below, followed by relaxation in the direction normal to the slip plane. The E_{usf} and E_{sf} can then be extracted from the GSFE curves obtained. A 96-atom supercell with 12 layers and a vacuum gap of 15 Å between periodically repeated slabs was used in the calculation. In order to study the effect of solute pairs on SFE, different configurations of solutes were shown in Fig. 2. And the solute site distributions were shown in Table 1. Additional benchmark calculations were also performed to confirm that the dimensions of the supercell are sufficient. A cut-off energy of 400 eV and a $7 \times 7 \times 1$ Monkhorst–Pack *k*-point mesh were used in our calculations. Internal coordinates of atoms within the supercells as well as the shape of the supercells were fully relaxed by the first-order Methfessel-Paxton smearing method with the width = 0.1 eV. The energy and force convergence criteria were set as 10^{-5} eV and 10^{-2} eV/Å, respectively.

Results

Binding Energy

To determine whether two solutes could form a solute pair, the binding energy between each two solutes was calculated using the formula below:

$$E_b^{XY} = E[\mathrm{Mg}_{n-2}XY] - E[\mathrm{Mg}_{n-1}X] - E[\mathrm{Mg}_{n-1}Y] + E[\mathrm{Mg}_n] \quad (1)$$

where $E[\mathrm{Mg}_{n-2}XY]$ is the energy of the bulk with solute *X* and *Y* at first near neighbor (1NN) or second near neighbor (2NN), and $E[\mathrm{Mg}_{n-1}X]$ and $E[\mathrm{Mg}_{n-1}Y]$ are the energy of a bulk with single solute *X* or *Y*, respectively. And $E[\mathrm{Mg}_n]$ is

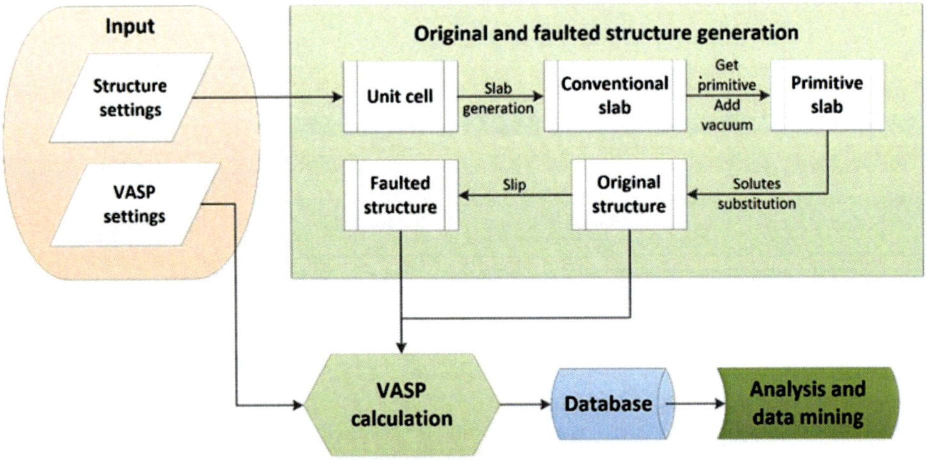

Fig. 1 Workflow to calculate SFE

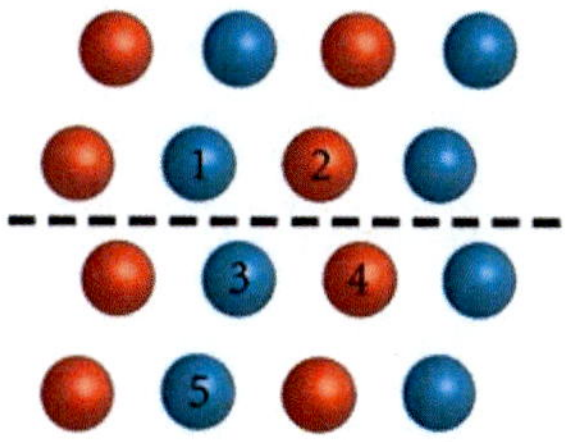

Fig. 2 Configurations of solute pairs

the energy of pure Mg. When the binding energy is minus, the two solutes attract each other and may form a solute pair. If the binding energy is positive, the solutes repel each other. Part of the calculated binding energy is shown in Fig. 3. We could find that if both of the two solutes are larger than Mg, the 1NN binding energy is usually positive, such as Ca with rare earth elements. On the contrary, if one of the solute is larger than Mg, and the other one is smaller than Mg, then the binding energy of 1NN becomes minus.

Effect of Single Solute on SFE

To probe the effects of alloying elements on the I2 SFEs of basal plane, we plot the SFE values in a periodic table of elements (as shown in Fig. 4) and divide it into three zones: Zone I (alkali and alkaline earth metals), Zone II (transition metals), and Zone III (rare earth elements). The calculated SFEs in this work show a good agreement with previous DFT calculation results. It can be noticed that the SFEs show a gradient decrease with the numbers of electron shells in Zone I and a gradient increase with the atomic number in Zone III. We deduce the changes in SFEs derive from the size effect of solutes. Placed substitutional solutes in the lattice introduce elastic strain considering the size misfit. In addition, alkali or alkaline metals possess same outermost electron number, and the atom size should be considered as a major influence factor of this tendency.

To quantitative measure the atomic size of each solute X, impurity volume V_{imp}^{X} induced by replacing a Mg atom with single solute X in pure Mg is calculated:

$$V_{\text{imp}}^{X} = V(\text{Mg}_{N-1}X) - V(\text{Mg}_{N}) \quad (2)$$

where $V(\text{Mg}_{N-1}X)$ and $V(\text{Mg}_{N})$ are the equilibrium volume of $\text{Mg}_{N-1}X$ system and pure Mg, respectively. Figure 5 shows the SFEs in terms of the calculated impurity volume. A significant linear correlation is found between SFEs and V_{imp}^{X}. Corresponding to the interaction energy between solute volume misfit and dislocation: $E_I = p\Delta V$, where p is the pressure at the site of the defect, ΔV is the total volume change, we have:

$$E_{sf}^{Mg_{N-1}-X} = E_{sf}^{Mg_N} + \lambda \cdot V_{\text{imp}}^{X} \quad (3)$$

where $E_{sf}^{Mg_N}$ and $E_{sf}^{Mg_{N-1}-X_1}$ are the SFEs of pure Mg and $\text{Mg}_{N-1}X$ system, and λ is a pressure-related coefficient (which can be considered as a constant in this model). The

Table 1 Distribution of solute pairs corresponding to Fig. 2

Config	Solute A	Solute B		
a	3	4	Same layer	Below slip
b	2	3	Different layer	Cross slip

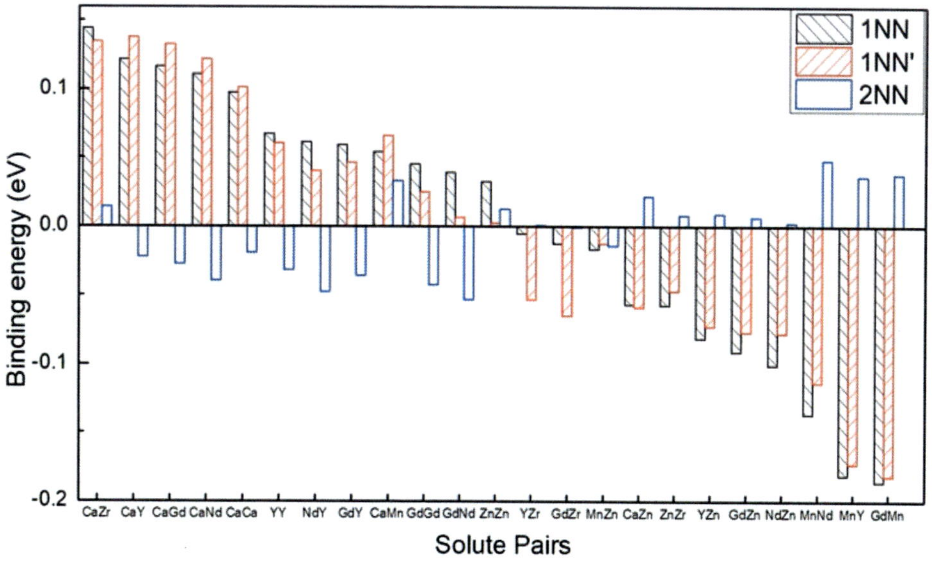

Fig. 3 Binding energy of selected solutes

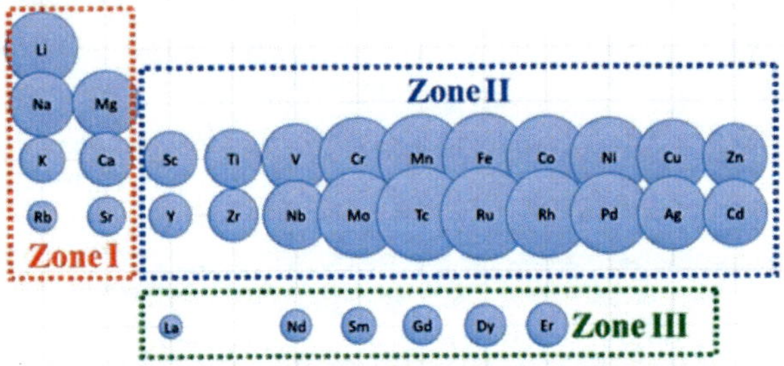

Fig. 4 Schematic showing the SFEs of $Mg_{N-1}X$ system in a sequence of periodic table (the area of the circle represents the SFE value)

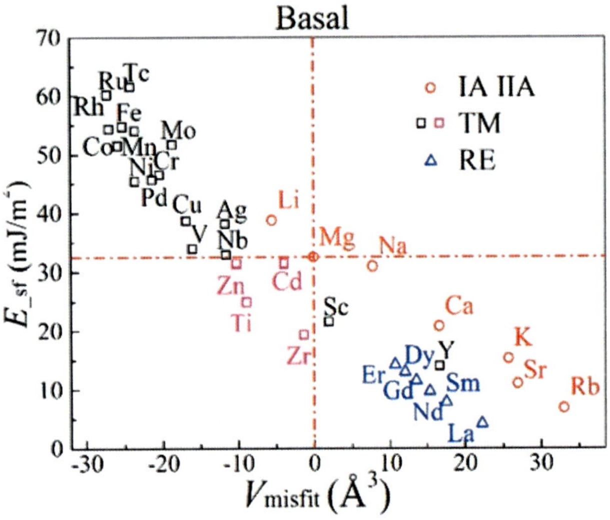

Fig. 5 Calculated SFEs of Mg doped with solutes versus impurity volume

SFEs decrease with the size of the solute. This correlation can be explained by a strain model. In the $Mg_{N-1}X$ system, a solute atom with larger volume can induce a larger lattice distortion and a significant strain, promoting the formation of SFs. Compared with direct calculations, this approximation of SFEs requires much fewer computing resources.

Effect of Solute Pairs on SFE

The effect of solute pairs on SFE under configuration a of Mg–Al and Mg–Zn alloys is shown in Fig. 6. In configuration a, the two solutes are at the same layer near the slip plane. The effect of the two solutes should be the same. Therefore, we first try to predict the effect of the solute pairs by adding the effect of each single solute using the formula below. And the predicted results were shown in the black point in Fig. 6.

$$\begin{aligned} E^{X-Z} &= E_{sf}^{\mathrm{Mg}} + (E_{sf}^{\mathrm{Mg}-X} - E_{sf}^{\mathrm{Mg}}) + (E_{sf}^{\mathrm{Mg}-Z} - E_{sf}^{\mathrm{Mg}}) \\ &= E_{sf}^{X} + E_{sf}^{Z} - E_{sf}^{\mathrm{Mg}} \end{aligned} \quad (4)$$

where E^{X-Z} is the SFE of $Mg_{N-2}XZ$ system.

The effect of solute pairs on SFE under configuration b of Mg–Al and Mg–Zn alloys is shown in Fig. 7. With machine learning, we obtain the formular below to predict the effect under this condition:

$$E_{sf}^{\mathrm{b}} = E_{sf}^{\mathrm{a}} + \mu * (E_{b-1NN} - E_{b-2NN}) \quad (5)$$

where E_{sf}^{b} and E_{sf}^{a} are the SFEs with solute pairs under configuration b and a, respectively. $E_{b-1\mathrm{NN}}$ and $E_{b-2\mathrm{NN}}$ are the binding energy of the solute pairs at 1NN and 2NN, respectively. We here used the basic regression. SFEs under different configuration, binding energy of solute pairs, and intrinsic properties of solutes, such as atom size and electronegativity were imported to the dataset. After normalization and regression, we found this obvious linear relationship between the SFEs under different configuration and binding energies.

Conclusion

The new model confirms the necessity of considering effects from both mechanic and electronic, and provides means to accurately quantify them. We also consider the solute effects on SFE under different configuration. The present study

Fig. 6 Calculated and predicted effect of solute pairs on SFE of Mg–Al and Mg–Zn alloys under configuration a

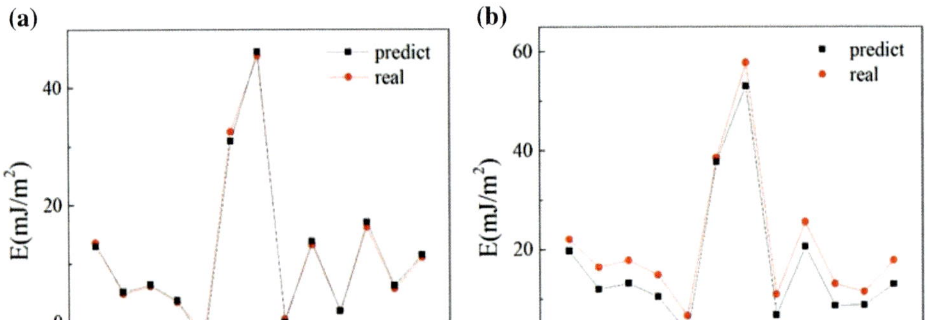

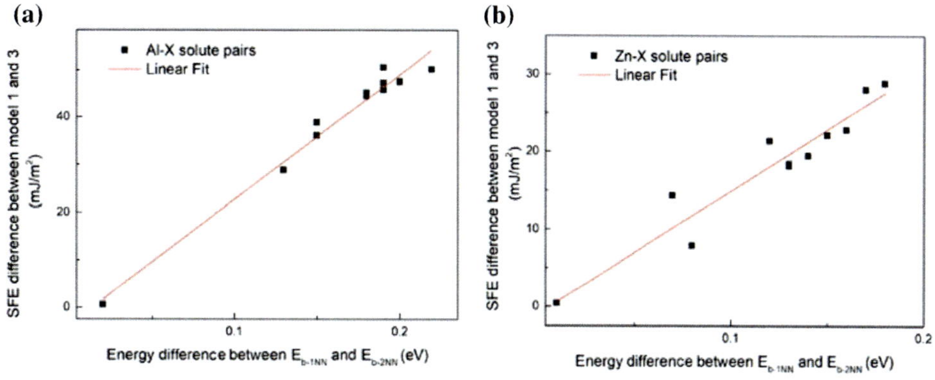

Fig. 7 Predicted effect of solute pairs on SFE of Mg–Al and Mg–Zn alloys under configuration b

provides the much-needed mechanistic insights and predictivity for understanding the role of solutes in modulating SFEs and subsequently plastic deformation behaviors of Mg alloys. It is expected to contribute critical knowledge towards rational and predictive Mg-based light alloy design for enhanced mechanical properties.

Acknowledgements We gratefully acknowledge the financial support from the Natural Sciences and Engineering Research Council of Canada (NSERC) Discovery grant (RGPIN-2017-05187) and China Scholarship Council (CSC). We also acknowledge Compute Canada for providing computing resources.

References

1. Z. Wu, W. Curtin, The origins of high hardening and low ductility in magnesium, Nature 526 (2015) 62–67.
2. T.M. Pollock, Weight loss with magnesium alloys, Science 328 (2010) 986–987.
3. S.R. Agnew, J.F. Nie, Preface to the viewpoint set on: the current state of magnesium alloy science and technology, Scr. Mater. 63 (2010) 671–673.
4. C.M. Byer, K.T. Ramesh, Effects of the initial dislocation density on size effects in single-crystal magnesium, Acta Mater. 61 (2013) 3808–3818.
5. Z. Wu, R. Ahmad, B. Yin, S. Sandlöbes, W.A. Curtin, Mechanistic origin and prediction of enhanced ductility in magnesium alloys, Science 359 (2018) 447–452.
6. J.P. Hirth, J. Lothe, Theory of dislocations, 2 ed., John Wiley & Sons (1982).
7. W.J. Kim, S.I. Hong, Y.H. Ki, Enhancement of the strain hardening ability in ultrafine grained Mg alloys with high strength, Scr. Mater. 67 (2012) 689–692.
8. Y. Chino, K. Sassa, M. Mabuchi, Enhancement of tensile ductility of magnesium alloy produced by torsion extrusion, Scr. Mater. 59 (2008) 399–402.
9. J. He, B. Jiang, Q. Yang, Y. Zeng, X. Xia, F. Pan, Improved ductility of magnesium alloy sheets by pre-hardening and annealing, Mater. Sci. Technol. 31 (2015) 1383–1387.
10. V. Vitek, Intrinsic stacking faults in body-centred cubic crystals, Philos. Mag. 18(154) (1968) 773–786.
11. J.R. Rice, Dislocation nucleation from a crack tip An analysis based on the Peierls concept, J. Mech. Phys. Solids 40(2) (1992) 239–70.
12. V. Vitek, Structure of dislocation cores in metallic materials and its impact on their plastic behaviour, Prog. in Mater. Sci. 36 (1992) 1–27.
13. B. Yin, Z. Wu, W.A. Curtin, First-principles calculations of stacking fault energies in Mg-Y, Mg-Al and Mg-Zn alloys and implications for <c+a> activity, Acta Mater. 136 (2017) 249–261.
14. T. Yonezawa, K. Suzuki, S. Ooki, A. Hashimoto, The effect of chemical composition and heat treatment conditions on stacking fault energy for Fe-Cr-Ni austenitic stainless steel, Metall. Mater. Trans. A Phys. Metall. Mater. Sci. 44A (2013) 5884–5896.
15. B. Yin, Z. Wu, W.A. Curtin, Comprehensive first-principles study of stable stacking faults in hcp metals, Acta Mater. 123 (2017) 223–234.
16. M. Muzy, Z. Pakieła, K.J. Kurzydłowski, Generalized stacking fault energies of aluminum alloys–density functional theory calculations, Metals. 8(10) (2018) 823–832.

Microstructure Characteristics of Nucleation and Growth for the $\{10\bar{1}1\}$ Twin in Mg Polycrystal via an Atomistic Simulation

Huicong Chen and Jun Song

Abstract

Using molecular dynamics (MD) simulations, microstructure evolution during nucleation and growth of the $\{10\bar{1}1\}$ twin in polycrystal Mg is studied. The results show that profuse basal stacking faults (BSFs) form inside the $\{10\bar{1}1\}$ twin embryo and disconnections with different step size are also characterized along the twinning boundary (TB) interface. The disconnections can be divided into two types according to their mobility, i.e., mobile steps with height of $2h_0$ and $4h_0$ and immobile steps with h_0 and $3h_0$. We find that the front line of mobile steps is divided into small segments due to the BSFs, and each segment will move as a unit at different velocity. When the mobile interfacial steps are consumed at grain boundaries, new steps are nucleated and glide on the TB interface. Such a process leads to twinning growth in the lateral and vertical directions. These insights contribute to the fundamental understanding of twinning mechanism in Mg.

Keywords

Magnesium • Molecular dynamics • $\{10\bar{1}1\}$ Twin • Disconnections

Introduction

Magnesium (Mg) and its alloys have been of great interest to researchers and industries because of their potential in light-weighting structural components [1]. However, the engineering applications of Mg alloys have been greatly limited by their poor ductility and formability at the room temperature. Such limitation is fundamentally attributed to their hexagonal close-packed (HCP) crystal structure which renders insufficient operative slip systems available to acquire homogenous deformation [2–6]. As such, twinning has been considered as a potential remedy to aid Mg alloys in enhancing their mechanical properties at finite temperature and numerous studies have been performed along this line to provide deep insight of twinning mechanism in Mg.

So far, the computational studies are limited to Mg single crystal which shows different microstructural and mechanical response comparing with simulations in Mg polycrystal where grain size, orientation, and shape play a crucial role. A very recent MD simulation on polycrystalline Mg [7] focuses on the identification of various of TBs using a new extended common-neighbor analysis (E-CNA) method, and the authors proposed that steps are formed at planar faults and can help the lateral twining growth. However, they did not provide a three-dimensional (3D) microstructure evolution which is more reliable to study TB interfaces including disconnections and steps. Similar simulations [8] were also performed on nano-twinned polycrystalline Mg, and the authors investigated effect of twinning lamella size and temperature on the migration of TBs, but they ignore the importance of twinning mechanism in a twin-free polycrystal. Because the 3D nature of twin domains [9], the morphology of local defects may differ viewed along different directions and it is thus crucial to reveal the characteristics of twins in 3D space. At the same time, it is essential to investigate twinning behavior starting from a twin-free Mg polycrystal.

To date, a great deal of effort has been devoted to understanding the twinning mechanism via focusing on $\{10\bar{1}2\}$ twin and work on $\{10\bar{1}1\}$ twinning is inadequate. In this study, MD simulation was performed on a polycrystalline Mg and the characteristic of microstructure evolution of $\{10\bar{1}1\}$ twin was carefully examined via a

H. Chen · J. Song (✉)
Department of Mining and Materials Engineering, McGill University, 3610 Rue University, Montréal, QC H3A 0C5, Canada
e-mail: jun.song2@mcgill.ca

H. Chen
e-mail: huicong.chen@mail.mcgill.ca

S. Barela et al. (eds.), *Magnesium Technology 2023*, The Minerals, Metals & Materials Series,
https://doi.org/10.1007/978-3-031-22645-8_7

three-dimensional visualization. Then, disconnection assisted twinning mechanism was proposed.

Methodology

The initial polycrystal structure with a total of 15 randomly oriented grains was constructed using the Voronoi tessellation method [10]. Then, MD simulation was performed using Large-scale Atomic/Molecular Massively Parallel Simulator (LAMMPS) [11]. The interatomic interactions are prescribed by the embedded-atom method (EAM) [12] potential developed by Wilson and Mendelev [13]. The model was first relaxed using conjugate gradient minimization method [14]. Afterwards the model was equilibrated at 300 K for 100 ps using the isothermal–isobaric (NPT) ensemble [15] to ensure zero pressures along all three periodic directions. Following the equilibration, the sample was deformed under uniaxial tensile stress loading along X axis to a total strain of 20% at a constant strain rate of $1 \times 10^9\ s^{-1}$. The integration time-step is fixed at 2 fs. Ovito [16] is used to visualize the results. The common-neighbor analysis (CNA) method [17] is used to identify the local defects and orientation mapping method [18] is used to color each atom.

Results and Discussion

Twinning Modes

As shown in Fig. 1, profuse twins are observed to be activated with straining. Figure 1(a) shows primarily activated twin T1 in grain G5 and twin T2 in grain G1 are corresponding to contraction twinning mode $\{10\bar{1}1\}$. Another two twin lamellas T3 and T4 in G5 (Fig. 1b) and T8 in grain G2 (Fig. 1f) are also identified as $\{10\bar{1}1\}$ twin. Besides, twin T6 and twin T7, identified as extension twinning mode $\{10\bar{1}2\}$ and $\{11\bar{2}1\}$, are characterized in G5 and G6 at a strain of 6.2% (Fig. 1c). T6 is activated due to high local stress produced by the twin-twin interaction [19] of T1, T3, and T4 in G5.

Evolution of Twin Structures

To analyse the twinning process of $\{10\bar{1}1\}$ twinning, grain G9 is selected from the polycrystal, and as shown in Fig. 2, its microstructural evolution is characterized by using a series of snapshots at different strains. The $\{10\bar{1}1\}$ twin nucleates as an initial embryo with the formation of BSFs as shown in Fig. 2a. However, the $\{10\bar{1}1\}$ twin grows into a plate shape (see Fig. 2b) which is surrounded by a pair of planar faults and the tip of the twin propagates until it reaches the grain boundary at the other end of the grain as shown in Fig. 2c. Then, further straining widens the width of the twin fault, leading to growth of twin in the normal direction (see Fig. 2d–e). Besides, another $\{10\bar{1}1\}$ twin is already formed at a strain of 10% and interacts with the first twin variant as shown in Fig. 2f, which provides a barrier for the migration of the twin fault and restricts the growth of the initial twin variant as shown in Fig. 2g–h. Obviously, as shown in Fig. 2d, the twin fault is disconnected by profuse steps at twin-matrix interfaces among which the mobile disconnections may migrate with the growth of twin fault. Figure 2b illustrates the BSFs in twinned region and immobile steps/disconnections at twin-matrix interfaces where the BSFs interact with twin faults.

Twinning Interface

To investigate the migration of steps and disconnections at $\{10\bar{1}1\}$ TB interface, a series of snapshots of grain G9 deformed under the uniaxial tensile stress loading are captured at different timesteps as shown in Fig. 3, where the matrix atoms have been deleted and atoms in GB, TB interface and twinned region are kept. There are two types of steps, i.e., step A and step B (see Fig. 3a). Steps A, consisting of two or four layers of $\{10\bar{1}1\}$ atoms, exist at the front of the migration line and move continuously along the interface (see Fig. 3b) until they merge into GB when reaching the other side of GB. Steps B are left after steps A interact with BSFs. Steps B are formed due to the interaction between BSFs and twinning interface as shown in Fig. 3a. Unlike steps A, steps B propagate conservatively and shift vertically as the twin lamella is thickening as shown in Fig. 3d. Interestingly, as shown in Fig. 3c, the front the migration line which consists of steps A is also disconnected due to the formation of BSFs in twinned region and divided into several segments. Each segment is observed to migrate at different velocity, and the center segment moves at the highest velocity, which could be the reason why the twin lamella is usually characterized as a lenticular shape. Besides, the newly nucleated steps A at GB often consist of two layers of $\{10\bar{1}1\}$ atoms (see Fig. 3a) because this type of step has the lowest formation energy among four possible ones [20, 21]. These steps can also catch up with the previous ones, then merging into new steps with four layers of atoms. Therefore, basal/prismatic interfaces may form at the disconnection sites as shown in Fig. 3c, where the interfaces are parallel to either both the basal plane of the matrix and prismatic plane of the twin (i.e., $\mathrm{basal_M/primatic_T}$ interface) or to both the basal plane from the twin and the prismatic

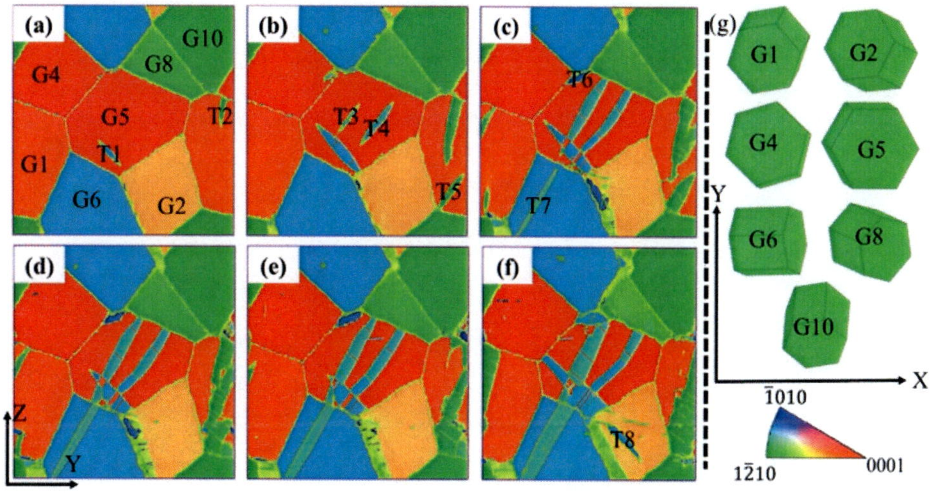

Fig. 1 Representative overview of twinning evolution at a strain of **a** 4.6%, **b** 5.2%, **c** 6.2%, **d** 6.8%, **e** 7.6%, **f** 12%; **g** geometrical orientation of each grain in the sample coordinate, where Gi ($i = 1 \sim 15$) represent each grain in the current polycrystal

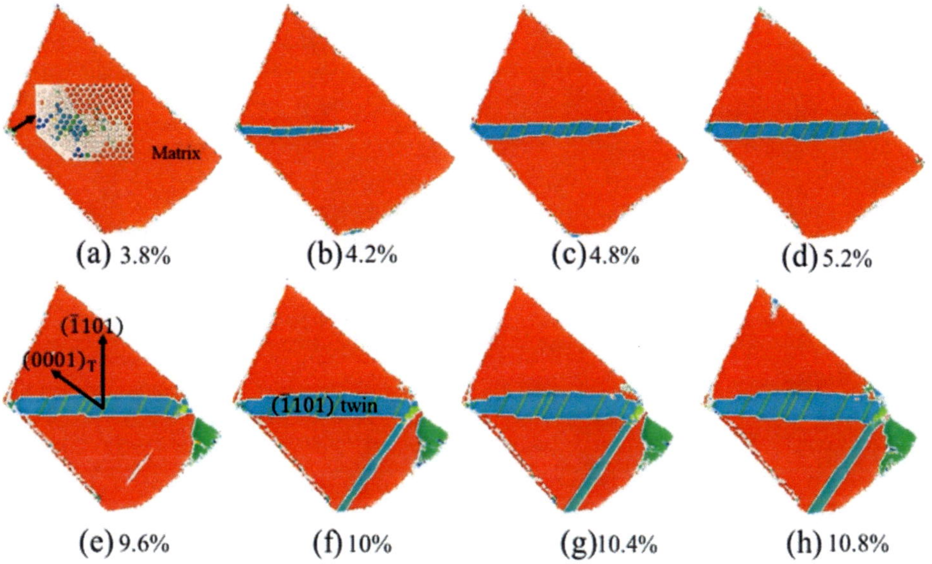

Fig. 2 Microstructural evolution of the $\{10\bar{1}1\}$ contraction twin in a selected grain deformed at different strain levels

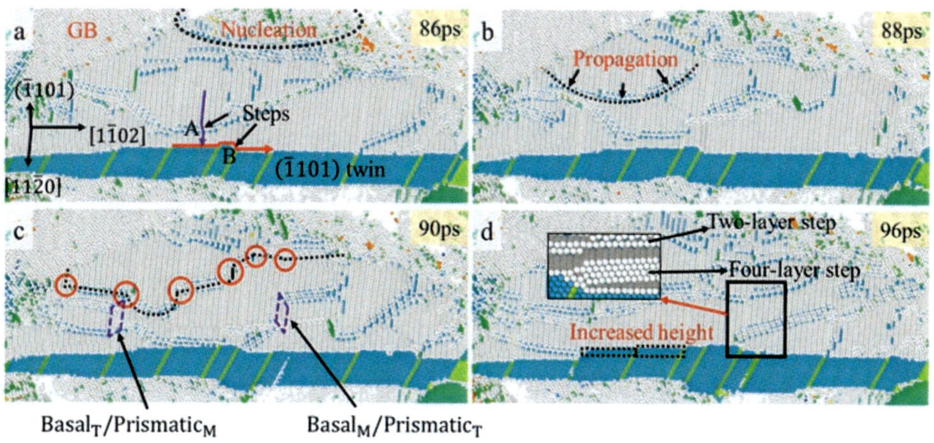

Fig. 3 Illustration of the growth of the $\{10\bar{1}1\}$ twin by movement of steps along the twin boundary interface deformed at different timesteps

plane from the matrix (i.e., $basal_T/primatic_M$ interface). The transformation and subsequent twinning growth are accompanied by nucleation and migration of disconnections (steps) and growth of stacking faults. The BSFs only grow along stacking fault planar direction and do not migrate along twinning direction, i.e., the distance between different BSFs keeps constant.

There are four kinds of disconnections ($\boldsymbol{b}_i$, h_i) (i = 1, 2, 3, 4), where $h_i = ih_0$ and $h_0 = d_{(10\bar{1}1)}$. Two steps are observed in the translation from matrix sites to twin sites and magnitudes of their height are h_2 ($2h_0$) and h_4 ($4h_0$). This observation agrees with topological analysis in a bicystal model [21, 22], which reveals that the two steps have minimal shuffles. Besides steps with h_2 and h_4, another two steps which have height of h_1 and h_3 are also observed. They exist near the interaction site between BSFs and TB interface, which are not the same case with h_2- and h_4-steps. For example, h_2-step exists in between two BSFs and it is mobile along twinning direction. For h_1- and h_3-steps, they cannot move along twinning direction (immobile), and they only migrate along $[11\bar{2}0]$ direction, leading to propagation of BSFs. Considering h_1- and h_3-steps having much higher formation energy and lower mobility along twinning direction than steps with height of h_2 and h_4, their formation is related to dislocation interaction between BSFs and h_2-step.

Conclusions

In this work, MD simulation is performed to investigate mechanisms of growth of $\{10\bar{1}1\}$ twins in polycrystalline Mg at the atomic scale. It is found that the nucleation and growth of $\{10\bar{1}1\}$ twins are accompanied by the formation and propagation of basal stacking faults. Due to the formation of BSFs inside twinned region, the front line of mobile steps is divided different segments and each segment will move as a unit at different velocity. When the mobile interfacial steps are consumed during twin growth, new steps are nucleated and glide on the TB interface, leading to twin growth in the lateral and vertical directions. Furthermore, two types of disconnections are observed in $\{10\bar{1}1\}$ twin. The first type of seps with height of $2h_0$ and $4h_0$ are mobile while the other type of steps with h_0 and $3h_0$ are immobile. The formation of h_0- and $3h_0$-steps may be due to the interaction between BSF and $2h_0$-step.

Acknowledgements The authors acknowledge financial support from Natural Sciences and Engineering Research Council of Canada (NSERC), McGill Engineering Doctoral Award (MEDA), and China Scholarship Council (201706050149), and thank Compute Canada for providing computing resources.

References

1. T.M. Pollock, Weight loss with magnesium alloys, Science 328 (2010) 986–987.
2. S.K. Sahoo, S. Biswas, L.S. Toth, P.C. Gautam, B. Beausir, Strain hardening, twinning and texture evolution in magnesium alloy using the all twin variant polycrystal modelling approach, International Journal of Plasticity 128 (2020).
3. Q. Peng, Y. Sun, J. Wang, Q. Zu, M. Yang, H. Fu, Structural characteristics of {101¯1} contraction twin-twin interaction in magnesium, Acta Materialia 192 (2020) 60–66.
4. S. Kweon, D.S. Raja, Investigation of the mechanical response of single crystal magnesium considering slip and twin, International Journal of Plasticity 112 (2019) 1–17.
5. Z. Huang, J.-F. Nie, Interaction between hydrogen and solute atoms in {101¯2} twin boundary and its impact on boundary cohesion in magnesium, Acta Materialia 214 (2021).
6. H. Fallahi, C. Davies, Evolution of twinning during cyclic loading of magnesium alloy examined by quasi-in-situ EBSD, Materials Science and Engineering: A 820 (2021).
7. G. Agarwal, A.M. Dongare, Deformation Twinning in Polycrystalline Mg Microstructures at High Strain Rates at the Atomic Scales, Scientific Reports 9(1) (2019) 3550.
8. C. Xu, L. Yuan, D. Shan, B. Guo, The influence of lamellar twins on deformation mechanism in nanocrystalline magnesium under uniaxial compression, Journal of Materials Science 54(19) (2019) 12623–12642.
9. Y. Liu, P.Z. Tang, M.Y. Gong, R.J. McCabe, J. Wang, C.N. Tomé, Three-dimensional character of the deformation twin in magnesium, Nature communications 10(1) (2019) 3308.
10. P.M. Derlet, H. Van Swygenhoven, Atomic positional disorder in fcc metal nanocrystalline grain boundaries, Physical Review B 67 (1) (2003).
11. S. Plimpton, Fast Parallel Algorithms for Short-Range Molecular Dynamics, Journal of Computational Physics 117 (1995) 1–19.
12. M.S. Daw, S.M. Foiles, M.I. Baskes, The embedded-atom method: a review of theory and applications, Materials Science Reports 9(7) (1993) 251–310.
13. S.R. Wilson, M.I. Mendelev, A unified relation for the solid-liquid interface free energy of pure FCC, BCC, and HCP metals, The Journal of Chemical Physics 144(14) (2016) 144707.
14. A.V. Ivanov, V.M. Uzdin, H. Jónsson, Fast and robust algorithm for energy minimization of spin systems applied in an analysis of high temperature spin configurations in terms of skyrmion density, Computer Physics Communications 260 (2021).
15. M.E. Tuckerman, J. Alejandre, R. López-Rendón, A.L. Jochim, G. J. Martyna, A Liouville-operator derived measure-preserving integrator for molecular dynamics simulations in the isothermal–isobaric ensemble, Journal of Physics A: Mathematical and General 39(19) (2006) 5629–5651.
16. A. Stukowski, Visualization and analysis of atomistic simulation data with OVITO–the Open Visualization Tool, Modelling and Simulation in Materials Science and Engineering 18(1) (2009) 015012.
17. H. Tsuzuki, P.S. Branicio, J.P. Rino, Structural characterization of deformed crystals by analysis of common atomic neighborhood, Computer Physics Communications 177(6) (2007) 518–523.
18. J.C. E, M.X. Tang, D. Fan, L. Wang, S.N. Luo, Deformation of metals under dynamic loading: Characterization via atomic-scale orientation mapping, Computational Materials Science 153 (2018) 338–347.
19. Q. Yu, J. Wang, Y. Jiang, R.J. McCabe, N. Li, C.N. Tomé, Twin–twin interactions in magnesium, Acta Materialia 77 (2014) 28–42.

20. O. MacKain, M. Cottura, D. Rodney, E. Clouet, Atomic-scale modeling of twinning disconnections in zirconium, Physical Review B 95(13) (2017) 134102.
21. J. Wang, I.J. Beyerlein, J.P. Hirth, Nucleation of elementary {10-11} and {-1013} twinning dislocations at a twin boundary in hexagonal close-packed crystals, Modelling and Simulation in Materials Science and Engineering 20(2) (2012) 024001.
22. A. Serra, R.C. Pond, D.J. Bacon, Computer simulation of the structure and mobility of twinning disclocations in H.C.P. Metals, Acta Metallurgica et Materialia 39(7) (1991) 1469–1480.

Precipitation Behavior in Low-alloyed Mg–Ca–Zn Alloys

Z. H. Li, T. T. Sasaki, D. Cheng, K. Wang, B. C. Zhou, A. Uedono, T. Ohkubo, and K. Hono

Abstract

Low-alloyed Mg–Ca–Zn alloys are promising as a base system of heat-treatable wrought alloys because of their rapid age-hardenability. The trace addition of Zn plays a critical role in accelerating the age-hardening response and following precipitation behavior of Mg–Ca binary alloys. This study reports the role of Zn on the rapid age-hardening and precipitation sequence in a Mg–0.3Ca–0.6Zn (at.%) alloy during isothermal aging at 200 °C using positron annihilation lifetime spectroscopy (PALS), scanning transmission electron microscopy (STEM), and atom probe tomography (APT). PALS analysis indicates the absence of excess quenched-in vacancies in the as-quenched condition. Instead, the smaller trapping sites, i.e., open spaces, can facilitate the formation of Ca–Zn co-clusters even in the as-quenched condition. APT analysis reveals that the number density of Ca–Zn co-clusters in the Mg–0.3Ca–0.6Zn alloy increases in the early stage of aging, while that of Ca clusters tends to decrease in the Mg–0.3Ca alloy. These results indicate that the rapid age-hardening is attributed to the formation of a large number of Ca–Zn co-clusters. Microstructure analysis using aberration-corrected STEM provides further insights into the precipitation process of the Mg–0.3Ca–0.6Zn alloy. The atomic structures and stability of precipitates are identified by first-principles calculations. A precise precipitation sequence is established as: S.S.S.S $\rightarrow$ G.P. zones $\rightarrow \eta'' \rightarrow \eta' \rightarrow \eta'$ pairs and stacks/$\eta_1 \rightarrow \eta$.

Keywords

Magnesium alloys • Precipitation • Age hardening • HAADF-STEM • Atom probe tomography

Z. H. Li (✉) · T. T. Sasaki (✉) · T. Ohkubo · K. Hono
National Institute for Materials Science, Tsukuba, 305-0047, Japan
e-mail: LI.Zehao@nims.go.jp

T. T. Sasaki
e-mail: SASAKI.Taisuke@nims.go.jp

T. T. Sasaki
Center for Elements Strategy Initiative for Structural Materials (ESISM), Kyoto University, Yoshida Honmachi, Sakyo, Kyoto, 606-8501, Japan

D. Cheng · K. Wang · B. C. Zhou
Department of Materials Science and Engineering, University of Virginia, Charlottesville, VA 22904, USA

A. Uedono
Division of Applied Physics, Faculty of Pure and Applied Science, University of Tsukuba, Tsukuba, 305-8573, Japan

Extended Abstract

The application of lightweight magnesium (Mg) alloys has attracted broad interest as an approach for weight-reduction toward the improvement of energy efficiency of vehicles [1, 2]. However, the practical use of commercial Mg alloys is mostly limited to cast products, i.e., there are very few applications of the wrought alloys such as Mg–3Al–1Zn (wt.%, AZ31) and Mg-6Al-1Zn (wt.%, AZ61) alloys owing to their poor room temperature (RT) formability or low strength [3, 4]. Among recently developed wrought Mg alloys, low-alloyed age-hardenable Mg–Ca–Zn(–Zr) alloy would be an important system for developing wrought Mg alloys because they provide the base for the first development of bake-hardenable Mg sheet alloys with comparable strength and RT formability with to those of 6XXX series aluminum alloys [5–7].

In the Mg–Ca–Zn alloys, the microalloyed Zn plays a critical role in accelerating the age-hardening response and the following precipitation behavior. An in-depth understanding of the mechanism for the rapid age-hardening [8, 9] and following precipitate microstructure evolution is expected to propose strategies for further improvement of mechanical properties [10–12]. This work investigates the origin of rapid age-hardening and precipitation process in a

S. Barela et al. (eds.), *Magnesium Technology 2023*, The Minerals, Metals & Materials Series,
https://doi.org/10.1007/978-3-031-22645-8_8

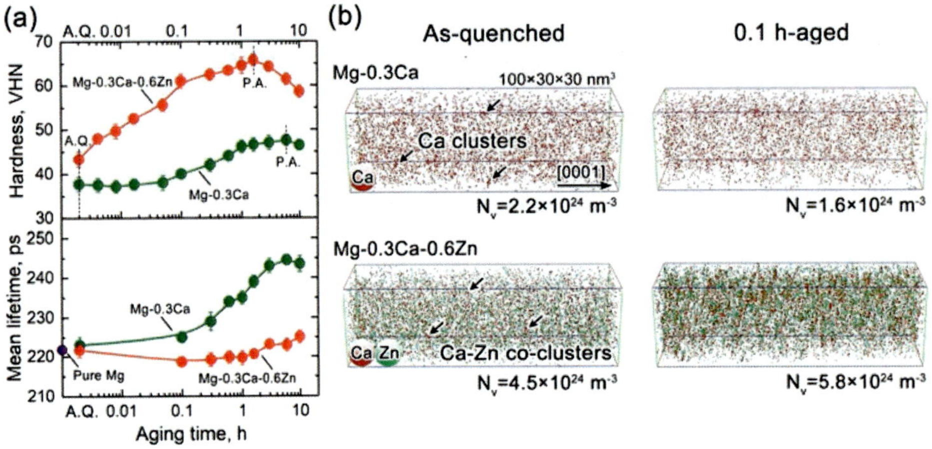

Fig. 1 **a** Variations in Vickers hardness and average positron lifetime as functions of aging time for Mg–0.3Ca and Mg–0.3Ca-0.6Zn alloys during isothermal aged at 200 °C. **b** 3D atom maps of Ca (red) and Zn (green) obtained from as-quenched and 0.1 h-aged Mg–0.3Ca and Mg–0.3Ca–0.6Zn alloys, respectively. Note that the data was analyzed along the $[0001]_{\alpha}$. (reproduce from Ref. [13] with permission from Elsevier)

Mg–0.3Ca–0.6Zn (at.%) alloy, using a combination of positron annihilation lifetime spectroscopy (PALS), scanning transmission electron microscopy (STEM), and 3D atom probe (3DAP) analyses.

Figure 1a shows variations in the Vickers hardness and mean positron lifetime during artificial aging at 200 °C. The Mg–0.3Ca alloy shows a Vickers hardness of 37.8 ± 1.6 HV in the as-quenched condition, and the hardness slightly increases to a peak hardness of 47.6 ± 1.3 HV at 6 h. The addition of trace Zn accelerates the age-hardening response and substantially increases the peak hardness. The hardness of the Mg–0.3Ca–0.6Zn alloy rapidly increases from 43.3 ± 1.7 HV to the peak hardness of 65.8 ± 1.3 HV in only 1.6 h. The mean positron lifetime is 223 and 221 ps for as-quenched Mg–0.3Ca and Mg–0.3Ca–0.6Zn alloys, respectively, which are much shorter than that for the monovacancy in pure magnesium, ~255 ps. That indicates the absence of excess quenched-in open spaces in the as-quenched condition. Subsequently, the mean positron lifetime of the Mg–0.3Ca alloy gradually increases with the aging time and reaches a maximum value of 242 ps at 6 h. In contrast, the Mg–0.3Ca–0.6Zn alloy shows only a slight decrease in mean positron lifetime until the peak hardness, 1.6 h, followed by a slight increase after peak aging. Figure 1b shows 3D atom maps obtained from the as-quenched and 0.1 h-aged Mg–0.3Ca and Mg–0.3Ca–0.6Zn alloys. Some Ca atoms form Ca clusters in the as-quenched Mg–0.3Ca sample. The number density of Ca clusters decreases from $\sim 2.2 \times 10^{24}$ m^{-3} to $\sim 1.57 \times 10^{24}$ m^{-3} after 0.1 h-aging. Unlike the Mg–0.3Ca alloy, the Mg–0.3Ca–0.6Zn alloy shows an enhanced clustering tendency upon aging. The number density of Ca–Zn co-clusters increases from $\sim 4.49 \times 10^{24}$ m^{-3} in the as-quenched condition to $\sim 5.83 \times 10^{24}$ m^{-3} at 0.1 h. Therefore, we can conclude that the origin of the rapid age-hardening in the Mg–0.3Ca–0.6Zn alloy is due to the increased number density of Ca–Zn co-clusters in the early stage of aging. Furthermore, we thoroughly investigated the precipitation process and the structures of precipitates in the Mg–0.3Ca–0.6Zn (at%) alloy during isothermal aging at 200 °C using aberration-corrected STEM and 3DAP. Based on the experimental results, first-principles calculations were performed to analyze the atomic structures and stabilities of precipitates and provide the thermodynamic rationale for the precipitation sequence. A precise precipitation sequence is established as: S.S.S.S → G.P. zones → η'' → η' → η' pairs and stacks/η_1 → η.

References

1. L.A. Dobrzanski, G.E. Totten, M. Bamberger, Magnesium and Its Alloys: Technology and Applications CRC Press, Taylor & Francis Group, Florida, 2020.
2. S. You, Y. Huang, K.U. Kainer, N. Hort, Recent research and developments on wrought magnesium alloys, J. Mag. Alloys 5 (2017) 239–253.
3. S.B. Yi, J. Bohlen, F. Heinemann, D. Letzig, Mechanical anisotropy and deep drawing behaviour of AZ31 and ZE10 magnesium alloy sheets, Acta Mater. 58 (2010) 592–605.
4. X. Huang, K. Suzuki, Y. Chino, M. Mabuchi, Texture and stretch formability of AZ61 and AM60 magnesium alloy sheets processed by high-temperature rolling, J. Alloy Compd. 632 (2015) 94–102.
5. M.Z. Bian, T.T. Sasaki, T. Nakata, Y. Yoshida, N. Kawabe, S. Kamado, K. Hono, Bake-hardenable Mg–Al–Zn–Mn–Ca sheet alloy processed by twin-roll casting, Acta Mater. 158 (2018) 278–288.
6. D. Klaumünzer, J. Victoria-Hernandez, S. Yi, D. Letzig, S.H. Kim, J.J. Kim, M.H. Seo, K. Ahn, Magnesium Process and Alloy Development for Applications in the Automotive Industry Magnesium Technology 2019, TMS, 2019, pp. 15–20.

7. Z.H. Li, T.T. Sasaki, T. Shiroyama, A. Miura, K. Uchida, K. Hono, Simultaneous achievement of high thermal conductivity, high strength and formability in Mg-Zn-Ca-Zr sheet alloy, Mater. Res. Lett. 8 (2020) 335–340.
8. Y. Ortega, J. del Río, Study of Mg–Ca alloys by positron annihilation technique, Scr. Mater. 52 (2005) 181–186.
9. Y. Ortega, M.A. Monge, R. Pareja, The precipitation process in Mg–Ca–(Zn) alloys investigated by positron annihilation spectroscopy, J. Alloys Comp. 463 (2008) 62–66.
10. J.F. Nie, Precipitation and hardening in magnesium alloys, Metall. Mater. Trans. A 43A (2012) 3891–3939.
11. K. Oh-ishi, R. Watanabe, C.L. Mendis, K. Hono, Age-hardening response of Mg–0.3 at.% Ca alloys with different Zn contents, Mater. Sci. Eng. A 526 (2009) 177–184.
12. R.E. Schäublin, M. Becker, M. Cihova, S.S.A. Gerstl, D. Deiana, C. Hébert, S. Pogatscher, P.J. Uggowitzer, J.F. Löffler, Precipitation in lean Mg–Zn–Ca alloys, Acta Mater. 239 (2022) 118223.
13. Z.H. Li, T.T. Sasaki, T. Shiroyama, A. Miura, K. Uchida, K. Hono, Role of Zn on the rapid age-hardening in Mg-Ca-Zn alloys, Scr. Mater. 216 (2022) 114735.

Multiscale, Multimodal Characterization of Recrystallized and Non-recrystallized Grains During Recrystallization in a Hot-Compressed Mg–3.2Zn–0.1Ca wt.% Alloy

Sangwon Lee, Tracy Berman, Can Yildirim, Carsten Detlefs, John Allison, and Ashley Bucsek

Abstract

High-strength, lightweight magnesium (Mg) alloys have substantial potential for reducing the weight of automobiles and other transportation systems and thus for improving fuel economy and reducing emissions. However, the strong crystallographic texture of rolled Mg sheet leads to poor formability and anisotropy. In specific non-rare earth Mg alloys, annealing can be used to desirably weaken the texture. Here, we present a multiscale in-situ study on the recovery and recrystallization of an 80% hot-compressed Mg–3.2Zn–0.1Ca wt% (ZX30) alloy using high-resolution 3D X-ray diffraction microscopy (HR-3DXRD) and dark field X-ray microscopy (DFXM). We track more than 8000 non-recrystallized grains during annealing. Relative changes in crystallographic orientation and volume of each recrystallized and non-recrystallized grain are measured as a function of annealing time. Finally, local strain and orientation are measured in the interior of the specific grains with a spatial resolution of 77 nm.

Keywords

Magnesium • Characterization • X-ray diffraction

S. Lee · A. Bucsek (✉)
Mechanical Engineering, University of Michigan, Ann Arbor, MI, USA
e-mail: abucsek@umich.edu

T. Berman · J. Allison
Material Science Engineering, University of Michigan, Ann Arbor, MI, USA

C. Yildirim · C. Detlefs
European Synchrotron Radiation Facility, Grenoble, France

Introduction

This work discusses the use of synchrotron-based X-ray diffraction microscopy techniques to investigate the recovery and recrystallization of an 80% hot-compressed Mg–3.2Zn–0.1Ca wt.% (ZX30) alloy in situ, in 3D, and across several orders of magnitude in length scale. High-strength lightweight magnesium (Mg) alloys have substantial potential for reducing the weight of automobiles and other transportation systems to improve fuel economy and reduce greenhouse gas emissions [1]. A major barrier, however, is the strong crystallographic texture of rolled Mg alloy sheet which can lead to anisotropy, brittleness, and poor machinability. There are several challenges to characterizing the microstructure evolution associated with recovery, recrystallization, and grain growth. Because the microstructure is constantly evolving, it is often necessary to characterize the microstructure evolution in situ. However, the relevant length scale (e.g., grain/subgrain size) and the degree of deformation vary significantly in the as-deformed versus recrystallized states, making it difficult to resolve the microstructure and its key features with a single technique. Temporal resolution is an inherent challenge to diffraction techniques that require a full 360° rotation because of (especially scintillation-based) detector readout times. Finally, the ability to apply these established techniques to highly deformed materials is not quite achievable yet, in part due to spatial resolution limitations and in part due to the necessary peak-finding and peak-fitting procedures.

To address these challenges, here, we use a multiscale, multimodal approach by combining two X-ray diffraction techniques: high-resolution X-ray diffraction (HR-XRD) statistical analysis [2] and dark field X-ray microscopy (DFXM) [3, 4]. For HR-XRD, we used a high-resolution detector to zoom into a subset of Bragg reflections appearing in one particular Debye–Scherrer ring. Using HR-XRD statistical analysis, we track more than 8000 sub-surface non-recrystallized and recrystallized grains during the

S. Barela et al. (eds.), *Magnesium Technology 2023*, The Minerals, Metals & Materials Series,
https://doi.org/10.1007/978-3-031-22645-8_9

recovery and recrystallization processes during in-situ annealing. At several points during the annealing process, we "zoom in" to individual recrystallized and non-recrystallized grains using DFXM. The results show small intragranular variations in elastic lattice strain and misorientation with a spatial resolution of ~100 nm. This combination of HR-XRD statistical analysis and DFXM enables a multiscale, multimodal diffraction microstructure imaging that is particularly useful for small and/or highly deformed grains that are often difficult to resolve using more standard 3D X-ray diffraction techniques.

Experimental

An 80% hot-compressed ZX30 alloy was used in this study. The deformed sample was machined into 1 × 1 × 5 mm^3 rectangular bars using electron discharge machining. Both HR-XRD and DFXM characterization techniques were performed during a single experiment on ID06-HXM [5] at the European Synchrotron Radiation Facility using an X-ray energy of 17 keV and a beam size of 200 μm × 200 μm^2. We annealed the sample for a total of 514 min, during which time we collected 33 h-XRD measurements. The sample was continuously heated at a rate of 10 °C/min for the first 129 min. Once the sample temperature reached 266 °C at 129 min, the sample temperature was maintained at 266 °C degrees for an additional 385 min (with a total heating time of 514 min). To calibrate the sample temperature, we used the thermal expansion of the {101} Debye–Scherrer ring. The multimodal, multiscale X-ray diffraction microstructure imaging using far-field high-energy diffraction microscopy (ff-HEDM), HR-XRD, and DFXM are shown in Fig. 1.

Results and Discussion

Results are shown in Fig. 2. The results of the HR-XRD statistical analysis are presented in Fig. 2. Figure 2a shows five of the 33 h-XRD statistical analysis measurements (summed over the 6° sample rotation) in order of increasing annealing time and temperature. In the first frame, 1,326 grains exist within our field of view. As annealing proceeds, the number of grains can be seen decreasing, and select grains become very large, as can be observed by the high relative intensity of their Bragg reflections. Toward the end of annealing, there are only 153 grains remain. Two additional sequences are provided in Fig. 2, where each measured Bragg reflection is represented as a circle. The location of each circle corresponds to the location of the Bragg reflection, the size of each circle corresponds to the relative volume of the grain, and the color of each circle corresponds to the relative grain volume in Fig. 2b, and ω (i.e., orientation) position in Fig. 2c. These statistical results show that only seven non-recrystallized grains exist out of 153 total grains at the end of annealing, these seven non-recrystallized grains account for 34% of the total volume [7]. These statistical and other results show the power of these "zoom in" and "zoom out" techniques for tracking microstructure evolution in situ from the as-deformed state during annealing.

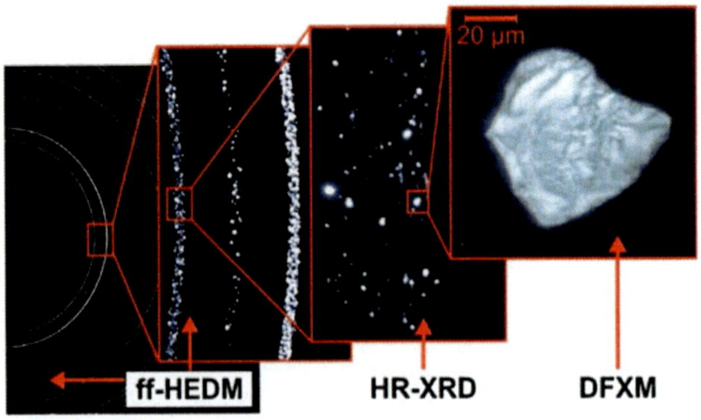

Fig. 1 Multimodal, multiscale X-ray diffraction microstructure imaging using ff-HEDM, HR-XRD, and DFXM. The present work includes only the HR-XRD and DFXM results [7]. The ff-HEDM results are published in [6]

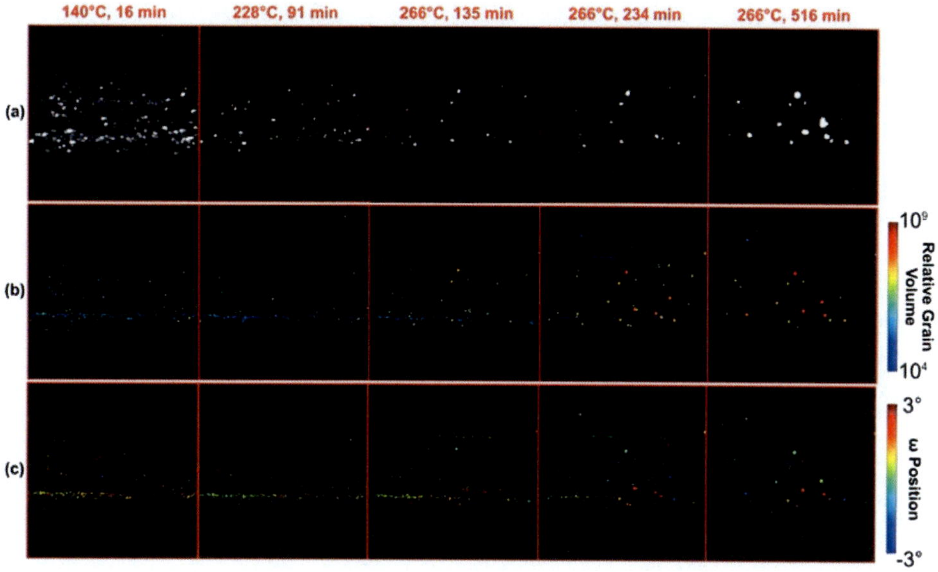

Fig. 2 HR-XRD statistical analysis results **a** raw detector images, **b** relative grain volume, **c** ω position [7]

References

1. M.K. Kulekci, Magnesium and its alloys applications in automotive industry, Int. J. Adv. Manuf. Technol. 39 (2008) 851–865, doi: https://doi.org/10.1007/s00170-007-1279-2.
2. Kutsal, M., Poulsen, H. F., Winther, G., Sørensen, H. O., & Detlefs, C. (2022). High-resolution 3D X-ray diffraction microscopy: 3D mapping of deformed metal microstructures. Journal of Applied Crystallography, 55(5). https://doi.org/10.1107/S1600576722007361
3. H. Simons, A. King, W. Ludwig, C. Detlefs, W. Pantleon, S. Schmidt, F. Stöhr, I. Snigireva, A. Snigirev, H.F. Poulsen, Dark-field X-ray microscopy for multiscale structural characterization, Nature Communications. 6 (2015) 6098.
4. H. Simons, A.C. Jakobsen, S.R. Ahl, C. Detlefs, H.F. Poulsen, Multiscale 3D characterization with dark-field x-ray microscopy, MRS Bulletin. 41 (2016) 454–459. https://doi.org/10.1557/mrs.2016.114.
5. Kutsal, M., Bernard, P., Berruyer, G., Cook, P. K., & Hino, R. (2019). The ESRF dark-field x-ray microscope at ID06 The ESRF dark-field x-ray microscope at ID06. https://doi.org/10.1088/1757-899X/580/1/012007
6. R. Roumina, S. Lee, T. Berman, K. Shanks, J. Allison, A. Bucsek, The dynamics of recrystallized grains during static recrystallization in a hot-compressed Mg-3.2Zn-0.1Ca wt.% alloy using in-situ far field high-energy diffraction microscopy, Acta Materialia. 234 (2022) 118039. doi: https://doi.org/10.1016/j.actamat.2022.118039.
7. S. Lee, T. Berman, C. Yildirim, C. Detlefs, J. Allison, A. Bucsek, Multiscale, and multimodal characterization of recrystallized and non-recrystallized grains during static recrystallization in a hot-compressed Mg-3.2Zn-0.1Ca wt.% alloy, (under review).

Combined Effect of Alloying and Grain Size on the Deformation Behavior of Magnesium Alloys

M. Arul Kumar, M. Wronski, and I. J. Beyerlein

Abstract

Hexagonal close-packed (HCP) magnesium metals are widely used in different industries due to their low density and high specific strength. Their applicability is restricted due to poor formability and pronounced plastic anisotropy. Commonly, the formability is improved by altering the chemistry (adding rare-earth elements like Y) or modulating the microstructure (e.g., grain refinement). Grain refinement alone cannot yield the desired ductility, and the scarcity of rare-earth elements also limits the alloying addition. In this work, using the crystal plasticity framework, the combined effect of Y-content and grain size on the mechanical responses of Mg alloy is studied. The influence of alloying is represented by varying the activation stress and hardening responses of basal, prismatic, pyramidal slip, and tensile twin systems. This detailed study provides a map of strength and tension–compression asymmetry for a wide range of Y-content and grain sizes. This work provides a pathway to optimize the microstructure and chemistry to achieve excellent structural properties.

Keywords

Magnesium • Modeling and simulation • Mechanical properties

M. Arul Kumar (✉)
Los Alamos National Laboratory, Los Alamos, NM 87545, USA
e-mail: marulkr@gmail.com

M. Wronski
AGH University of Science and Technology, Mickiewicza 30, 30-059 Krakow, Poland

I. J. Beyerlein
University of California Santa Barbara, Santa Barbara, CA 93106, USA

Extended Abstract

Hexagonal close-packed (HCP) magnesium (Mg) and derivative alloys are widely used for lightweight transportation due to their low density and high specific strength [1]. However, poor formability at room temperature limits their widespread use in many applications. The low formability of Mg is a consequence of its pronounced plastic anisotropy [2, 3]. The plastic deformation in Mg and its alloys is accommodated by a combination of basal $\langle a \rangle$, prismatic $\langle a \rangle$, and pyramidal $\langle c+a \rangle$ slip, along with deformation twinning [4]. The activation barrier or critical resolved shear stresses (CRSS) for each slip dislocation mode is significantly different. In turn, a crystal oriented to activate one set of slip systems can have a different plastic response than another crystal oriented to activate another set of slip systems. Further, the deformation twinning is directionally dependent, so the deformation twins accommodate either c-axis contraction or extension but not both. The anisotropy in CRSS values and directional dependency leads to a pronounced tension–compression asymmetry, which limits the material's formability.

Grain refinement, texture control, and alloying are commonly considered potential strategies to improve the formability of Mg by reducing the dissimilarities in the CRSS among slip modes and suppressing direction-dependent twinning. Reducing the grain size has been found to suppress twinning activity and increase formability [2, 5]. The crystallographic texture is modified, specifically the texture weakening, via alternative metal processing techniques, such as asymmetric rolling and equal channel angular pressing to improve the formability [6, 7]. Finally, the additions of rare earth elements (Y, Ce, Nd, La, etc.) have been reported to lower plastic anisotropy significantly [8–11]. Among these three strategies, the alloying addition is known to be a viable way to improve formability. However, these alloying elements are not abundant and thus limit the applicability of these alloys. This work hypothesizes that this

S. Barela et al. (eds.), *Magnesium Technology 2023*, The Minerals, Metals & Materials Series,
https://doi.org/10.1007/978-3-031-22645-8_10

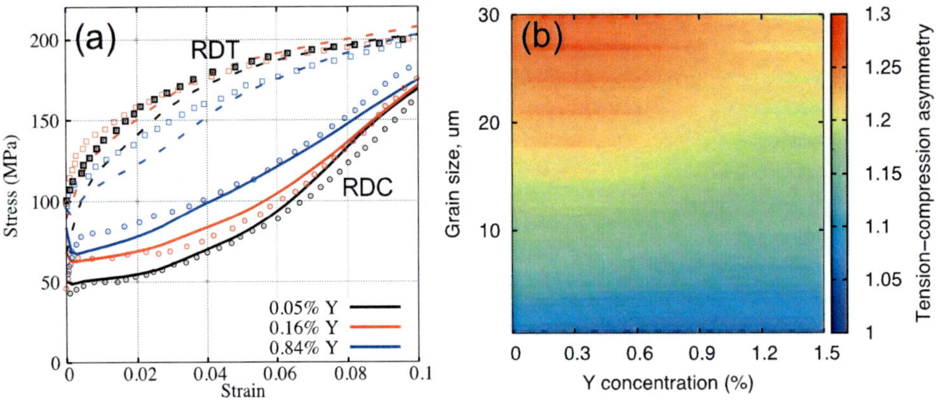

Fig. 1 a Model predicted (solid and dashed lines) stress–strain response along with the experimental values (symbols) for rolling direction tension (RDT) and compression (RDC) for three different Y content. b Distribution of tension–compression asymmetry in the grain size and Y-content space

limitation can be overcome by combining the alloying method with the grain refinement strategy. That is, the minimum amount of rare-earth elements needed to improve the formability of Mg alloys can be reduced by properly tuning the grain size.

To capture the effect of grain size and Y-content on mechanical responses, a visco-plastic self-consistent (VPSC) crystal plasticity model is employed [12]. In the calculation, basal $\langle a \rangle$, prismatic $\langle a \rangle$, and pyramidal $\langle c + a \rangle$ slip modes and $\{10\bar{1}2\}$ tensile twinning are assumed to accommodate the plastic deformation. The dislocation density-based hardening law is used [13]. The effect of grain sizes on slip and twinning is modeled through directional-dependent micro-Hall-Petch equations [14]. Following the classical works of [15–17], the alloying concentration is directly related to the strength terms for each slip and twinning mode. It allows us to directly capture the role of alloying elements on dislocation slip motion and deformation twinning. The stress–strain responses and twin volume fraction values reported in [18] for different Y content are used to calibrate and validate the model. The model-predicted mechanical responses for both rolling direction tension (RDT) and compression (RDC), along with the experimental values, are shown in Fig. 1a. The solid and dashed lines correspond to the model-predicted stress–strain response for RDC and RDT loadings, whereas the symbols refer to experiments. Note that the grain size is different for different Y-content cases. Overall, the developed correctly captures the twinning signature in the RDC loading and the complex dependence of Y content on slip and twinning activity for both RDC and RDT. The increase in Y-content lowers the tension–compression asymmetry at yield and also later in the strain hardening regime.

The developed model is further used to simulate the mechanical responses under tension and compression for a wide range of grain sizes and Y-content. The calculated tension–compression asymmetry at yield is plotted in the grain size and Y-content, see Fig. 1b. As one expects, the tension–compression asymmetry decreases with an increase in Y-content and a reduction in grain sizes. Interestingly, the distribution shown in Fig. 1b suggests that the Y-content can be reduced significantly with a slight to moderate decrease in grain sizes. For instance, the required reduction in grain size to lower the Y-content by 1% to achieve similar tension–compression asymmetry is only ~5 micron, which is feasible. Overall, this work establishes a plausible framework to optimize the microstructure (grain size) and Y-content to achieve desired structural properties such as strength and asymmetry.

References

1. M.K. Kulekci, International Journal of Advanced Manufacturing Technology 39(9–10) (2008) 851–865.
2. S.R. Agnew, O. Duygulu, International Journal of Plasticity 21(6) (2005) 1161–1193.
3. M.A. Kumar, I.J. Beyerlein, R.A. Lebensohn, C.N. Tome, Materials Science and Engineering a-Structural Materials Properties Microstructure and Processing 706 (2017) 295–303.
4. P.G. Partridge, Metallurgical Reviews 12 (1967) 169–194.
5. Q. Yu, L. Qi, R.K. Mishra, J. Li, A.M. Minor, Proceedings of the National Academy of Sciences 110(33) (2013) 13289–13293.
6. W. Kim, S. Hong, Y. Kim, S. Min, H. Jeong, J. Lee, Acta materialia 51(11) (2003) 3293–3307.
7. W. Kim, J. Lee, W. Kim, H. Jeong, H. Jeong, Scripta Materialia 56 (4) (2007) 309–312.
8. T. Al-Samman, X. Li, Materials Science and Engineering: A 528 (10–11) (2011) 3809–3822.
9. Chakkedath, J. Bohlen, S. Yi, D. Letzig, Z. Chen, C.J. Boehlert, Metallurgical and Materials Transactions A 45(8) (2014) 3254–3274.
10. Y. Chino, K. Sassa, M. Mabuchi, Materials Science and Engineering: A 513 (2009) 394–400.
11. H. Ding, X. Shi, Y. Wang, G. Cheng, S. Kamado, Materials Science and Engineering: A 645 (2015) 196–204.

12. R.A. Lebensohn, C.N. Tome, Acta Metallurgica Et Materialia 41 (9) (1993) 2611–2624.
13. I.J. Beyerlein, C.N. Tome, International Journal of Plasticity 24(5) (2008) 867–895.
14. G. Proust, C.N. Tome, A. Jain, S.R. Agnew, International Journal of Plasticity 25(5) (2009) 861–880.
15. R. Labusch, physica status solidi (b) 41(2) (1970) 659–669.
16. R.L. Fleisgher, Acta metallurgica 9(11) (1961) 996–1000.
17. F. Nabarro, Philosophical magazine 35(3) (1977) 613–622.
18. J. Wang, X. Wang, K. Yu, T.J. Rupert, S. Mahajan, E.J. Lavernia, J.M. Schoenung, I.J. Beyerlein, Materials Science and Engineering: A 817 (2021) 141373.

Quantifying the Role of Coarse Intermetallic Particles on Twinning Behavior

Benjamin Anthony and Victoria Miller

Abstract

Deformation twinning is a mechanism of critical interest in magnesium alloys and other HCP metals, both due to its ability to accommodate strain and its tendency to contribute to failure by providing a preferential crack pathway along twin boundaries. This deleterious behavior is worsened by instances of twin transmission, where a twin impinging on a grain boundary nucleates an adjacent, connected twin in the neighboring grain due to intense local stresses. Many commercial Mg alloys feature coarse grain boundary intermetallic particles in their as-produced state which potentially impede or exacerbate the localized stresses that play a role in both twin transmission and twinning behavior. Combined EDS-EBSD is used to analyze grain boundary particles, deformation twins, and transmission events to determine how particle morphology, position, and grain orientation modify twinning behavior and transmission likelihood and how these findings compare to computational results from Crystal Plasticity—Fast Fourier Transform modeling.

Keywords

Deformation twinning • Twin transmission • Intermetallic phases

Extended Abstract

Deformation twinning is a mechanism of critical interest in magnesium alloys and other HCP metals as it provides both beneficial and deleterious effects. Twinning enables magnesium alloys to accommodate strain in crystallographic directions that are relatively difficult to achieve through slip alone such as along the **c**-axis [1]. At the same time, twin boundaries can serve as a preferential crack pathway which can contribute to premature failure of Mg parts during fabrication or service [2, 3]. Understanding and control of twinning behavior are thus crucial for producing tough and formable parts, as well as for broadening the application space of magnesium alloys.

This deleterious behavior of twinning is made worse by instances of twin transmission, where the intense local stresses caused by a twin impinging upon a grain boundary induce the nucleation of a new twin in the neighboring grain that can then propagate, grow, and potentially transmit across yet another grain boundary. Transmission results in a pair of adjoined twins that form a continuous twin boundary across the grain boundary, providing a longer pathway for a crack to follow. As twin transmission is more favorable across grain boundaries with low misorientation [4, 5], highly textured material is particularly susceptible to transmission events and the potential formation of long chains of twins across many grains, leading to a long, continuous twin boundary which can greatly contribute to premature brittle failure [6].

Many commercial Mg alloys feature intermetallic particles in their as-produced state, such as the β-$Mg_{17}Al_{12}$ phase in Mg–Al alloys [1, 7]. While the effects of the nanoscale intragranular precipitates on twinning behavior have been studied extensively [8], little research has been done on the coarse micron-scale particles that typically form at the grain boundaries in these alloys during solidification [9]. These coarse intermetallic particles are elastically harder than the matrix, and the backstresses due to mechanical mismatch can impede or exacerbate the localized stresses caused by twin impingement that play a role in both twin transmission and twinning behavior.

Previous computational studies by our group using Crystal Plasticity—Fast Fourier Transform (CP-FFT) modeling [10] indicated that certain microstructural cases, such

B. Anthony (✉) · V. Miller
University of Florida, Gainesville, FL, USA
e-mail: benjamin.anthony@ufl.edu

S. Barela et al. (eds.), *Magnesium Technology 2023*, The Minerals, Metals & Materials Series,
https://doi.org/10.1007/978-3-031-22645-8_11

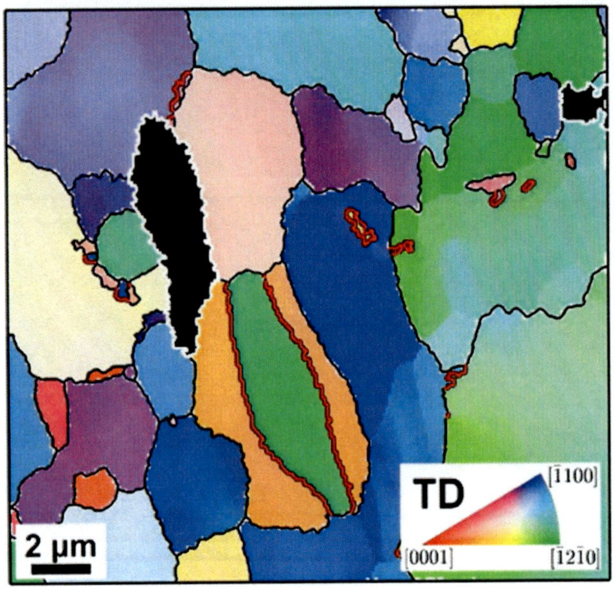

Fig. 1 Inverse pole figure map from combined EDS-EBSD of strained AM60 Mg alloy. Grain boundaries are highlighted in black, twin boundaries are highlighted in red, and particle phases are colored in black with white borders

as instances of a twin directly impinging upon an intermetallic particle of sufficient size, would reduce the likelihood of nucleating a new twin in the neighboring grain. However, other microstructures—such as a twin impinging upon the grain boundary nearby the particle without directly making contact—resulted in an increased likelihood of transmission. Additionally, the modification of stress states in the parent and neighboring grains due to the presence, size, and position of the particle was associated with changes in the twin thickness in the parent grain and the potential variant selection in the secondary grain. These simulation results for single twins and particles can help to interpret the statistical occurrence of twinning in real particle-containing magnesium microstructures.

Toward that goal, this work uses electron dispersive spectroscopy (EDS) and electron backscatter diffraction (EBSD) in tandem to simultaneously analyze grain boundary particles, deformation twins, and transmission events over a broad region of the microstructure. As shown in Fig. 1, the combined EDS-EBSD data is processed using automated algorithms designed to work with the MTEX [11] open-source toolbox for MATLAB to identify both intergranular particles and twin boundaries. Chemical information provided by EDS is combined with band contrast values from EBSD to segment particles, while orientation information from EBSD is used to automatically identify twin boundaries. Twinning statistics such as area fraction, thickness, and variant selection—including Schmid and non-Schmid variant identification—can be automatically calculated, and instances of adjoined twin pairs across grain boundaries determined. Similarly, instances of twin-particle contact can be identified or the distance between a twin and particle along a grain boundary (such as in Fig. 1) can be calculated. These automated segmentation and analytical methods using combined EDS-EBSD enable a statistical study of how particle morphology, position, and grain orientation modify twinning behavior and transmission likelihood, and allow for the findings to be compared to computational results from CP-FFT modeling.

References

1. J.F. Nie, K.S. Shin, Z.R. Zeng, Microstructure, Deformation, and Property of Wrought Magnesium Alloys, Springer US, 2020, https://doi.org/10.1007/s11661-020-05974-z
2. D. Ando, J. Koike, Y. Sutou, The role of deformation twinning in the fracture behavior and mechanism of basal textured magnesium alloys, Mater. Sci. Eng. A 600 (2014) 145–152, https://doi.org/10.1016/j.msea.2014.02.010
3. D.A. Basha, H. Somekawa, A. Singh, Crack propagation along grain boundaries and twins in Mg and Mg–0.3 at.%Y alloy during in-situ straining in transmission electron microscope, Scr. Mater. 142 (2018) 50–54, https://doi.org/10.1016/j.scriptamat.2017.08.023
4. I.J. Beyerlein, L. Capolungo, P.E. Marshall, R.J. McCabe, C.N. Tome, Statistical analyses of deformation twinning in magnesium, Philos. Mag. 90 (16) (2010) 2161–2190, https://doi.org/10.1080/14786431003630835
5. M. Arul Kumar, I.J. Beyerlein, R.J. McCabe, C.N. Tomé, Grain neighbour effects on twin transmission in hexagonal close-packed materials, Nat. Commun. 7 (May) (2016) 1–9, https://doi.org/10.1038/ncomms13826
6. G. Liu, R. Xin, F. Liu, Q. Liu, Twinning characteristic in tension of magnesium alloys and its effect on mechanical properties, Mater. Des. 107 (2016) 503–510, https://doi.org/10.1016/j.matdes.2016.06.073
7. A .A . Luo, P. Fu, L. Peng, X. Kang, Z. Li, T. Zhu, Solidification microstructure and mechanical properties of cast magnesium-aluminum-tin alloys, Metall. Mater. Trans. A 43 (1) (2012) 360–368, https://doi.org/10.1007/s11661-011-0820-y
8. J.D. Robson, N. Stanford, M.R. Barnett, Effect of precipitate shape and habit on mechanical asymmetry in magnesium alloys, Metall. Mater. Trans. A 44 (7) (2013) 2984–2995, https://doi.org/10.1007/s11661-012-1466-0
9. A.K. Dahle, Y.C. Lee, M.D. Nave, P.L. Schaffer, D.H. StJohn, Development of the as-cast microstructure in magnesium-aluminium alloys, J. Light Metals 1 (1) (2001) 61–72, https://doi.org/10.1016/S1471-5317(00)00007-9
10. B. Anthony, B. Leu, I.J. Beyerlein, V.M. Miller, Deformation twin interactions with grain boundary particles in multi-phase magnesium alloys, Acta Mater. 219 (2021) 117225, https://doi.org/10.1016/j.actamat.2021.117225
11. F. Bachmann, R. Hielscher, H. Schaeben, Texture Analysis with MTEX – Free and Open Source Software Toolbox, Solid State Phenomena 160 (2010) 63–68, https://doi.org/10.4028/www.scientific.net/SSP.160.63

Optimization of the Microstructure and Performance of Aluminum Alloy Cold Spray Coatings on Magnesium Alloy Substrates

Sridhar Niverty, Rajib Kalsar, Anthony Naccarelli, Timothy Eden, Glenn Grant, and Vineet Joshi

Abstract

The widespread implementation of Magnesium alloys for the automotive industry warrants an improvement in their corrosion performance and their ability to withstand dissimilar material contact under corrosive environments. Solid-phase coating methods offer opportunities to precisely tune the coating microstructure, thickness, and coating-substrate interface. We report on the optimization and characterization of cold spray deposited AA6061 aluminum alloy coatings on Magnesium alloy corrosion and automotive components. Coatings with low porosity, high thickness, hardness, and excellent corrosion properties were achieved using a bond coating composed of pre-treated CP-Al powders. The dynamic nature of the cold spray process enabled the formation of a metallurgical bond at the coating-substrate interface leading to high adhesion strengths. Furthermore, the ability of cold spray coatings to repair manufacturing defects was explored, and the microstructural mechanism for the same was investigated.

Keywords

Magnesium • Corrosion • Cold spray coatings

Extended Abstract

Magnesium alloys are favorable candidates for use in automotive applications due to their low density (1.7 g/cm^3) and high specific strength. Recently, there has been a growing interest in utilizing these alloys for automotive components, such as strut tower base [1]. However, Mg alloys are susceptible to corrosion attack (filiform, pitting, and general corrosion) when exposed to harsh corrosive automotive environments [2]. Furthermore, Mg alloys can also be exposed to dissimilar material contact (Example-with Al alloys and galvanized steel), which under corrosive environments can lead to galvanic and crevice corrosion [2, 3]. Thus, there is an urgent need to improve the surface properties of Mg alloys with the aim of improving their corrosion resistance and enabling dissimilar material contact. To that end, several surface coating methods have been developed to improve their corrosion resistance [4–7]. Among these, spray coating methods enable layer by layer deposition of metallic powders sprayed at supersonic velocities. In particular, cold spray technology is a solid-phase coating method that minimizes undesirable intermetallic formation at the coating-substrate interface and minimizes oxidation of powders. It also enables precise coating thickness control, can be optimized for minimum porosity formation, and offers better scalability and lower costs compared to other coating methods [7, 8].

In this study, we have utilized cold spray coating technology to coat Aluminum 6061 alloy powders on to AZ91 Magnesium alloy substrates. Al6061 powders were utilized for the coating because Mg alloys are often in contact with Al6061 plates or Al alloy based isolators in automotive components [1]. Nitrogen gas was used as a carrier gas. Al6061 powders were coated onto the AZ91 substrates using higher carrier gas temperature and pressure than those utilized in the existing literature (Fig. 1a) as an increase in gas temperature directly correlates with improved deposition efficiency [7, 8]. Over several trials, the deposition of the Al6061 powders was optimized by introducing a bond coating of commercially pure Aluminum (CP-Al). Furthermore, the size distribution of both the Al and Al6061 powders was restricted to between 20 and 53 μm. While the Al6061 powders were utilized as received, the CP-Al

S. Niverty · R. Kalsar · G. Grant · V. Joshi (✉)
Energy and Environment Directorate, Pacific Northwest National Laboratory, Richland, WA 99354, USA
e-mail: Vineet.joshi@pnnl.gov

A. Naccarelli · T. Eden
Applied Research Laboratory, Pennsylvania State University, University Park, PA 16802, USA

S. Barela et al. (eds.), *Magnesium Technology 2023*, The Minerals, Metals & Materials Series,
https://doi.org/10.1007/978-3-031-22645-8_12

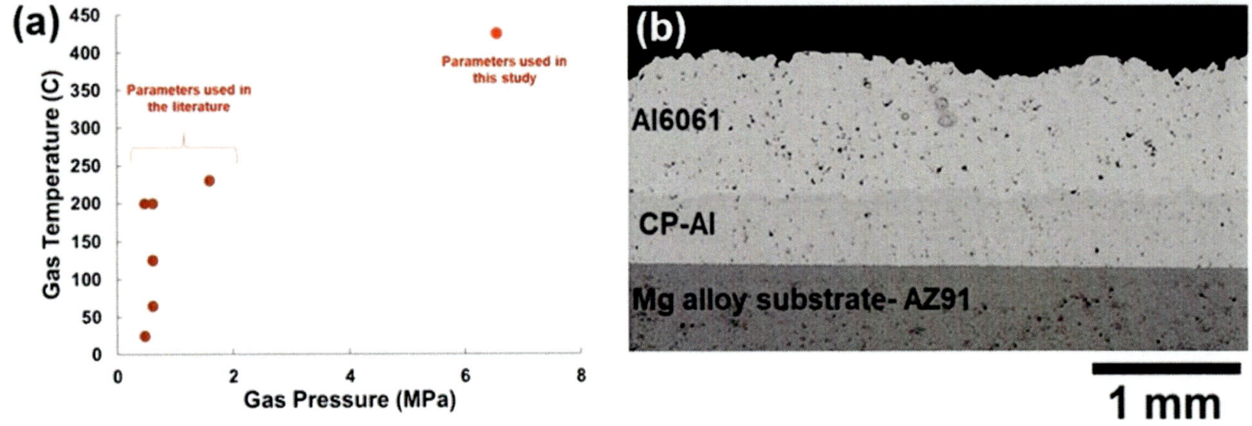

Fig. 1 a Plot showing a comparison of spray parameters used in the literature versus those used in our study, **b** cross section SEM micrograph showing the AZ91 substrate, the CP-Al bond coating and Al6061 top coating

powders were softened by annealing at 220–240 °C in an Argon atmosphere for 1 h. The microstructure and performance of the case with CP-Al bond coat and Al6061 top coat were compared with the case with only Al6061 coating. Powder material and process changes led to a substantial increase in coating thickness (from 398 ± 153 μm to greater than 1.4 mm), decrease in coating porosity (from 3.4 ± 0.3% to less than 0.5%), and decrease in maximum peak roughness (from 594 ± 3 μm to 290 ± 3 μm). While Scanning Electron Microscopy (SEM) and Energy Dispersive Spectroscopy (EDS) showed no metallurgical bond formation between the coating and substrates, Transmission Electron Microscopy (TEM) imaging showed the presence of a thin layer of $Mg_{17}Al_{12}$ at the interface between the substrate and coating. Due to the high strain rate deformation experienced by the powders upon impact, Electron Back Scatter Diffraction (EBSD) showed dynamic recrystallization at the interface between adjacent powder particles and substrate-particle interface. This led to the formation of small grains (< 5 μm in size) decorating powder particle borders. Additionally, a region of compressive residual stress (30–50 μm) was observed on the surface of the substrate due to the impact of the CP-Al powders.

The coating brought about a notable improvement in surface properties of the alloy that was characterized by mechanical and corrosion property testing. Vickers hardness testing performed across the cross section of coated samples showed the bond coat to have a substantially lower hardness (40 Hv) compared to the substrate (Average = 70 Hv) or top coating (Average = 75 Hv). The softer CP-Al matrix aided in improving deposition of the Al6061 powders. Coating adhesion testing (ASTM D4541) showed a promising adhesion strength of 16.1 ± 2.8 MPa that is further being improved through cold spray trials at higher temperatures. Furthermore, the presence of the coatings decreased the wear rate of the coated alloy (1E−3 mm^3/N m) by greater than 25% as compared to the substrate (1.345E−3 mm^3/N m). Potentiodynamic polarization electrochemical testing showed a drastic increase in exchange current density of the coating due to the presence of the coating (from − 1530 mV without the coating to − 706 mV with the coatings). Finally, localized Electrochemical Impedance Spectroscopy (EIS) showed a sixfold increase in the resistance to corrosion with the presence of coating. These metrics show that the CP-Al bond coating + Al6061 top coating combination creates nearly pore free microstructures with favorable improvements in mechanical and corrosion properties.

In summary, coatings of Al6061 were deposited on top of AZ91 Mg alloy and optimized using a CP-Al bond coat. These coatings yielded notable improvements in surface properties that were substantiated through a variety of mechanical and corrosion tests. The widespread use of the alloys used in this study also aids in end-of-life recycling, thereby making these coatings favorable for use in the automotive industry.

References

1. K. Lazarz, J. Cahill, T.J. Ciccone, K. Redlin, S. Simko, Corrosion Performance of a Magnesium Tower Brace, SAE Tech. Pap. (2021) 1–7. https://doi.org/10.4271/2021-01-0276.
2. M. Esmaily, J.E. Svensson, S. Fajardo, N. Birbilis, G.S. Frankel, S. Virtanen, R. Arrabal, S. Thomas, L.G. Johansson, Fundamentals and Advances in Magnesium Alloy Corrosion, Prog. Mater. Sci. 89 (2017) 92–193. https://doi.org/10.1016/j.pmatsci.2017.04.011.
3. G. Song, B. Johannesson, S. Hapugoda, D. StJohn, Galvanic corrosion of magnesium alloy AZ91D in contact with an aluminium alloy, steel and zinc, Corros. Sci. 46 (2004) 955–977. https://doi.org/10.1016/S0010-938X(03)00190-2.
4. Y. Wang, H. Yu, C. Chen, Z. Zhao, Review of the biocompatibility of micro-arc oxidation coated titanium alloys, Mater. Des. 85 (2015) 640–652. https://doi.org/10.1016/J.MATDES.2015.07.086.
5. R. Hu, Y. Su, H. Liu, Deposition behaviour of nickel phosphorus coating on magnesium alloy in a weak corrosive electroless nickel

plating bath, J. Alloys Compd. 658 (2016) 555–560. https://doi.org/10.1016/J.JALLCOM.2015.10.300.
6. C. Taltavull, A.J. Lopez, B. Torres, A. Atrens, J. Rams, Optimisation of the high velocity oxygen fuel (HVOF) parameters to produce effective corrosion control coatings on AZ91 magnesium alloy, Mater. Corros. 66 (2015) 423–433. https://doi.org/10.1002/MACO.201407982.
7. A. Moridi, S.M. Hassani-Gangaraj, M. Guagliano, M. Dao, Cold spray coating: Review of material systems and future perspectives, Surf. Eng. 30 (2014) 369–395. https://doi.org/10.1179/1743294414Y.0000000270.
8. F. Gärtner, T. Stoltenhoff, T. Schmidt, H. Kreye, The cold spray process and its potential for industrial applications, J. Therm. Spray Technol. 15 (2006) 223–232. https://doi.org/10.1361/105996306X108110.

Part III

Corrosion and Coatings

Open-Air Plasma Assisted Si–O–C Layer Deposition on AZ91D Mg Alloy for Corrosion Mitigation

Jiheon Jun, Yong Chae Lim, Yi-Feng Su, Andrew Sy, Ryan Robinson, and Daphne Pappas

Abstract

Use of Mg alloy structural parts can enable lightweight automobiles for improved energy efficiency. High corrosion susceptibility of Mg alloys, however, remains as a technical challenge against their application in vehicle-structures. This work investigates open-air plasma assisted Si–O–C coating as a corrosion barrier for AZ91D Mg alloy. The open-air plasma coating does not involve wet chemical process and is amenable to the industries experienced in other plasma-based processes. As-deposited coatings on AZ91D substrates were characterized by advanced microscopic characterization techniques, including SEM, STEM, and EDS. Corrosion evaluation was performed using electrochemical impedance spectroscopy, polarization, and H_2 collection measurements in 3.5 wt.% NaCl solution. Post-immersion AZ91D samples, that were uncoated or Si–O–C coated condition, were also characterized by SEM, STEM, and EDS. The results indicate that plasma assisted Si–O–C coating delayed the initiation of corrosion and the progression of corrosion attack.

Keywords

Magnesium • Environmental effects • Process technology

J. Jun (✉) · Y. Chae Lim · Y.-F. Su
Materials Science and Technology Division, Oak Ridge National Laboratory, One Bethel Valley Road, Oak Ridge, TN, USA
e-mail: junj@ornl.gov

A. Sy · R. Robinson · D. Pappas
Plasmatreat USA, Inc, 30695 Huntwood Ave, Hayward, CA 94544, USA

Extended Abstract

Use of Mg alloy structural parts can enable lighter-weight automobiles for improved energy efficiency [1, 2]. High corrosion susceptibility of Mg alloys, however, remains as a technical challenge against their application in vehicle-structures. This work investigated open-air plasma assisted Si–O–C (silane-based) coatings as corrosion barriers for AZ91D Mg alloys. This open-air plasma coating process does not involve wet chemical processes and is amenable to industries already experienced in other plasma-based processes.

High pressure diecast AZ91D Mg alloy was obtained from a commercial vendor and finished with 600 grit SiC paper prior to the plasma coating process or corrosion exposure. Figure 1 describes the two-step plasma process for Si–O–C coating deposition on an AZ91D substrate. An example of Si–O–C coated AZ91D sample is shown in Fig. 2. This plasma assisted Si–O–C coating is referred to as P01 coating according to the plasma process ID used to deposit the barrier layer. The coating thickness, estimated by light reflection measurement, was ~ 220 nm.

ASTM B117 exposure test was used for initial corrosion evaluation of uncoated and P01 coated AZ91D samples. After 24 h exposure, the center of P01-coated AZ91D sample remained visually intact while the entire surface of uncoated AZ91D developed corrosion products (see Fig. 2). This result highlights the corrosion protection/mitigation effect from P01 coating for AZ91D. As another method of corrosion exposure, uncoated and P01 coated AZ91D samples were immersed in 3.5% NaCl solution open to air at room temperature. The appearance of pre-defined sample surfaces before and after NaCl solution exposure is presented in Fig. 3. It was observed that the formation of corrosion ditches was suppressed when P01 coating was applied on AZ91D after 47 h immersion in 3.5% NaCl solution.

As-deposited coatings on AZ91D substrates were characterized by advanced microscopic characterization

S. Barela et al. (eds.), *Magnesium Technology 2023*, The Minerals, Metals & Materials Series,
https://doi.org/10.1007/978-3-031-22645-8_13

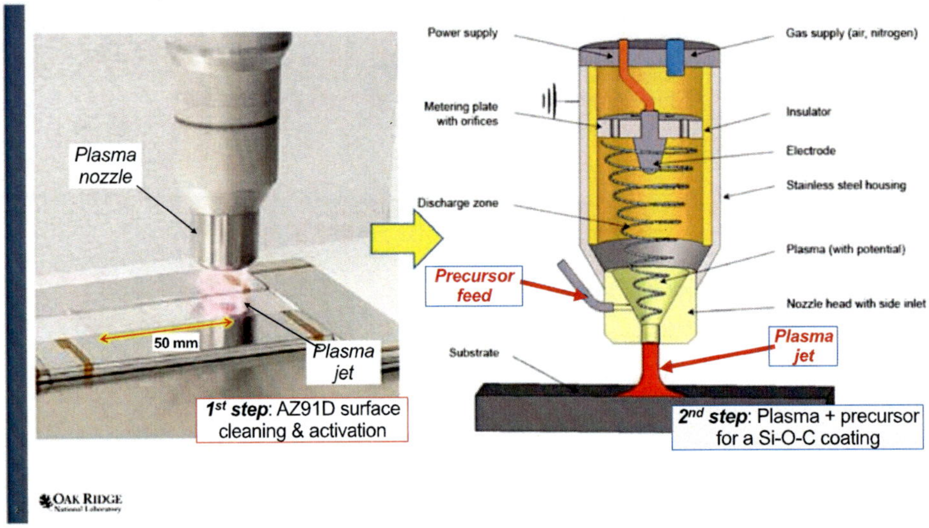

Fig. 1 Description of two-step plasma process for Si–O–C coating deposition on AZ91D Mg alloy coupon. For further details, refer to the following article: https://www.degruyter.com/document/doi/10.1515/cdbme-2016-0013/html

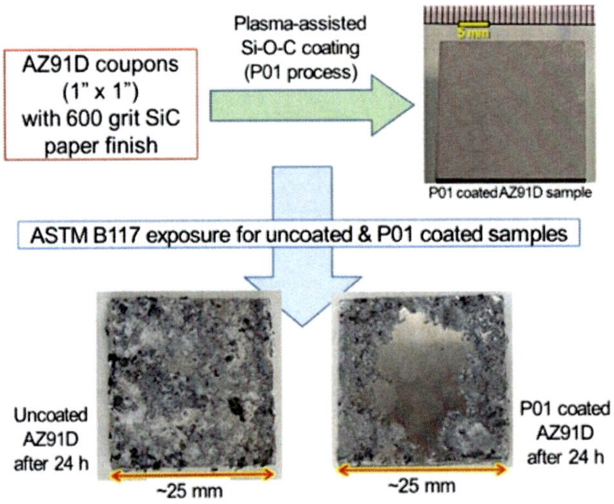

Fig. 2 P01 coated AZ91D coupon (top right) and the appearance of uncoated and P01 coated AZ91D samples after B117 salt spray exposure for 24 h (bottom). Note that P01 coating was not applied on ~ 3 mm marginal area on each corner of the sample, so corrosion attack on the P01 coated sample was initiated from the uncoated area

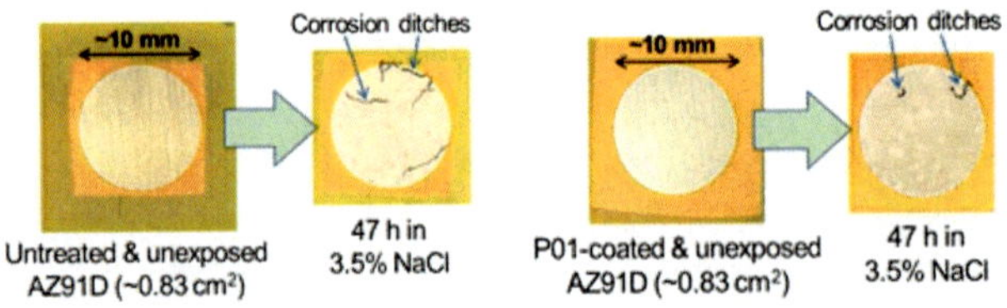

Fig. 3 Comparison of uncoated and P01 coated AZ91D sample surfaces before and after 3.5% NaCl immersion (open to air at room temperature). Immersion time: 47 h

techniques, including SEM, STEM, EDS, and EELS. From the EDS characterization results, it was confirmed that as-deposited P01 coating is ~ 200 nm thick and composed of ~ 45 wt.% Si, ~ 36 wt.% O, and ~ 17 wt.% C. Characterization of post-exposure P01 coated and uncoated AZ91D samples was also performed. An example of post-exposure STEM characterization is presented in Fig. 4 for P01-coated AZ91D sample cross-sectioned after 166 h immersion in 3.5% NaCl solution. In this specific section, Mg and O rich corrosion product layer was observed instead of Si–O–C coating, which implies that Si–O–C

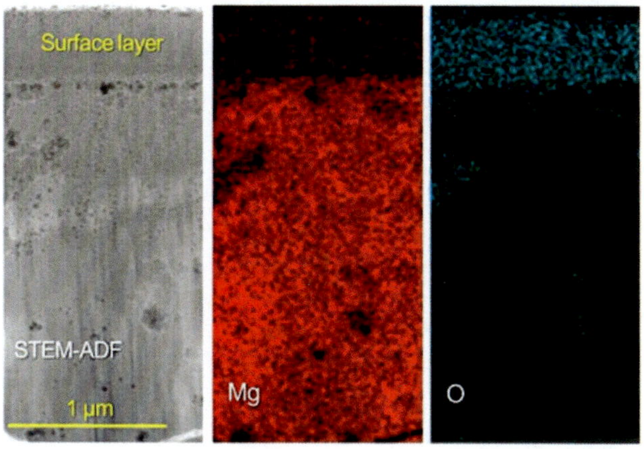

Fig. 4 STEM and EDS maps of a post-exposure P01 coated AZ91D sample. The exposure condition was 3.5% NaCl immersion for 166 h. Annular dark field mode was used for imaging this TEM lamella

coating in this location was removed during NaCl solution immersion.

Corrosion evaluation was performed using electrochemical impedance spectroscopy, polarization, and H_2 collection measurements in 3.5 wt.% NaCl solution. Post-immersion AZ91D samples, that were uncoated or Si–O–C coated condition, were also characterized by SEM, STEM, and EDS. The results, which will be discussed in detail during the presentation, indicated that plasma assisted Si–O–C coating delayed the initiation of corrosion and the progression of corrosion attack.

Acknowledgements This work was supported by the Lightweight Metals Core Program, Vehicle Technologies Office, U.S. Department of Energy. This manuscript has been authored by UT-Battelle, LLC, under contract DE-AC05-00OR22725 with the US Department of Energy (DOE). The US government retains and the publisher, by accepting the article for publication, acknowledges that the US government retains a nonexclusive, paid-up, irrevocable, worldwide license to publish or reproduce the published form of this manuscript or allow others to do so, for US government purposes. DOE will provide public access to these results of federally sponsored research in accordance with the DOE Public Access Plan (http://energy.gov/downloads/doe-public-access-plan).

References

1. W. J. Joost, P. E. Krajewski, Towards magnesium alloys for high-volume automotive applications. *Scripta Materialia* **128**, 107–112 (2017).
2. S. Kleinbaum, C. Jiang, S. Logan, Enabling sustainable transportation through joining of dissimilar lightweight materials. *MRS Bulletin* **44**, 608–612 (2019).

First-Principles Investigations into the Electrochemical Behavior of Mg-Based Intermetallics

Pragyandipta Mishra, Pranav Kumar, Lakshman Neelakantan, and Ilaksh Adlakha

Abstract

Magnesium alloys have drawn considerable attention for several engineering applications, owing to their excellent properties like low density and high specific strength. The room temperature ductility and mechanical properties of Mg are usually enhanced by alloying additions. Based on the thermomechanical processing, the presence of critical concentration of alloying element typically leads to the formation of stable binary intermetallic phases with Mg, thereby distinctly altering the microscopic electrochemical properties of the alloy. However, the secondary intermetallic phases in Mg alloys are typically of sub-micron size; thus, accurate electrochemical characterization is a challenging issue. Using first-principles calculations, the electrochemical behavior of various Mg intermetallics was comprehensively quantified. The electrochemical polarization behavior of the intermetallics was strongly dependent on surface-mediated properties and chemical bonding characteristics. Finally, the computational framework provides an accurate screening tool that can assist in alloy design and development of coatings.

Keywords

First-principles • Mg alloys • Mg intermetallics • Electrochemical behavior

P. Mishra (✉) · P. Kumar · I. Adlakha (✉)
Department of Applied Mechanics, Indian Institute of Technology Madras, Chennai, India
e-mail: pragyandipta.m@gmail.com

I. Adlakha
e-mail: ilaksh.adlakha@iitm.ac.in

L. Neelakantan
Department of Metallurgical and Materials Engineering, Indian Institute of Technology Madras, Chennai, India

Extended Abstract

The last few decades have witnessed a rising interest in magnesium (Mg) alloys in many engineering industries owing to their excellent properties like high strength and stiffness, low density, and high ductility [1–3]. These properties make them especially useful in the transportation sector as a lightweight material which subsequently reduces energy consumption and CO_2 emission [4, 5]. However, their utilization is still restricted because of its poor corrosion resistance arising due to its inability to form a stable oxide film and high electrochemical activity of Mg [6, 7]. Further, the presence of microstructural and chemical heterogeneities in the form of intermetallics promotes the formation of localized galvanic cells leading to accelerated corrosion behavior [7]. Therefore, accurate characterization of the electrochemical behavior of the various secondary intermetallic phases is essential for a better understanding of the properties of Mg alloys.

Although investigations into the electrochemical behavior of Mg alloys have made great progress, due to the involvement of a large number of local and environmental factors, accurate prediction of corrosion-related degradation still remains as one of the most challenging issues in material science [7–10]. The investigations into the electrochemical behavior of Mg alloys have been largely experimental, due to the unavailability of a consistently proven computational model to predict the same. This was especially true in the case of first-principles calculations, as no methodology existed to correlate the microscale quantities derived from first-principles calculations and the macroscale electrochemical behavior [8]. Although these experimental characterizations often yield useful results with respect to feasibility and kinetics of the electrochemical process, they give little to no information on the local chemistry as these results are more focused on the average effects over large surface areas.

S. Barela et al. (eds.), *Magnesium Technology 2023*, The Minerals, Metals & Materials Series,
https://doi.org/10.1007/978-3-031-22645-8_14

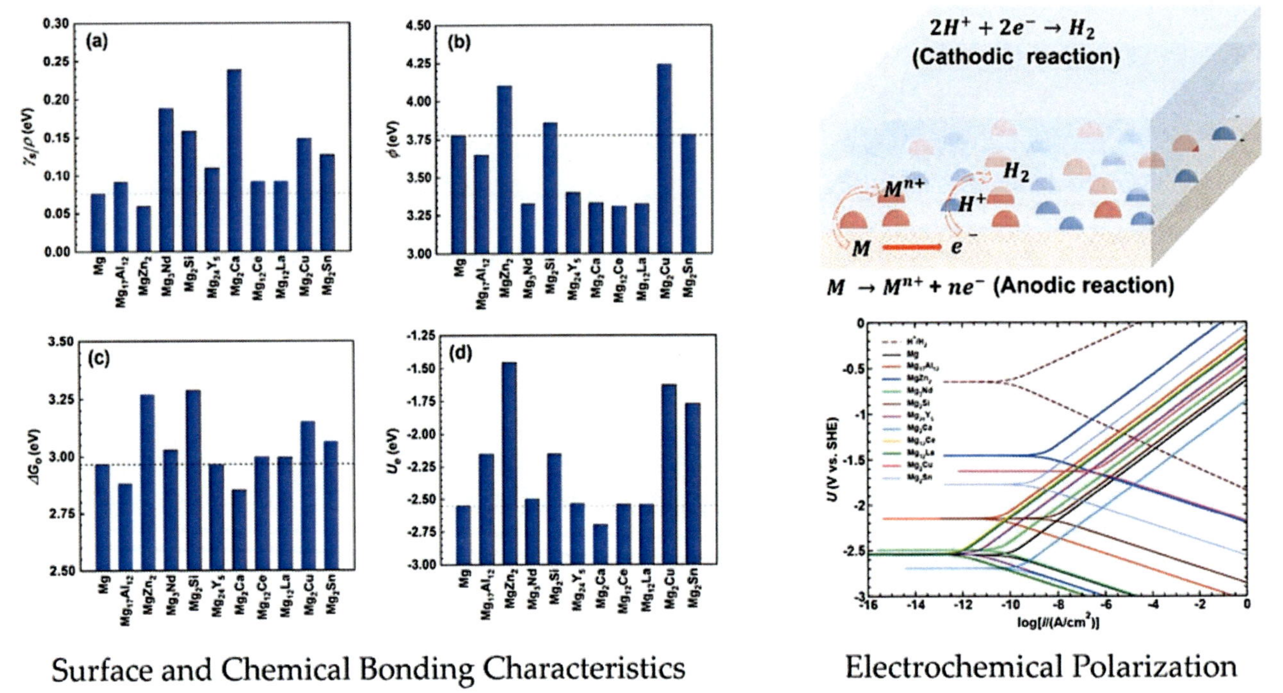

Fig. 1 Influence of surface and chemical bonding characteristics on electrochemical behavior. Reproduced from Mishra et al. [7]

The electrochemical behavior of a crystalline surface is strongly dependent on surface properties (surface energy density and work function) and chemical bonding characteristics (bulk contribution towards Gibbs free energy and equilibrium potential) [7]. The surface energy density is the amount of energy required to create the surface, and therefore, a crystalline surface that has a higher surface energy density when compared to another would be more electrochemically active [11, 12]. On the other hand, work function quantifies the energy required to remove an electron through a surface; thus, a smaller value of work function indicates a surface that is more likely to lose electrons, which in turn makes the surface more reactive when compared to a surface with higher work function [8]. The bulk contribution towards Gibbs free energy (ΔG_0) provides a measure of the energy required to form ions by breaking the underlying chemical bonds while equilibrium potential is defined as the electrochemical potential at which the oxidation and reduction reactions occur at the same rate for a given electrolyte [7].

In the present study, DFT method was employed to comprehensively characterize the electrochemical properties of Mg-based binary intermetallics commonly found in Mg alloys ($Mg_{17}Al_{12}$, $MgZn_2$, Mg_3Nd, Mg_2Si, $Mg_{24}Y_5$, Mg_2Ca, $Mg_{12}Ce$, $Mg_{12}La$, Mg_2Cu, and Mg_2Sn). A robust first-principles based thermodynamic framework was developed to predict the polarization behavior of Mg-based intermetallics. The surface-mediated properties (surface energy and work function) and the intrinsic chemical characteristics of the intermetallics were calculated by a series of first-principles based calculations and used as inputs to the thermodynamic framework (Fig. 1). Finally, the findings were interpreted with electronic insights using the Bader charge analysis and density of states (DOS).

References

1. B. Smola, I. Stulíková, F. von Buch, B.L. Mordike, Structural aspects of high performance Mg alloys design, Mater. Sci. Eng. A. 324 (2002) 113–117. https://doi.org/10.1016/S0921-5093(01)01291-6.
2. M. Bamberger, G. Dehm, Trends in the Development of New Mg Alloys, Annu. Rev. Mater. Res. 38 (2008) 505–533. https://doi.org/10.1146/annurev.matsci.020408.133717.
3. B.L. Mordike, T. Ebert, Magnesium: Properties — applications — potential, Mater. Sci. Eng. A. 302 (2001) 37–45. https://doi.org/10.1016/S0921-5093(00)01351-4.
4. X. Zhang, W. Jiang, Elastic, lattice dynamical, thermal stabilities and thermodynamic properties of BiF3-type Mg3RE compounds from first-principles calculations, J. Alloys Compd. 663 (2016) 565–573. https://doi.org/10.1016/j.jallcom.2015.12.190.
5. E. Aghion, B. Bronfin, D. Eliezer, The role of the magnesium industry in protecting the environment, J. Mater. Process. Technol. 117 (2001) 381–385. https://doi.org/10.1016/S0924-0136(01)00779-8.
6. J. Xie, J. Zhang, Z. You, S. Liu, K. Guan, R. Wu, J. Wang, J. Feng, Towards developing Mg alloys with simultaneously improved strength and corrosion resistance via RE alloying, J. Magnes. Alloys. 9 (2021) 41–56. https://doi.org/10.1016/j.jma.2020.08.016.

7. P. Mishra, P. Kumar, L. Neelakantan, I. Adlakha, First-principles prediction of electrochemical polarization and mechanical behavior in Mg based intermetallics, Comput. Mater. Sci. 214 (2022) 111667. https://doi.org/10.1016/j.commatsci.2022.111667.
8. H. Ma, X.-Q. Chen, R. Li, S. Wang, J. Dong, W. Ke, First-principles modeling of anisotropic anodic dissolution of metals and alloys in corrosive environments, Acta Mater. 130 (2017) 137–146. https://doi.org/10.1016/j.actamat.2017.03.027.
9. S.P. Knight, N. Birbilis, B.C. Muddle, A.R. Trueman, S.P. Lynch, Correlations between intergranular stress corrosion cracking, grain-boundary microchemistry, and grain-boundary electrochemistry for Al–Zn–Mg–Cu alloys, Corros. Sci. 52 (2010) 4073–4080. https://doi.org/10.1016/j.corsci.2010.08.024.
10. G.L. Song, A. Atrens, Corrosion Mechanisms of Magnesium Alloys, Adv. Eng. Mater. 1 (1999) 11–33. https://doi.org/10.1002/(SICI)1527-2648(199909)1:1<11::AID-ADEM11>3.0.CO;2-N.
11. G.-L. Song, R. Mishra, Z. Xu, Crystallographic orientation and electrochemical activity of AZ31 Mg alloy, Electrochem. Commun. 12 (2010) 1009–1012. https://doi.org/10.1016/j.elecom.2010.05.011.
12. R. Xin, Y. Luo, A. Zuo, J. Gao, Q. Liu, Texture effect on corrosion behavior of AZ31 Mg alloy in simulated physiological environment, Mater. Lett. 72 (2012) 1–4. https://doi.org/10.1016/j.matlet.2011.11.032.

Integrating Multimodal Corrosion with Correlative Microscopy Across Multiple Length Scales

Sridhar Niverty, Rajib Kalsar, Lyndi Strange, Venkateshkumar Prabhakaran, and Vineet V. Joshi

Abstract

The corrosion behavior of light metal alloys is complex due to the simultaneous interaction of variables such as processing history, microstructure, corrosive environment, and temperature. Each of these factors contributes to corrosion damage at multiple length scales. This talk will focus on the combination of multimodal corrosion with multiscale imaging to probe the corrosion behavior of light metal alloys. This unique approach involves the simultaneous application of electrochemical methods such as Scanning Electrochemical Cell Microscopy (SECCM), in situ/ex situ imaging, and hydrogen collection. It is particularly beneficial to the study of complex heterogeneous microstructures (varying grain size, chemical composition, porosity, nature of oxide film) obtained during processes such as Friction Stir Processing and surface plasma treatment. Employing this approach correlatively over multiple length scales has aided in the identification of microstructural features that contribute to local and global corrosion damage.

Keywords

Corrosion • Scanning Electrochemical Cell Microscopy • Hydrogen collection

Extended Abstract

Corrosion of materials is a complex phenomenon dictated by several factors such as chemical composition, processing history, presence of intermetallic particles with differing electrochemical properties, corrosive environment, and temperature. Several methods have been developed to study corrosion initiation and progression. Among these, immersion into corrosive fluid, alternate immersion, electrochemical methods, hydrogen collection, imaging (2D and 3D), and spray methods are most often used. Electrochemical methods yield several quantifiable corrosion parameters such as corrosion potential, corrosion current, and corrosion resistance that can be used to rapidly compare materials in a variety of environments [1–3]. In particular, polarization enables the study of different stages of corrosion by monitoring corrosion current at constant or dynamically varying imposed potentials. Interpreting the changes seen in the corrosion, current behavior without accompanying correlative microstructural information can be challenging. To that end, we have developed a multimodal corrosion measurement setup that can be used to perform bulk corrosion, sample imaging (2D), and evolved gas collection simultaneously. We have studied several materials under open circuit potential as well as under accelerated corrosion conditions using potentiodynamic polarization. This has enabled the observation of the onset and progression of corrosion damage during the anodic polarization while also quantifying hydrogen bubble evolution as a function of corrosion potential and time.

In this study, we describe the corrosion behavior of AZ91 magnesium alloy. Open circuit potential (OCP) and potentiodynamic polarization (PDP) tests were performed on an AZ91 sample mounted in an epoxy mount. The surface of the AZ91 was polished to a 1 μm diamond finish. OCP was performed for a duration of 1800s following which PDP was performed from 200 mV below OCP up to 500 mV above OCP (Fig. 1). A saturated calomel electrode was used as the reference electrode, and graphite rods were used as counter electrodes. Figure 1 also shows locations on the OCP and PDP curves where snippets of the sample video are included. These locations include (a) At the beginning of OCP, (b) End of OCP, (c) At E_{corr}, (d) At a current density of

S. Niverty · R. Kalsar · L. Strange · V. Prabhakaran · V. V. Joshi (✉)
Energy and Environment Directorate, Pacific Northwest National Laboratory, Richland, WA 99354, USA
e-mail: Vineet.joshi@pnnl.gov

S. Barela et al. (eds.), *Magnesium Technology 2023*, The Minerals, Metals & Materials Series,
https://doi.org/10.1007/978-3-031-22645-8_15

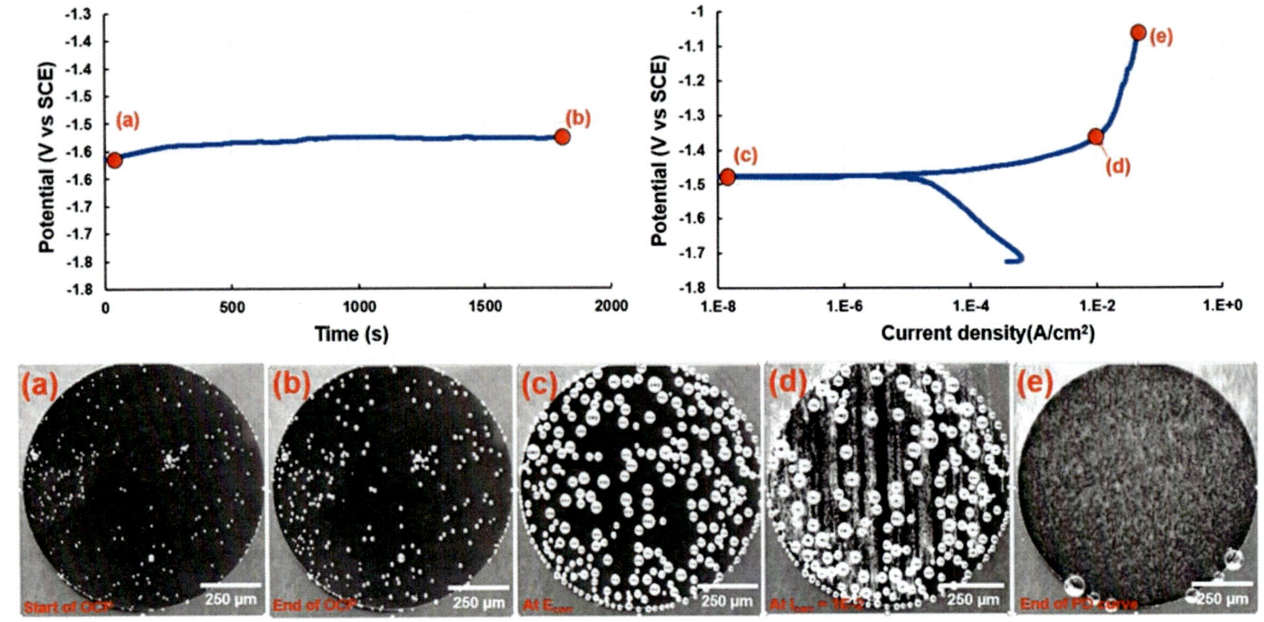

Fig. 1 OCP and PDP curves of AZ91 alloy. Annotations highlight locations where images of the sample surface are shown, i.e. **a** At the beginning of OCP, **b** End of OCP, **c** At E_{corr}, **d** At a current density of 1E-02 A/cm^2, and **e** At a potential of −1.05 V. Images show the increase in corrosion as well as hydrogen bubble evolution as a function of imposed potential

1E-02 A/cm^2, and (e) At a potential of −1.05 V or the potential at which the PDP test was terminated. The reactions involved during the corrosion of magnesium are as follows [4]:

$$\text{Anodic Reaction}: \ Mg \rightarrow Mg^{2+} + 2e^-$$

$$\text{Cathodic Reaction}: \ 2H_2O + 2e^- \rightarrow 2OH^- + H_2 \uparrow$$

$$\text{Net Redox Reaction}: \ Mg + 2H_2O \rightarrow Mg(OH)_2 + H_2 \uparrow$$

Inset images (a) and (b) show the corrosion of AZ91 during 1800s under OCP. Hydrogen bubble evolution is an excellent indicator of the location and magnitude of localized corrosion reaction. During OCP, the magnitude of corrosion taking place is very minimal, as indicated by the minor increase in size of corrosion bubbles on the surface of the sample. Comparing inset images (b) and (c) shows an increase in size and number of bubbles seen on the surface during the cathodic polarization curve. The magnitude of corrosion increases drastically on increasing the potential into the anodic polarization curve. The number of corrosion sites and evolution of hydrogen are observed to increase drastically with applied potential. The locations of the sample with a large magnitude of surface corrosion (such as pits) are also the locations that show a greater magnitude of hydrogen bubble release. Larger bubbles are usually seen where the surface corrosion rate is slow. With increase in potential, the entire sample surface is observed to undergo rapid corrosion as shown in inset image (e) where small and fast evolving hydrogen bubbles encompass the entire surface of the sample. Such visual insight is quintessential to improving our understanding of microstructural changes taking place during the corrosion of Mg alloys.

In summary, through the use of the multimodal corrosion setup, we gained microstructural insights into the locations and the magnitude of evolution of hydrogen during the corrosion of AZ91 magnesium alloy. We expect that the real time observation capability will support experiments such as comparison of the mechanism of corrosion attack in different materials, as well as aid correlative microscopy to observe the role of microstructural constituents on corrosion.

References

1. Fontana, M. Corrosion Engineering (Materials Science & Engineering). (1987).
2. Frankel, G. S. & Rohwerder, M. Electrochemical Techniques for Corrosion. *Encycl. Electrochem.* (2007) https://doi.org/10.1002/9783527610426.BARD040007.
3. Prabhakaran, V. *et al.* Understanding Localized Corrosion on Metal Surfaces Using Scanning Electrochemical Cell Impedance Microscopy (SECCIM). *J. Phys. Chem. C* **126**, 12519–12526 (2022).
4. Frankel, G. S., Samaniego, A. & Birbilis, N. Evolution of hydrogen at dissolving magnesium surfaces. *Corros. Sci.* **70**, 104–111 (2013).

Protective Micro-Arc Oxidation Surface Coating on AZ80 Forged Magnesium Alloy

Xin Pang, Yuna Xue, and Hamid Jahed

Abstract

Magnesium alloys, with a high weight-to-strength ratio, have attracted considerable attention for lightweight applications, especially in automotive industry. However, their inferior corrosion properties when in contact with aqueous solutions, salt, and other metals have become a significant hindrance to more widespread use of magnesium alloys. Surface coating has been considered an economic and effective strategy for corrosion protection of magnesium alloys. In this work, the corrosion behavior of uncoated and micro-arc oxidation (MAO) coated cast, extruded, and forged AZ80 magnesium alloys has been studied using salt fog corrosion chamber test and microscopic characterization techniques. The influence of the forging process and MAO surface treatment on the corrosion properties of the AZ80 magnesium alloys was investigated. The experimental results showed that the low forging temperature of 250 °C led to fine grain size and uniform distribution of secondary phase in the magnesium alloy, resulting in higher corrosion resistance. The MAO coating provided significant corrosion protection of the AZ80 magnesium alloy substrates in the aggressive continuous salt fog environment.

Keywords

Magnesium alloy • Corrosion protection • Forging • Micro-arc oxidation • Surface coating

Introduction

Automotive manufacturers are facing huge challenges for vehicle weight saving to improve fuel efficiency and reduce greenhouse gas emissions [1]. Magnesium and magnesium alloys, due to their excellent specific strength, have attracted considerable attention for lightweight applications and are being considered for external components like the lower front-end control arm [2–4]. In service, these components are subjected to both corrosive environments like the road salt and cyclic mechanical loading. Mg is known for its inferior corrosion properties in such service environments [5, 6], and it is of crucial importance to develop effective corrosion protection strategies for Mg alloy components.

Surface coatings are considered very effective for corrosion protection of Mg alloys [7]. Recently, the micro-arc oxidation (MAO) surface coating technique has attracted significant attention for enhancing the corrosion resistance of Mg alloys and most of the researches focused on the optimization of MAO processing parameters or electrolyte composition [8–12]. However, studies on the influence of manufacturing processes such as forging on the corrosion properties and subsequent MAO treatment of wrought Mg alloys are limited.

In this work, the corrosion behavior of uncoated and micro-arc oxidation (MAO) coated AZ80 forged magnesium alloy has been studied. The influence of forging process and MAO surface treatment parameters on the corrosion properties of the AZ80 wrought magnesium alloy was investigated using salt spray corrosion chamber test and microscopic characterization techniques. Here, AZ80 alloy is selected as a potential candidate alloy for load-bearing

X. Pang (✉)
CanmetMATERIALS, Natural Resources Canada, 183 Longwood Road South, Hamilton, ON L8P 0A5, Canada
e-mail: xin.pang@nrcan-rncan.gc.ca

Y. Xue
School of Materials Science and Engineering, Xi'an Shiyou University, Xi'an, Shaanxi, 710065, China

H. Jahed
Department of Mechanical and Mechatronics Engineering, University of Waterloo, 200 University Avenue West, Waterloo, ON N2L 3G1, Canada

S. Barela et al. (eds.), *Magnesium Technology 2023*, The Minerals, Metals & Materials Series,
https://doi.org/10.1007/978-3-031-22645-8_16

component applications due to their high strength and good forgeability.

Experimental

Materials and Forging Process

Commercially available AZ80 Mg alloy in the form of extrusion (8.2 Al, 0.42 Zn, 0.31 Mn, 0.10 Si, 0.05 Cu, 0.005 Ni, and 0.005 Fe, and other impurities ≤ 0.30 wt.% as per ASTM B91-12 standard [13]) was used as the starting material for forging in this investigation. Extruded rods (AZ80E) from Magnesium Elektron North America Inc. were forged at CanmetMATERIALS, Natural Resources Canada (Hamilton, Canada), using a hydraulic press of 500 tons with I-beam shaped molds. Prior to forging, the billet and tooling were heated to and kept at desired temperatures to reduce thermal gradients. The forging direction was perpendicular to the extrusion orientation of the AZ80E (Fig. 1). The forging process was conducted in a single step at a displacement rate of 20 mm/s at 250 °C, 300 °C, and 450 °C, respectively. Flat coupons of different dimensions were cut from the as-extruded AZ80E and extrusion-forged I-beam AZ80F alloys (Fig. 2), considering the size limitation of the materials (i.e. $50 \times 25 \times 3$ mm^3 for AZ80E, and $65 \times 25 \times 3$ mm^3, $65 \times 20 \times 3$ mm^3, and $65 \times 10 \times 3$ mm^3 for AZ80F).

Metallography and Scanning Electron Microscope (SEM) Analysis

For microstructure analysis via light optical microscope (LOM), the coupons were cold mounted in resin, ground using SiC sandpapers from 400 to 1200 grit, and then polished with diamond paste. The etching reagent used was an acetic-picral solution consisting of 10 mL acetic acid glacial, 4.2 g picric acid, 10 mL H_2O, and 70 mL ethanol (95%). Post etching, the specimens were washed in a stream of alcohol, dried with a blast of air, and stored in a desiccator for subsequent analysis. The microstructure and surface morphology of the MAO coated specimens were analyzed using a Philips XL30 SFEG SEM equipped with the energy dispersive X-ray spectrometry (EDS).

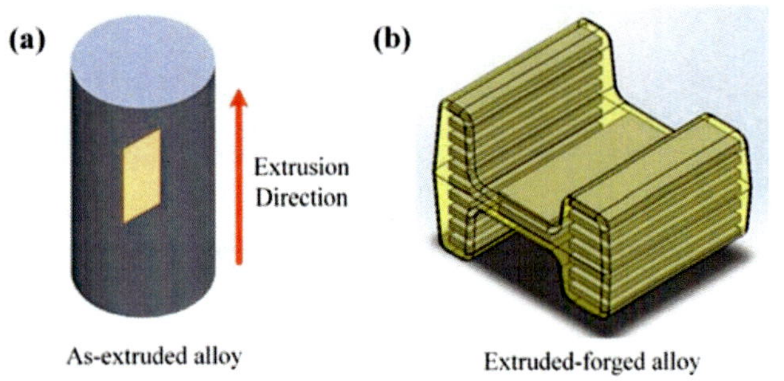

Fig. 2 Location and orientation of the coupons machined from **a** AZ80E and **b** AZ80F alloys

MAO Treatment

Prior to the MAO treatment, the specimen surfaces were polished using SiC sandpapers from 320 up to 1200 grit, degreased with acetone, rinsed with deionized water, blown dry, and stored in a desiccator immediately to prevent surface oxidation. The MAO process was carried out under a constant current mode with a pulsed current at a frequency of 500 Hz, using a single pulse direct current power supply developed by Xi'an University of Technology. The MAO electrolyte contained 0.065 mol/L Na_2SiO_3, 15 mol/L KF, and 0.18 mol/L KOH and was adjusted to pH 13 using a KOH solution. During the MAO process, the Mg alloy specimens served as the anode and a stainless steel plate was used as the cathode. The processing parameters optimized by our previous work [14] were adopted: a current density of 34 mA/cm^2, a pulse width of 80 μs, and coating time of 10 min. Post MAO, the specimens were first cleaned ultrasonically in trichloroethylene, rinsed with deionized water, air-dried, and then stored in a desiccator for further analysis.

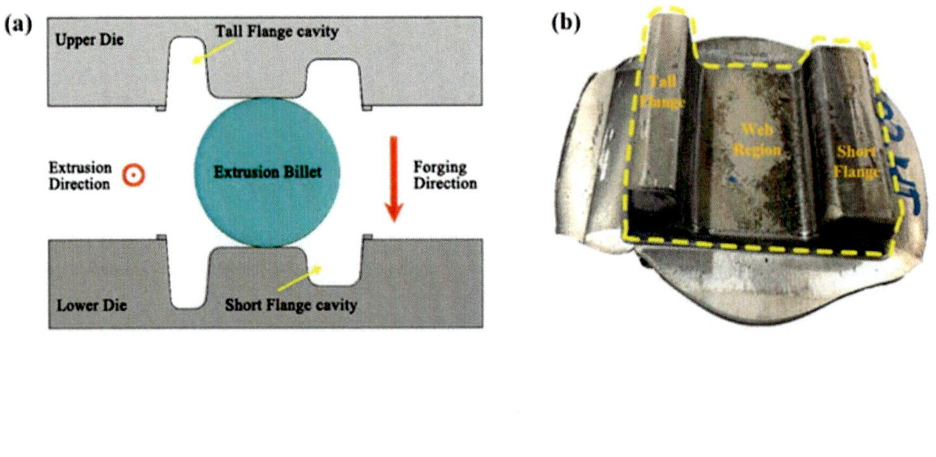

Fig. 1 **a** The schematic of the forging process and **b** photo of a forged I-Beam

Corrosion Testing

Continuous salt fog corrosion testing of the uncoated and MAO coated AZ80 alloy coupons was performed in a Singleton salt fog chamber, with a salt fog of 5 wt.% NaCl and 100% relative humidity at 35 °C solution according to ASTM B 117 standard [15] for 840 h (35 days). At least two repetitive samples of each alloy were used for mass loss measurement, and the average mass loss was presented. Prior to mass loss measurements, the corroded coupons were cleaned in a solution consisting of 200 g/L CrO_3, 10 g/L $AgNO_3$, and 20 g/L $Ba(NO_3)_2$ as per ASTM G1-03 standard [16].

Results and Discussion

Microstructure of Extruded and Forged AZ80 Alloys

Itt can be seen that the as-extruded AZ80E alloy is featured by α-Mg matrix with embedded secondary β-phase (Fig. 1). Due to its high aluminum content, the AZ80E alloy contains a high volume of β-$Mg_{17}Al_{12}$ intermetallic precipitates, forming a lamellar structure along the extrusion direction similar to what was reported in the literature [17]. In contrast, the AZ80 alloy forged at 250 °C alloy (AZ80F-250) showed a much finer microstructure (Fig. 4), which can be attributed to the dynamic recrystallization (DRX) during the forging processing at 250 °C [18]. The extrusion orientation is still evident. No significant differences in the microstructure were seen for different locations of the forged I-beam. The forging process at 250 °C induced severe plastic deformation, resulting in a more homogeneous distribution of the secondary β-$Mg_{17}Al_{12}$ phase of smaller size. This is consistent with the result reported by Wang et al. [19].

The AZ80 alloy forged at 300 °C (AZ80F-300) has a fine microstructure (Fig. 5a, b) similar to that of AZ80F-250, but with a larger DRX grain size. Pronounced grain refinement compared to the as-extruded AZ80E can be seen under higher magnification (Fig. 5b). For the AZ80 forged at 450 °C (AZ80F-450), the β-$Mg_{17}Al_{12}$ precipitates were dissolved in the α–Mg matrix forming a homogeneous microstructure with larger DRX grains than the AZ80F-300 alloy (Fig. 3).

It can be seen from Figs. 4 and 5 that the volume fraction of β-$Mg_{17}Al_{12}$ precipitates in the forged AZ80 alloys reduced with the increasing forging temperature, while the microstructure homogeneity enhanced. A smaller DRX grain size for lower forging temperature can be attributed to less thermal energy available to drive the growth of DRX grains and also to the pinning of the DRX grain boundaries by finely distributed β-$Mg_{17}Al_{12}$ particles [20–22]. Such phenomena were also observed by other researchers for AZ80 alloy. Li et al. [23] reported that similar eutectic morphologies were seen post hot deformation at 250 °C, and the β-phase precipitates would dissolve back into the α-Mg matrix at temperatures between 350 and 400 °C. The temperature range for dissolution of β-$Mg_{17}Al_{12}$ precipitates observed in our work agrees very well with the results from the literature [2, 23–24].

MAO Coated AZ80 Alloys

Corrosion protective MAO coatings were deposited on AZ80E and AZ80F alloys, and the SEM images of the surface and cross-sectional morphologies of AZ80E are presented in Fig. 6. Relatively uniform micro-pores and oxide particles are seen on the coating surface (Fig. 6a), which are typical features of MAO coatings on Mg alloys [11, 25] resulted from the throwing out of molten compound and gas bubbles from the micro-arc discharge channels during the MAO processing [26]. The ceramic MAO coating is around 15 μm thick and consists of a porous outer layer and a thin inner (barrier) layer of about 1 μm thickness at the coating/substrate interface (Fig. 6b). The relatively dense and compact barrier layer can block corrosive media from penetrating to the substrate and plays a crucial role in protecting the underneath Mg alloy against corrosion. In addition, because the MAO coating was formed under a high

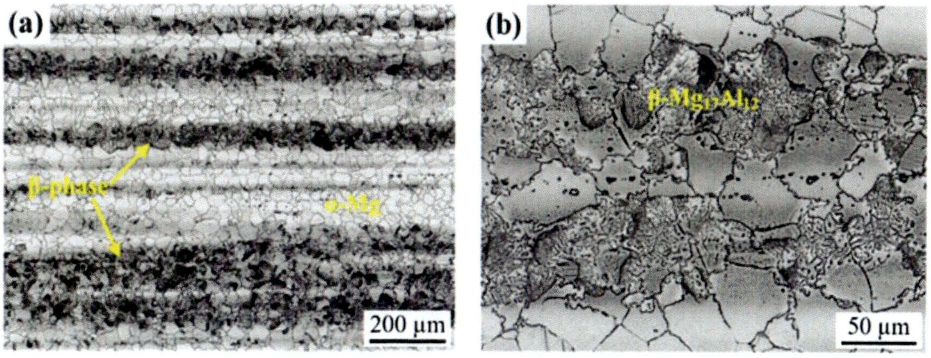

Fig. 3 Microstructure of as-extruded AZ80E alloy at different magnifications by LOM

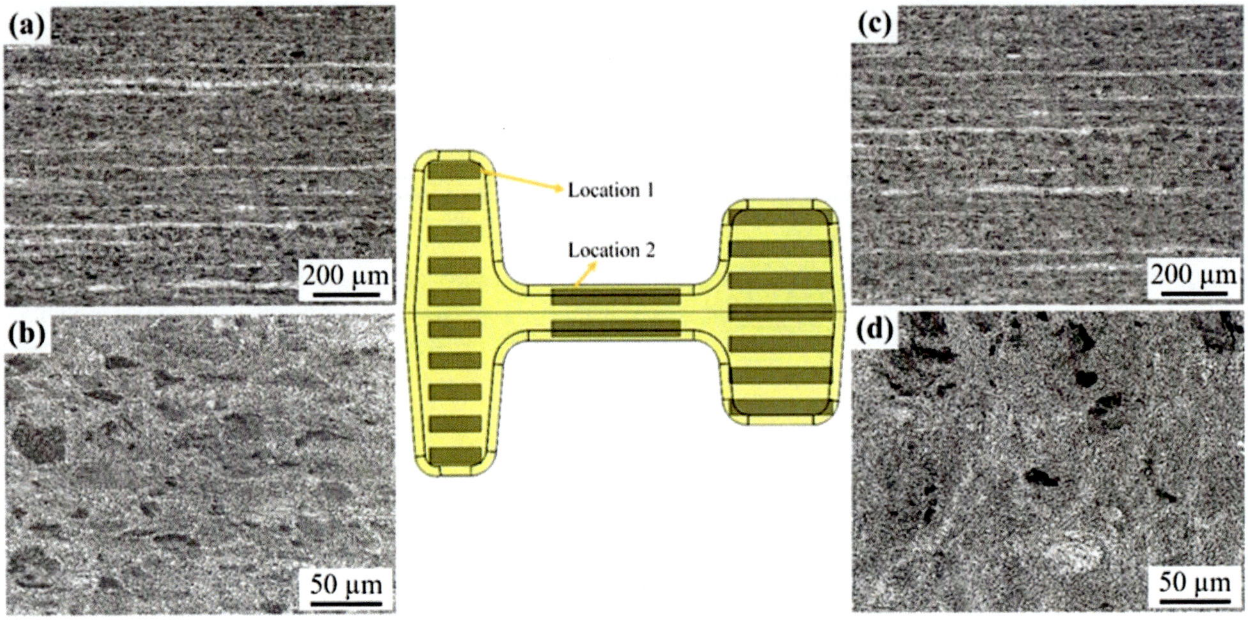

Fig. 4 Microstructure of AZ80F-250 alloy for locations **a, b** 1 and **c, d** 2 at different magnifications. The extrusion orientation is perpendicular to the cross-section shown in the center

Fig. 5 Microstructure of **a, b** AZ80F-300 and **c, d** AZ80F-450 alloys at different magnifications

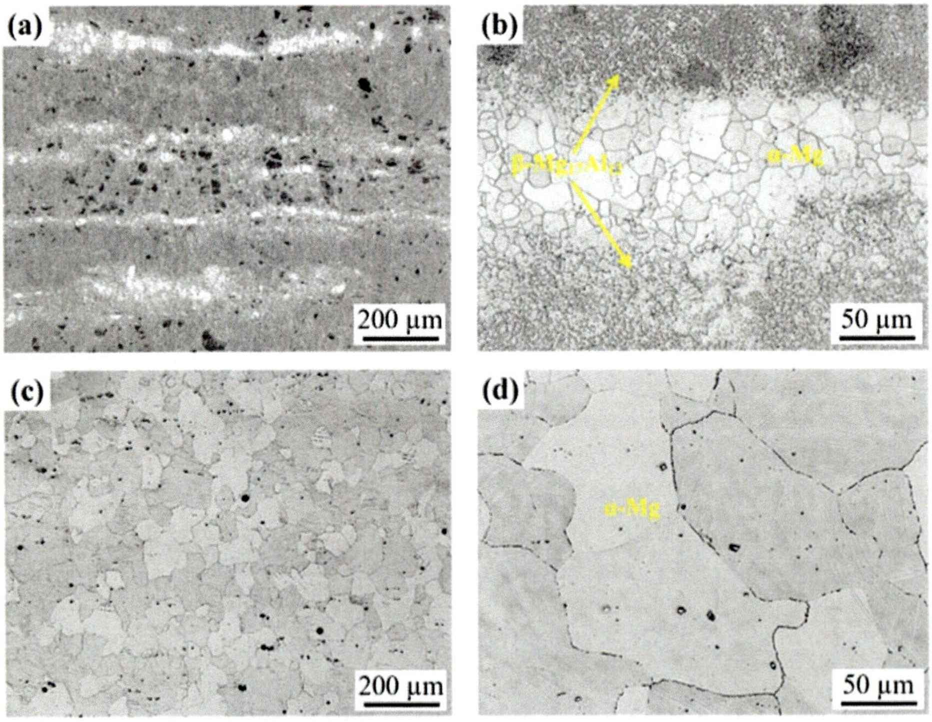

voltage plasma condition, it has a strong bonding with the Mg alloy substrate [8, 27], which confers further coating robustness and protectiveness. No significant differences in the microstructure, thickness, and surface morphology of the MAO coatings on extruded and forged AZ80 alloys were seen in the SEM analysis (not presented). Based on our previous study [28], the MAO coating is mainly composed of MgO, MgF_2, Mg_2SiO_4, and other complex silicate compounds.

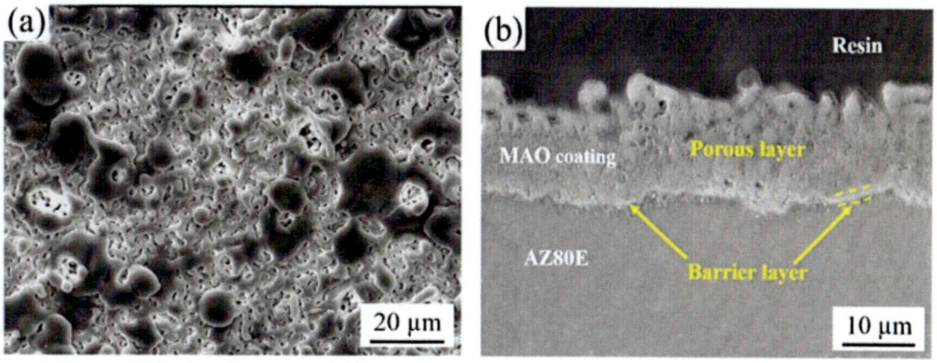

Fig. 6 The SEM images of the **a** surface and **b** cross-section of the MAO coated AZ80E alloy

Corrosion Testing of Uncoated and MAO Coated AZ80 Alloys

Specimens of uncoated and MAO coated AZ80 extruded and forged alloys were exposed to aggressive continuous salt fog environment as per ASTM B 117 standard [15]. Post 35 days, the corrosion rate calculated from mass loss for uncoated and MAO coated cast AZ80 (C), extruded AZ80 (E), and forged AZ80 (F) formed at different temperatures are presented in Fig. 7.

Among all the uncoated AZ80 specimens, cast AZ80 (C) showed the largest corrosion rate and was fully corroded before the end of 35 days. This can be ascribed to the severe micro-galvanic corrosion between the α-Mg matrix and large second-phase precipitates, and the inclusions and defects in the alloy inherited from the casting process as shown in our previous study [29]. It was reported that the distribution of secondary phase $Mg_{17}Al_{12}$ in Mg alloys significantly influences their corrosion resistance [5, 30]. The β-$Mg_{17}Al_{12}$ phase has a higher corrosion potential than the α-Mg phase and can either act as micro-cathodes accelerating the corrosion of α-Mg matrix or form a continuous network of corrosion barrier [5]. For the AZ80F-250 and AZ80F-300 alloys, the fine and more uniformly distributed second-phase particles acted as an almost continuous network of corrosion barrier network, resulting in a lower corrosion rate. For the AZ80F-450, the secondary phase dissolved into the α-Mg matrix and the corrosion barrier network was eliminated, which led to a lower corrosion resistance.

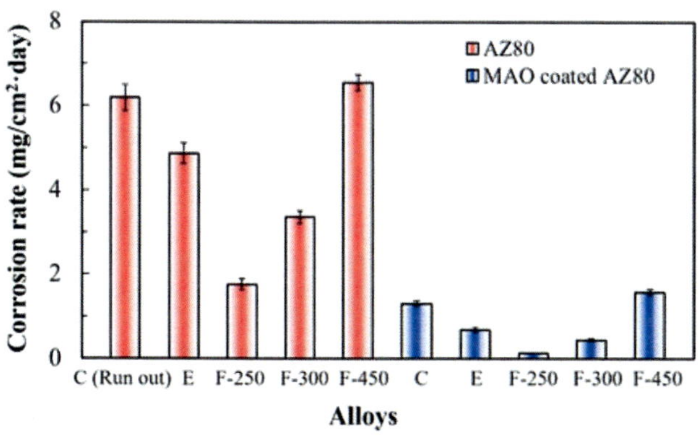

Fig. 7 The corrosion rate of uncoated and MAO coated cast AZ80 (C), extruded AZ80 (E), and forged AZ80 (F) formed at different temperatures post 35 days of salt fog test

Remarkably lower corrosion rate was seen for MAO coated AZ80 alloys compared to the uncoated alloys, showing significant corrosion protection of the Mg alloys by the hard ceramic MAO coating. Note that the MAO coating did not change the ranking of corrosion rate of the substrate Mg alloys. As was found in our previous studies [14, 28], the phase and chemical compositions of the MAO coating formed on magnesium alloys are mainly determined by the type of MAO treatment solution and chemical composition of the substrate. Therefore, MAO coatings of similar thickness and properties, offering a similar extent of corrosion protection, are expected to form on the AZ80 alloys of different processing history as they underwent the same MAO treatment process. It was observed that after almost the same length of time, i.e. around 6 days of salt fog testing, localized corrosion occurred on all MAO coated AZ80 alloys penetrating the coating. This is likely associated with the penetration of aggressive salt fog through micro-pores or micro-cracks on the MAO coatings of similar thickness on all alloys. The corrosion then propagated beneath the coating (see Fig. 8 for the example of AZ80E). Once the coating breakdown began the corrosion rate of the MAO coated specimens depended on the corrosion resistance of the substrate alloy, and thus, a very similar ranking of corrosion rate as for the uncoated AZ80 alloys is seen for the MAO coated AZ80 alloys (Fig. 7).

Conclusions

In this work, the corrosion properties of uncoated and micro-arc oxidation (MAO) coated AZ80 Mg alloys forged at different temperatures have been studied using continuous salt spray corrosion chamber test and microscopic characterization techniques. It was found that the forging

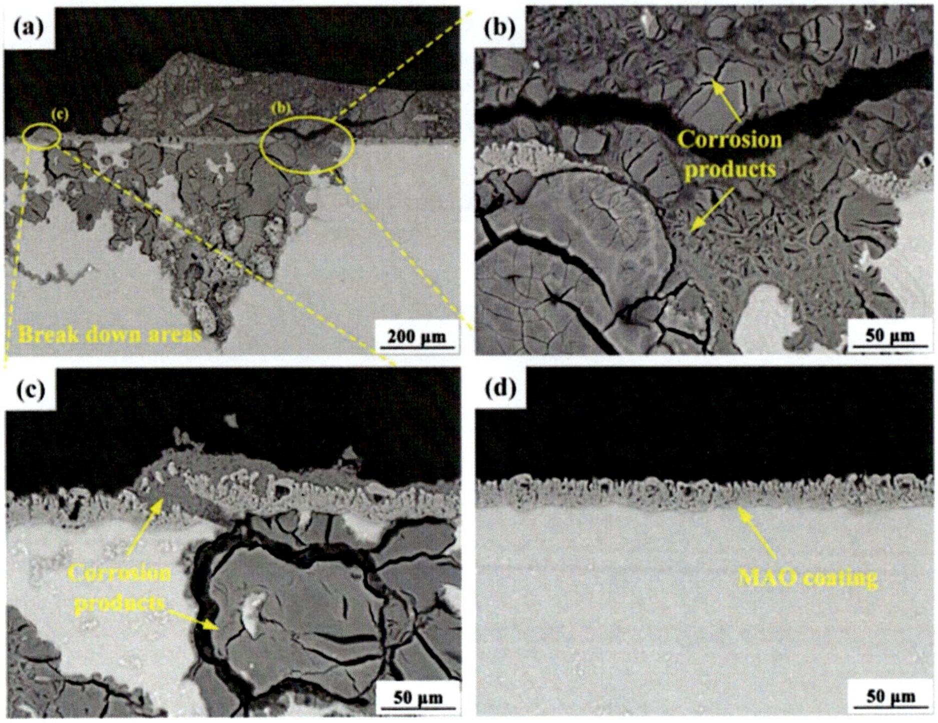

Fig. 8 Cross-sectional SEM images of the MAO coated AZ80E alloy post 6 days of salt fog test. The **a–c** coating breakdown region and **d** non-breakdown region (intact coating)

temperature significantly influenced the microstructure, in particular the volume fraction and distribution of secondary phase in AZ80 alloy, which in turn affected the corrosion rate of the alloy. The experimental results showed that the low forging temperature of 250 °C led to fine grain size and uniform distribution of secondary phase in the magnesium alloy, resulting in higher corrosion resistance. The MAO coating reduced the corrosion rate of all extruded and forged AZ80 alloy specimens and provided robust corrosion protection of the AZ80 substrates in the aggressive continuous salt fog environment.

Acknowledgements This work was financially supported by the Natural Sciences and Engineering Research Council of Canada through the Automotive Partnership Canada (APC) program under grant APCPJ 459269-13. The authors are grateful to Chao Shi and Renata Zavadil of CanmetMATERIALS, Natural Resources Canada, Dr. Sugrib Shaha, and Dr. Jie Wang of the University of Waterloo for their technical support.

References

1. D. S. Kumar, C. T. Sasanka, K. Ravindra and K. Suman, "Magnesium and its alloys in automotive applications - A Review," *Am. J. Mater. Sci. Technol.*, vol. 4, pp. 12-30, 2015.
2. A. Gryguc, S. B. Behravesh, S. K. Shaha, H. Jahed, M. Wells, B. Williams and X. Su, "Low-cycle fatigue characterization and texture induced ratcheting behaviour of forged AZ80 Mg alloys," *Int. J. Fatigue,* vol. 116, pp. 429-438, 2018.
3. A. Gryguc, S. K. Shaha, S. B. Behravesh, H. Jahed, M. Wells, B. Williams and X. Su, "Monotonic and cyclic behaviour of cast and cast-forged AZ80 Mg," *Int. J. Fatigue,* vol. 104, pp. 136-149, 2017.
4. D. Toscano, S. K. Shaha, B. Behravesh, H. Jahed and B. Williams, "Effect of forging on the low cycle fatigue behavior of cast AZ31B alloy," *Mater. Sci. Eng. A,* vol. 706, pp. 342-356, 2017.
5. G. L. Song and A. Atrens, "Corrosion mechanisms of magnesium alloys," *Advanced Engineering Materials,* vol. 1, pp. 11-31, 2000.
6. L. J. Liu and M. Schlesinger, "Corrosion of magnesium and its alloys," *Corrosion Science,* vol. 51, no. 8, pp. 1733-1737, 2009.
7. J. Wang, X. Pang and H. Jahed, "Surface protection of Mg alloysin automotive applications: a review," *AIMS Materials Science,* vol. 6, no. 4, pp. 567-600, 2019.
8. B. L. Jiang and Y. F. Ge, "Micro-arc oxidation (MAO) to improve the corrosion resistance of magnesium (Mg) alloys," in *Corrosion Prevention of Magnesium Alloys*, G. Song, Ed., Woodhead Publishing Limited, 2013, pp. 163–196.
9. B. V. Vladimirov, B. L. Krit, V. B. Lyudin, N. V. Morozova, A. D. Rossiiskaya, I. V. Suminov and A. V. Epel'feld, "Microarc oxidation of magnesium alloys: A review," *Surf. Eng. Appl. Electrochem.*, vol. 50, pp. 195-232, 2014.
10. P. Su, X. Wu, Y. Guo and Z. Jiang, "Effects of cathod current density on structure and corrosion resistance of plasma electrolytic oxidation coatings formed on ZK60 Mg alloy," *J. Alloys Compd.*, vol. 475, pp. 773-777, 2009.
11. V. Ezhilselvi, J. Nithin, J. N. Balaraju and S. Subramanian, "The influence of current density on the morphology and corrosion properties of MAO coatings on AZ31B magnesium alloy," *Surf. Coatings Technol.*, vol. 288, pp. 221-229, 2016.
12. R. F. Zhang, S. F. Zhang, J. H. Xiang, L. H. Zhang, Y. Q. Zhang and S. B. Guo, "Influence of sodium silicate concentration on properties of micro arc oxidation coatings formed on AZ91HP magnesium alloys," *Surf. Coatings. Technol.,* vol. 206, pp. 5072-5079, 2012.
13. ASTM B91–17, *Standard specification for magnesium-alloy forgings,* ASTM International, 2017.

14. Y. Xue, X. Pang, B. Jiang and H. Jahed, "Corrosion and corrosion fatigue performances of micro-arc oxidation coating on AZ31B cast magnesium alloy," *Materials and Corrosion*, vol. 70, no. 2, pp. 268-280, 2019.
15. ASTM B117–17, *Standard practice for operating salt spray (fog) apparatus*, ASTM International, 2017.
16. ASTM G1–03, *Standard practice for preparing, cleaning, and evaluating corrosion test specimens*, ASTM International, 2011.
17. R. C. Zeng, Y. B. Xu, W. Ke and E. H. Han, "Fatigue crack propagation behavior of an as-extruded magnesium alloy AZ80," *Mater. Sci. Eng. A*, vol. 509, no. 1-2, pp. 1-7, 2009.
18. S. Lee, J. Seok, S. Park, S. Hyuk and J. Yoon, "Evolution of tension and compression asymmetry of extruded Mg-Al-Sn-Zn alloy with respect to forming temperatures," *Mater. Des.*, vol. 110, pp. 510-518, 2016.
19. L. Wang, E. Mostaed, X. Cao, G. Huang, A. Fabrizi, F. Bonollo, C. Chi and M. Vedani, "Effects of texture and grain size on mechanical properties of AZ80 magnesium alloys at lower temperatures," *Mater. Des.*, vol. 89, pp. 1-8, 2016.
20. S. W. Xu, S. Kamado, N. Matsumoto, T. Honma and Y. Kojima, "Recrystallization mechanism of as-cast AZ91 magnesium alloy during hot compressive deformation," *Mater. Sci. Eng. A*, vol. 527, pp. 52-60, 2009.
21. P. Changizian, A. Zarei-Hanzaki and H. R. Abedi, "On the recrystallization behavior of homogenized AZ81 magnesium alloy: The effect of mechanical twins and r precipitates," *Mater. Sci. Eng. A*, vol. 558, pp. 44-51, 2012.
22. F. Guo, D. Zhang, X. Yang, L. Jiang and F. Pan, "Strain-induced dynamic precipitation of Mg17Al12 phases in Mg-8Al alloys sheets rolled at 748K," *Mater. Sci. Eng. A*, vol. 636, pp. 516-521, 2015.
23. H. Z. Li, X. Y. Wei, J. Ouyang, J. Jiang and Y. Li, "Hot deformation behavior of extruded AZ80 magnesium alloy," *Trans. Nonferrous Met. Soc. China (English Ed.)*, vol. 23, pp. 3180-3185, 2013.
24. A. Gryguc, S. B. Behravesh, S. K. Shaha, H. Jahed, M. Wells, B. Williams and X. Su, "Multiaxial cyclic behaviour of extruded and forged AZ80 Mg alloy," *Int. J. Fatigue*, vol. 127, pp. 324-337, 2019.
25. H. Duan, C. Yan and F. Wang, "Effect of electrolyte additives on performance of plasma electrolytic oxidation films formed on magnesium alloy AZ91D," *Electrochim. Acta*, vol. 52, pp. 3785-3793, 2007.
26. H. F. Guo and M. Z. An, "Growth of ceramic coatings on AZ91D magnesium alloys by micro-arc oxidation in aluminate-fluoride solutions and evaluation of corrosion resistance," *Appl. Surf. Sci.*, vol. 246, pp. 229-238, 2005.
27. S. V. Gnedenkov, O. A. Khrisanfova, A. G. Zavidnaya, S. L. Sinebrukhov, P. S. Gordienko, S. Iwatsubo and A. Matsui, "Composition and adhesion of protective coatings on aluminum," *Surf. Coatings Technol.*, vol. 145, pp. 146-151, 2001.
28. Y. Xue, X. Pang, B. Jiang, H. Jahed and D. Wang, "Characterization of the corrosion performances of as-cast Mg-Al and Mg-Zn magnesium alloys with micrarc oxidation coatings," *Materials and Corrosion*, vol. 71, no. 6, pp. 992-1006, 2020.
29. Y. Xue, X. Pang, B. Jiang and H. Jahed, "Corrosion performances of micro-arc oxidation coatings on AZ31B, AZ80, and ZK60 cast Mg alloys," in *Proceedings of the Canadian Society for Mechanical Engineering International Congress*, May 27–30, 2018, Toronto, Ontario, Canada.
30. M. Ben-Haroush, G. Ben-Hamu, D. Eliezer and L. Wagner, "The relation between microstructure and corrosion behavior of AZ80 Mg alloy following different extrusion temperatures," *Corrosion Science*, vol. 50, pp. 1766–1778, 2008.

Effect of Deformation Speed on Stress Corrosion and Fracture Toughness of Extruded Mg10Dy and Mg10Dy1Nd Using C-Ring Tests

Petra Maier, Easwar Pamidi, Benjamin Clausius, and Norbert Hort

Abstract

The influence of the deformation speed in C-ring tests in Ringer's solution on crack initiation and propagation of extruded Mg10Dy and Mg10Dy1Nd is investigated. Deformation speeds varying from 2 to 0.012 mm/min allow corrosion times from a few minutes to hours. Both the crack initiation force (higher for Mg10Dy1Nd) and displacement (higher for Mg10Dy) increase with decreasing deformation speed up to a corrosion time of 1 h and then decrease, more for Mg10Dy1Nd and slightly more for the displacement compared to the force at higher corrosion times. The decrease is associated with the higher corrosion times—corrosion pits become visible at a test time of 1 h on the tensile side. In Mg10Dy1Nd the fracture toughness increases with decreasing deformation speed, and no clear picture is seen for Mg10Dy. Sub-cracks often initiate at corrosion pits, which show a correlation to twinned grains at the tensile side—increasing fracture toughness.

Keywords

Magnesium • C-ring tests • Fracture toughness • Crack propagation

Introduction and Motivation

This study focuses on the alloy Mg−Dy and Mg−Dy−Nd, which together with Zn and Zr are the basis for Resoloy® [1], a bioresorbable Mg-alloy for scaffolds. Research by Yang et al. [2] showed that the binary cast Mg–Dy alloy has the best combination of mechanical and corrosive properties with 10 wt. % Dy when the Dy alloy is between 5 and 20 wt. %. Dy is interesting because of its high solubility in Mg, so mechanical and corrosion properties can be well tailored by heat treatment [3]. Nd has a low solid solubility in Mg and is of interest because it forms stable intermetallic phases mainly at grain boundaries via precipitation hardening. In addition, Dy improves the corrosion resistance of Mg alloys by minimizing the effects of galvanic corrosion between the secondary phases and the Mg-matrix and leading to improved passive layers [4–6].

Some Mg alloys are susceptible to stress corrosion cracking (SCC) in aqueous environments, which can lead to hydrogen embrittlement. AZ61 was found to be highly susceptible to SCC, and cracking initiation occurs already in the elastic region at ~50% of the ultimate tensile strength [7]. According to [8] strain rate is a critical variable in slow strain rate tensile testing. The SCC susceptibility for as-cast AZ80 and AZ31 was found to increase with a decrease in strain rate, and higher for AZ80 due to the higher volume fraction of $Mg_{17}Al_{12}$ particles [9]. Suitable alloy selection can increase the resistance to SCC [10]. According to [11], ZX50 suffers more stress corrosion cracking than WZ21 and WE43, which was attributed to its higher susceptibility towards localized corrosion/pitting. The SCC mechanism for these extruded alloys was concluded by the combined effect of hydrogen-assisted cracking and anodic dissolution. Study [12] showed increased resistance to SCC in both distilled water and 0.5 wt% NaCl solution for EV31A, known commercially as Elektron 21, compared to ZE41 and QE22. Rare earths, such as Nd, can improve the SCC resistance of EV31A. A slow strain rate test showed that EV31A was less susceptible to SCC than AZ91E [13]. Under a low-intensity stress situation using constant load, EV31A even proved to be resistant. The fractography of EV31A showed little evidence of hydrogen embrittlement. C-ring specimens are preferred for determining SCC susceptibility because they

P. Maier (✉) · E. Pamidi · B. Clausius
University of Applied Sciences Stralsund, Stralsund, Germany
e-mail: Petra.maier@hochschule-stralsund.de

N. Hort
Helmholtz-Zentrum Hereon, Geesthacht, Germany

N. Hort
Leuphana University Lüneburg, Lüneburg, Germany

S. Barela et al. (eds.), *Magnesium Technology 2023*, The Minerals, Metals & Materials Series,
https://doi.org/10.1007/978-3-031-22645-8_17

ensure a complex strain field with very high local tensile and compressive strains [14, 15]. For this study, Ringer's solution is used because it can simulate the corrosion behavior of biodegradable metals.

The motivation of this study is the effect of slowing down the strain rate (deformation speed) in C-ring tests on the stress corrosion susceptibility of Mg10Dy(1Nd), especially on fracture toughness. Previous research [16] has shown that the slow deformation rates (with corrosion times up to 24 h) also lead to crack initiation, and crack initiation proceeds from corrosion pits after some corrosion time (1 h). Strong pitting corrosion, as after immersion tests (7 and 20 days) in [17], did not occur. Whether fracture toughness and crack propagation, being influenced by microstructural feature, is dependent on the deformation speed will be evaluated here. The studies in [16, 17] have shown that stress corrosion can be inferred from parallel sub-cracks from the main crack and crack propagation is strongly influenced by twins in the microstructure or these are responsible for crack opening under tensile loading, at higher corrosion times due to the corrosion pits formed on the twins. Mg10Dy and Mg10Dy1Nd were cast and hot-extruded according to [17]. That study further shows that the grain size for Mg10Dy with 36.2 ± 10.2 μm is slightly higher than that for Mg10Dy1Nd with 33.6 ± 9.7 μm. The ternary alloy shows a less homogeneous microstructure, and the second phases are attached to grains of small grain size and aligned in extrusion direction. Mg10Dy shows a higher corrosion rate in immersion: 2.46 mm/year versus 1.53 mm/year for Mg10Dy1Nd. Static stress corrosion at 7 mm deformation, being in the plastic region, over 7 and 20 days shows significantly stronger corrosion attack for Mg10Dy. Large pitting best describes the corrosion morphology, significantly more pronounced in Mg10Dy. However, even after a corrosion duration of 20 days, the C-rings were intact despite severe corrosion and by that cross-section reduction—cracking at corrosion pits was not observed.

Experimental

C-ring specimens with an outer diameter of 34 mm and a width of 10 mm were milled from extruded tubes. The wall thickness was selected as 2 mm. C-ring tests in Ringer's solution were made at the following deformation speeds: 2, 1, 0.283, 0.142, 0.057, 0.028 , and 0.012 mm/min displacement controlled. Force–displacement curves were recorded on each of the 3 specimens to evaluate the force and the displacement at crack initiation, and the fracture toughness (via fracture energy of the force–displacement curve section starting from crack initiation). Table 1 shows the maximum corrosion times that would occur if the specimens were to hold up to the geometric restriction of 17 mm. The plastically deformed C-rings are ground in 4 planes analogous to [17] and prepared metallographically.

Figure 1 shows the crack propagation curve section, starting from crack initiation, of the force–displacement curves of the C-ring tests of Mg10Dy (Fig. 1a) and Mg10Dy1Nd (Fig. 1b) of the selected deformation rates. A clear influence of the deformation speed becomes visible, also that Mg10Dy1Nd reaches higher forces (up to 260 N), but also lower deformation paths (up to 7 mm).

The evaluation of the force–displacement curves shows that both, the force and the displacement until crack initiation (see here Fig. 2a), first increase slightly with decreasing deformation speed and then consistently decrease. Regarding the values of the force [16], it is higher for Mg10Dy1Nd compared to Mg10Dy, and the values for the displacement are higher for Mg10Dy. Studies [15, 16] attribute this to the more inhomogeneous microstructure of Mg10Dy1Nd. The bar graph of the displacement values at crack initiation in Fig. 2a shows that for the deformation speeds of 1 and 0.283 mm/min the specimens fail, if at all, only very close to the geometric restriction of 17 mm. Thus, the fracture toughness for these two speeds, see Fig. 2b, is not quite comparable and is lower than if deformation could exceed 17 mm. In the bar chart in Fig. 2b especially values for Mg10Dy were excluded for calculating the mean value due to the better ductility (later displacement at crack initiation). However, Fig. 2b makes clear that the lowest two deformation speeds (0.028 and 0.012 mm/min) lead to a higher fracture toughness in comparison with 2 and 0.142 mm/min.

For the discussion of the influence of the deformation speed on crack initiation and propagation, and stress corrosion, it will be useful to take the time needed for crack initiation in account. By converting the displacement at crack initiation into time for each deformation speed (this is possible due to the displacement control), it becomes clear, see Fig. 3, that the Mg10Dy1Nd alloy exhibits worse values compared to the Mg10Dy—lower values are seen, meaning less time. Up to deformation speeds of 0.142 mm/min the difference is not obvious, but above that, from 0.057 mm/min, it is clearly seen. Figure 3 shows in a different way to Fig. 2a the difference among the deformation speeds, especially at the slower ones, where for Mg10Dy at 0.057 mm/min the highest time of 300 min, see Table 1, is still almost reached, but Mg10Dy1Nd starts falling behind the maximum value.

For direct comparison of the two alloys and discussion of the extent of stress corrosion on the C-ring surfaces, and possible attack on the crack flanks during crack propagation, the specific corrosion times including the duration of crack propagation must be taken into account. Therefore Table 2 shows the mean time for crack propagation, calculated by the difference in displacement—this duration is then added to the time to crack initiation in Fig. 4. It can be seen that the

Table 1 Maximum corrosion time according to deformation speed

Deformation speed (mm/min)						
2	1	0.283	0.142	0.057	0.028	0.012
Maximum corrosion time for reaching 17 mm (s)						
8.5	17	60	120	300	600	1440

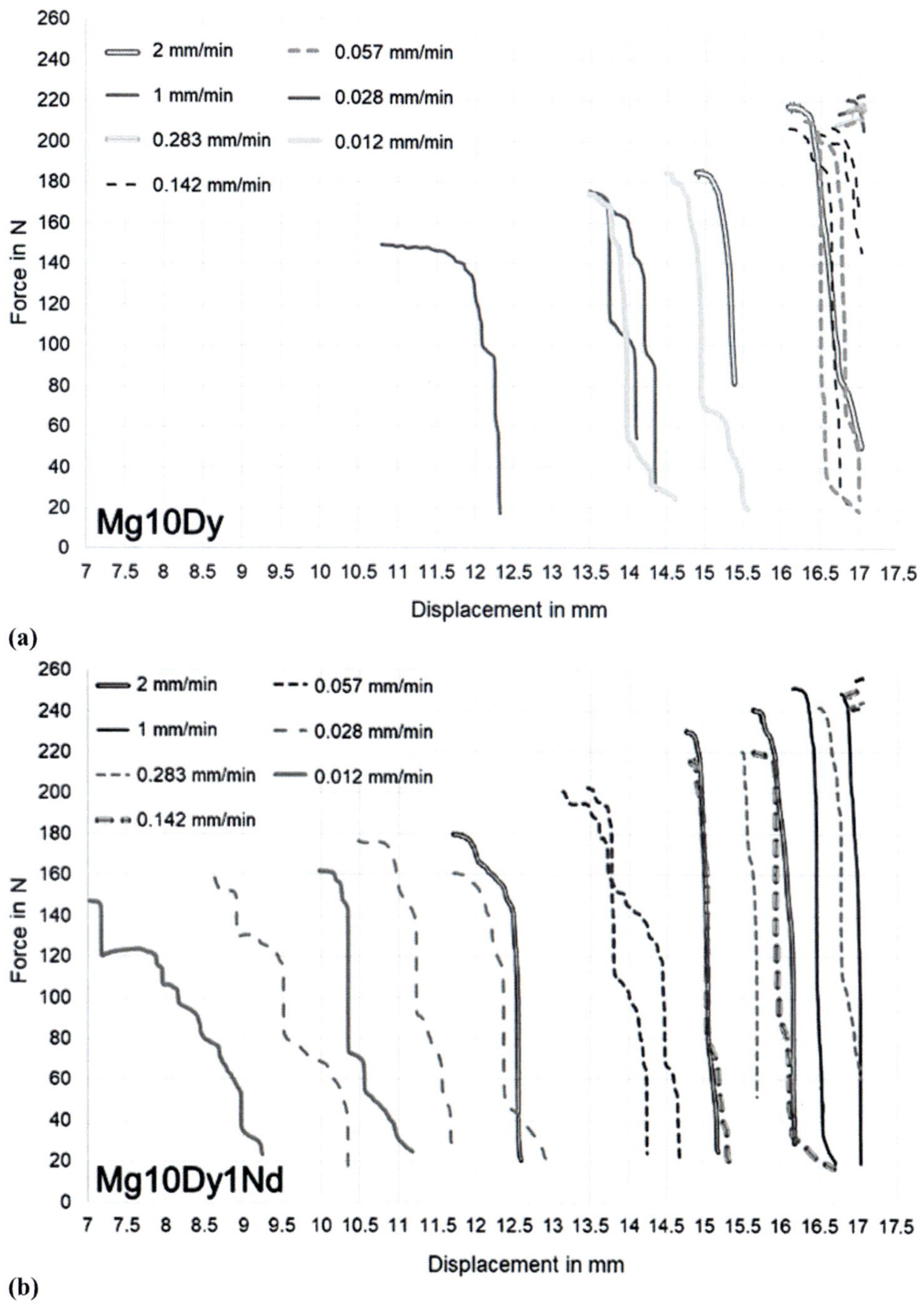

Fig. 1 Crack propagation curve section of the force–displacement curves of the C-ring tests of Mg10Dy (**a**) and Mg10Dy1Nd (**b**) of the selected deformation rates

duration of crack propagation increases with decreasing deformation speed.

The duration of crack propagation, being the time for corrosion of the crack flanks as well as a contribution to the fracture toughness, and additional time for surface corrosion, is for Mg10Dy1Nd, which showed less displacement for crack initiation, higher than for Mg10Dy. Like mentioned in the introduction, the study on the corrosion rate and

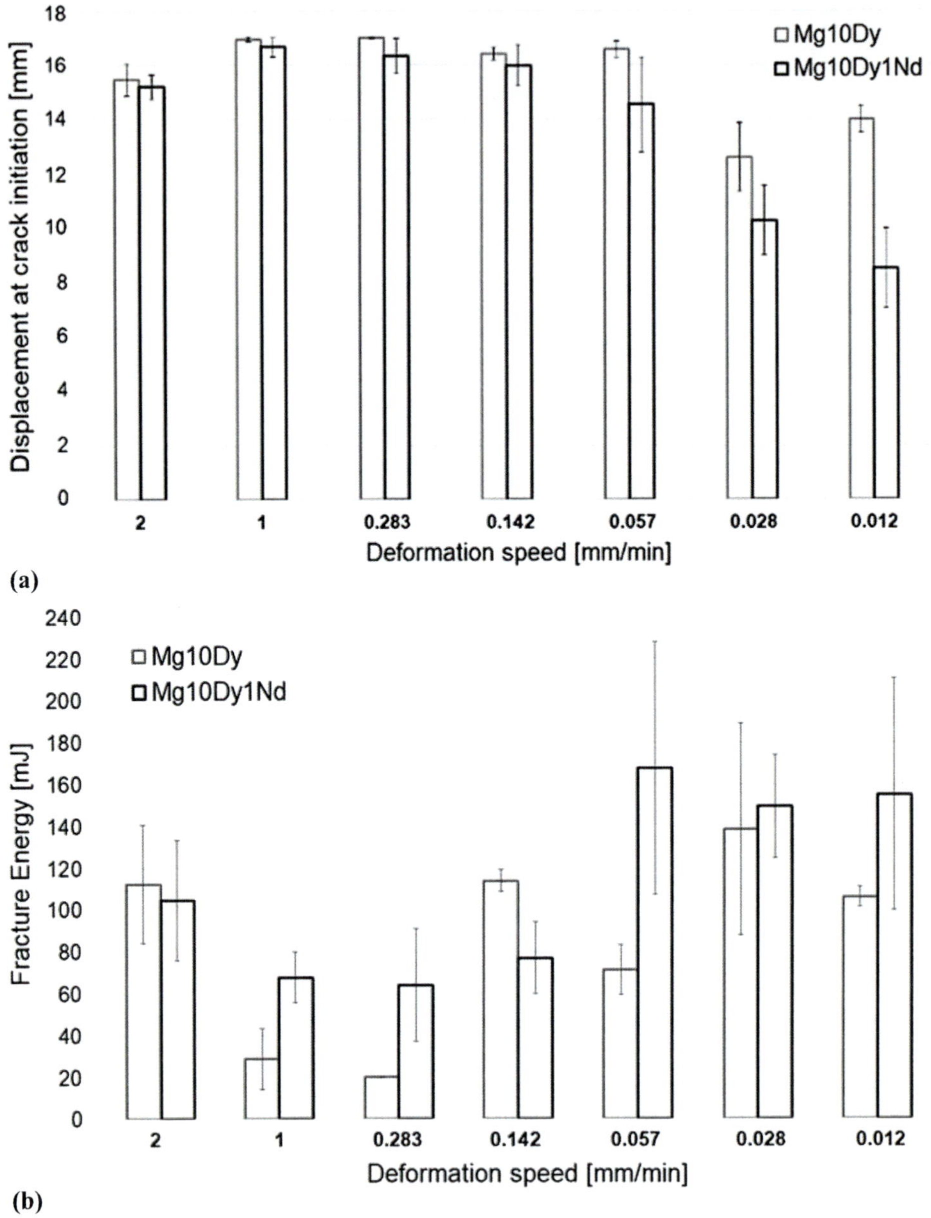

Fig. 2 Influence of deformation speed on displacement at crack initiation (**a**) and fracture energy (**b**) of Mg10Dy and Mg10Dy1Nd

corrosion morphology [17] showed that Mg10Dy has a higher corrosion rate—so the opposite behavior could have been expected.

Figures 5 and 6 show representative cross-sectional micrographs of the fractured C-rings of Mg10Dy and Mg10Dy1Nd of different deformation rates: 1 and 0.028 mm/min. Whereas the cracks mostly run through the neutral zone to the compressive side; in some samples, the crack stops in the tensile side (see Fig. 2a for samples, which almost reach 17 mm—mostly Mg10Dy at 1 and 0.283 mm/min), here shown for Mg10Dy at 1 mm/min.

The cross-sectional micrographs in Fig. 5 reveal that the number of secondary cracks is more pronounced in the compression-hardened region. Here twinning has already changed the microstructure and interferes with the crack growth. A difference in crack propagation among the different deformation speeds cannot be seen. Figure 6 shows the crack propagation in more detail: transgranular, cleavage crack propagation, which is mostly twin-initiated. Like described in [17] the twins will act as a potent stress concentrator in the nucleation and propagation of cracks. Here the deformation speed does not change the mechanism. Even at higher times for crack propagation and therefore for exposure of the crack flanks to the corrosion media, no sub-cracks form at corrosion pits along the crack flanks. The assumption, based on [18], was different—in that study on

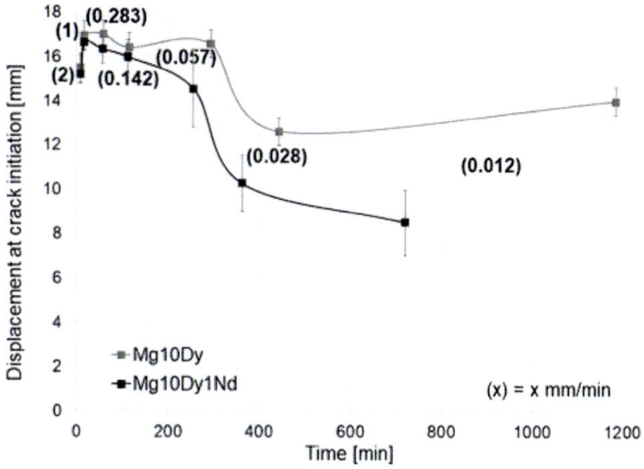

Fig. 3 Influence of deformation speed, converted to its corresponding time, on displacement at crack initiation of Mg10Dy and Mg10Dy1Nd

Table 2 Crack propagation time in min (mean value)

Deformation speed (mm/min)						
2	1	0.283	0.142	0.057	0.028	0.012
Crack propagation time (min)						
Mg10Dy						
0.26	0.14	0.33	4.21	11.44	35.65	88.44
Mg10Dy1Nd						
0.31	0.42	1.38	5.50	20.52	49.28	145.24

crack propagation under 3-point-bending in cast Mg10GdxNd-T4 alloys corrosion pits formed along the crack flanks.

Only some specimens, see here Mg10Dy at 0.012 mm/min, which have been exposed to the corrosion media all together for 21 h, show some corrosion attack along the crack flanks—mostly at the surface near region,

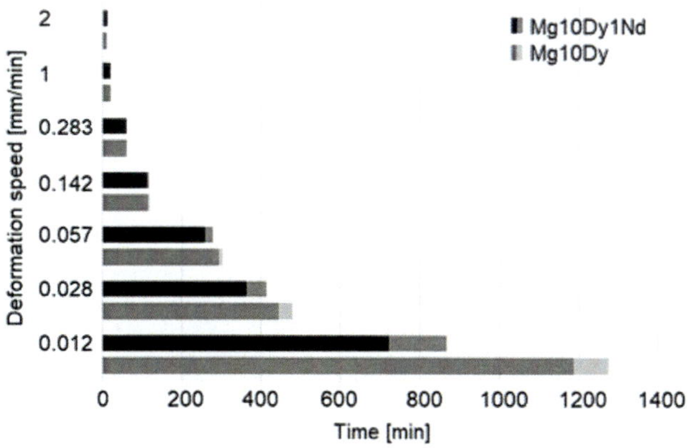

Fig. 4 Influence of deformation speed, converted to its corresponding time (mean value), on time to crack initiation and duration of Mg10Dy and Mg10Dy1Nd

see Fig. 7. Most likely corrosion took part during further crack opening. Also the surface of the tensile loaded side of the C-ring shows strong corrosion attack—the pits have a rounded shape, where the depth is almost the same as the width. The pits start overlapping.

Figure 8 shows surface cracks in cross-sectional micrographs of corroded C-ring surfaces at tensile side of Mg10Dy (Fig. 8 a, c, e, g) and Mg10Dy1Nd (Fig. 8b, d, f, h) at 2 mm/min (Fig. 8a, b), 0.283 mm/min (Fig. 8c, d), 0.028 mm/min (Fig. 8e, f), and 0.012 mm/min (Fig. 8g, h). It can be seen, more pronounced at higher deformation speeds, that the near surface region (up to 50 μm) twinned under the tensile load and that corrosion attack becomes visible starting from 0.283 mm/min. The corrosion pits and the surface cracks are influenced from the twin boundaries, very convincing in Fig. 8g, where the large grain shows 3 parallel twins to the crack formed. The corrosion pits have a critical pit shape to begin with, and it is obvious that the crack initiates from the pit—see Fig. 8c, and d, the corrosion pit widening the crack opening at the surface before the crack becomes thinner by growing. Figure 8h corresponds with Fig. 7 and shows that the pit shape, once the stress intensity has been released, grows into a half-round shape and starts overlapping.

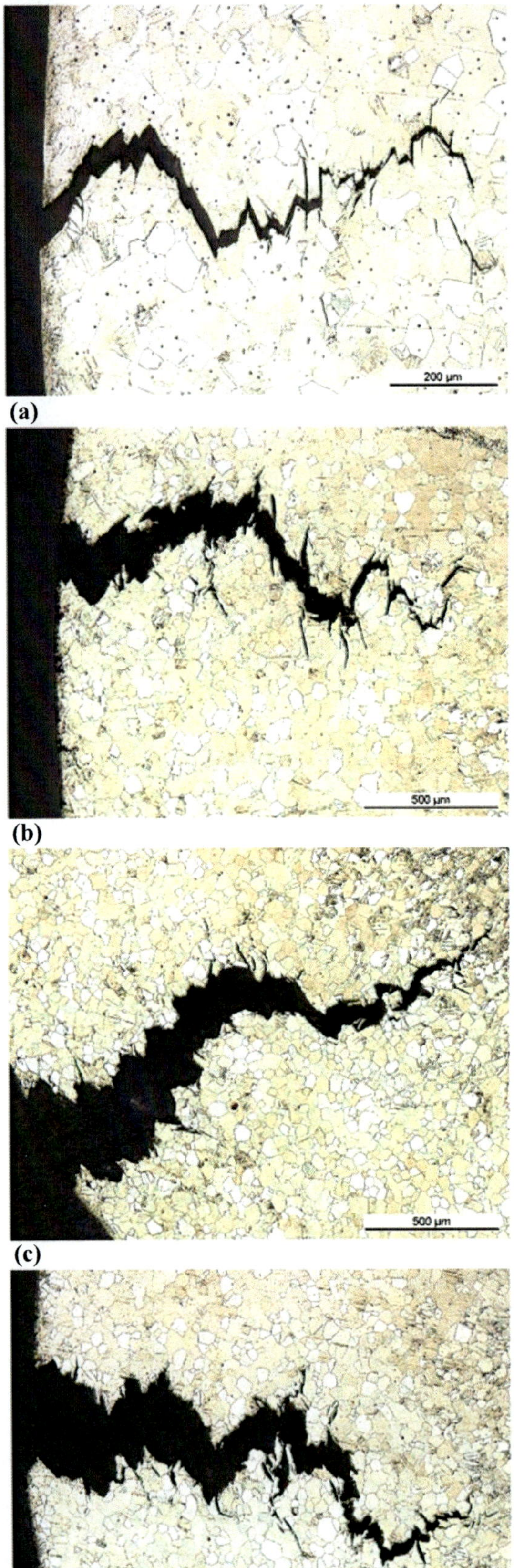

Fig. 5 Cross-sectional micrographs of crack path in Mg10Dy (**a**, **b**) and Mg10Dy1Nd (**c**, **d**) at 1 mm/min (**a**, **c**) and 0.028 mm/min (**b**, **d**), due to shorter crack in (**a**) the magnification is higher

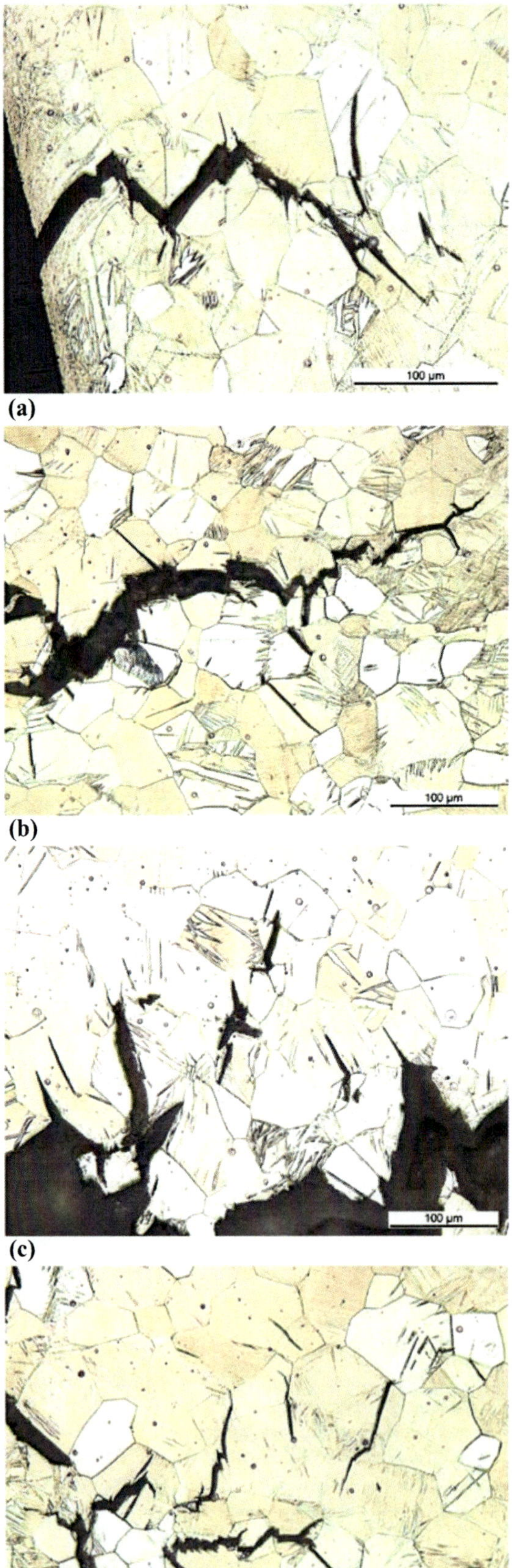

Fig. 6 Cross-sectional micrographs of crack propagation in Mg10Dy (**a**, **c**) and Mg10Dy1Nd (**b**, **d**) at 1 mm/min (**a**, **b**) and 0.028 mm/min (**c**, **d**)

Fig. 7 Cross-sectional micrograph of a second larger surface crack at the corroded C-ring surface at tensile side of Mg10Dy at 0.012 mm/min

The micrographs in Fig. 8 allow the conclusion that the corrosion pits are regions for crack initiation and influence the fracture toughness in following way: at all corrosion pits the stress intensity increases depending size and shape, more than one crack forms at the surface at the critical corrosion pits, twin boundaries will support crack initiation, all allow to absorb energy, and the most critical microcrack growth into the macrocrack being responsible for the final failure. Apart from the crack in Fig. 8c, which is the main crack in that specimen, all the cracks shown are in the 500 µm region of the main crack. That the combination of twin boundaries, corrosion pits, and tensile stress is reducing the stress corrosion resistivity is supported by the fact that also the microstructure of the compression-loaded side of the C-ring is twinned, but none of these conspicuous corrosion pits as crack nucleation sites are seen. Only after corrosion times of more than a few hours shallow, very wide pits become evident.

The increase of fracture toughness during the C-ring tests is indirectly influenced by slowing down the deformation speed: lower deformation speeds lead to more corrosion pits, increasing stress intensity, causing a plastic deformation zone, and eventually nucleating micro-cracks. Every energy delaying final fracture by plastic deformation increases the toughness account—here multiple surface cracks act beneficial. Of course also the strength and ductility of the material are responsible for the fracture toughness. At lower deformation speeds, Mg10Dy1Nd shows higher fracture values, agreeing with a longer time of crack propagation and starting off at higher forces at crack initiation. However, the ductility was reduced by adding Nd, which in the first place caused earlier crack initiation.

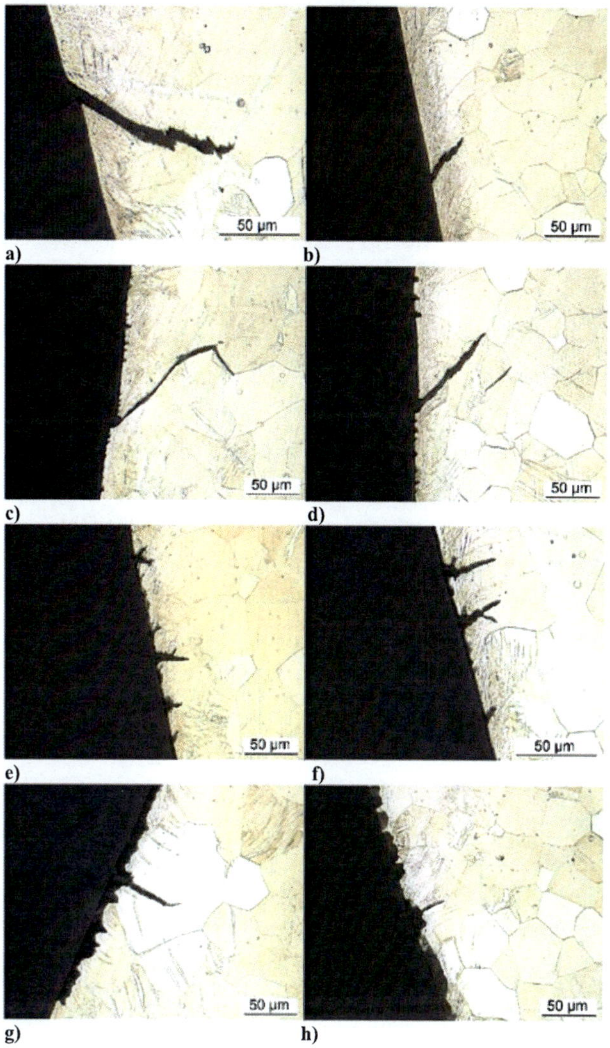

Fig. 8 Cross-sectional micrographs of corroded C-ring surface at tensile side of Mg10Dy (**a**, **c**, **e**, **g**) and Mg10Dy1Nd (**b**, **d**, **f**, **h**) at 2 mm/min (**a**, **b**), 0.283 mm/min (**c**, **d**), 0.028 mm/min (**e**, **f**), and 0.012 mm/min (**g**, **h**)

Summary

Reducing the deformation speed to 0.012 mm/min in C-ring test in Ringer solution is still resulting in stress corrosion cracking in extruded Mg10Dy and Mg10Dy1Nd. The corrosion times in this study ranged from minutes to 20 h. The fracture toughness increases towards lower deformation speeds. Here multiple surface cracking at the tensile side of the C-ring acts beneficial. The lower deformations speeds do not cause large pitting corrosion at the surface and the surface of the crack flanks is also free of corrosion pits. Cross-sectional micrographs do not show a difference in crack propagation among the different test speeds—the interaction with twin boundaries is a typical appearance in these alloys. That the corrosion pits are in a clear context with the twins formed at the surface at the tensile side of the C-ring becomes very obvious. The crack-opening load conditions transform the pits into micro-cracks. Mg10Dy1Nd fails earlier due to its less ductility, but reveals a higher fracture toughness, with is mostly based on the higher strength. Mg10Dy was defeated with longer test times.

References

1. US patents: 9,566,367 B2 & 9,522,219 B2, EU patents: 2744531 & 2744532
2. L. Yang et al. Mater. Sci. Eng. B, 176, 1827–1834 (2011)
3. D. Tolnai et al. Magnesium Technology 2016, 17–21 (2016)
4. W. Jin et al. Corros. Sci. 94, 142–155 (2015)
5. R. Arrabal et al. Corros. Sci. 55 (2012), S. 301–312
6. R. Eifler et al. J. Mater. Sci. 27, 25 (2016)
7. K. Törne et al. Acta Biomater, 48, 541–550 (2017)
8. L. Choudhary and R.K. Singh Raman, Eng. Fracture Mechanics, 103, 94–102 (2013)
9. D. Dubey et al. Mater. Sci. Eng. A, 792, 139793 (2020)
10. S. Jafari, S.E. Harandi, R.K. Singh Raman, JOM, 67 1143–1153 (2015)
11. L. Choudhary et al. Mater. Sci. Eng. C 42 629–636 (2014)
12. M.B. Kannan et al. Mater. Sci. Eng. A 480, 529–539 (2007)
13. B.S. Padekar et al. Corros. Sci. 71, 1–9 (2013)
14. R. Singh, NML: Jamshedpur, India, 48–67 (2007)
15. ASTM International. Standard Practice for Making and Using C-Ring Stress-Corrosion Test Specimens; ASTM G38–01(2013); West Conshohocken, PA, USA (2013)
16. P. Maier et al. Proc. of 40. Vortrags- und Diskussionstagung Werkstoffprüfung 2022, Dresden, Germany (2022)
17. P. Maier et al. Metals 10, 5, 584 (2020)
18. P. Maier et al. Magnesium Technology 2014, 77–82 (2014)

A Comparative Study About Hydroxyapatite Coated AZ31 and AZ91 Mg Alloys

S. Baslayici, M. Bugdayci, K. Benzesik, O. Coban, O. Yucel, and Ercan Acma

Abstract

Magnesium alloys are potential candidates for hard tissue replacements due to their structural and mechanical properties close to bone. Unlike conventional metallic implants, the corrosion rate of magnesium is quite high. This will be an advantage if magnesium-based materials are used as biodegradable. In this study, Magnesium-based AZ31 and AZ91 magnesium alloys were coated with hydroxyapatite by plasma spray and electrostatic spray methods and their corrosion rates were compared.

Keywords

Mg alloys • Corrosion • Hydroxyapatite • Coatings

Introduction

Nowadays, interest in the use of magnesium alloys as biomaterials has increased. The near-bone mechanical properties of magnesium alloys make them important candidates for use in hard tissue replacements. Conventional bio alloys are desired to be bio-inert and not interact with surrounding tissues. Mg and its alloys are biodegradable. Due to its low corrosion resistance and dissolution in the body, Mg and its alloys have been considered insufficient as bio alloys for many years. However, the idea of using biodegradable implants as a result of increasing corrosion resistance by alloying and surface modification has recently been adopted by researchers and attracted attention [1–10]. In this study, we compared the corrosion rates of magnesium alloys by coating hydroxyapatite with 2 different methods.

Experimental

AZ31 and AZ91 Mg alloys are coated with Hydroxyapatite by Atmospheric Plasma Spray and Electrostatic Spray methods. Afterwards, corrosion tests were carried out and the corrosion rates were found for each alloy separately for the coated and uncoated samples, and the results were compared.

While making atmospheric plasma spray coatings, hydroxyapatite particles were sprayed on the substrate material from the plasma spray gun with powder feed.

While making electrostatic coating, there must be electrostatic attraction that can overcome the gravitational force at least so that Hydroxyapatite powders can accumulate on Mg alloys. During the experiments, spraying was carried out from 20, 40 and 60 cm distances in a closed cabinet. Then the samples were subjected to heat treatment.

The heat treatment was tried at 400, 500 and 600 °C and it was observed that the base material burned at 600 °C but did not take place at 400 °C. Then it was tried at 460 °C, but traces of burning were observed on the substrate. For this reason, all heat treatment experiments were carried out at 450 °C.

The general flow chart of atmospheric plasma spray coating is given in Figs. 1 and 2, the general flow chart of the electrostatic coating is seen.

After the coating tests, corrosion tests were started. The process was performed according to ASTM G31-72 Standard Practice for Laboratory Immersion Corrosion Testing

S. Baslayici (✉) · M. Bugdayci
Construction Technology Department, Vocational School, Istanbul Medipol University, 34810 Beykoz, Istanbul, Turkey
e-mail: sbaslayici@medipol.edu.tr

M. Bugdayci
Chemical Engineering Department, Faculty of Engineering, Yalova University, 77100 Yalova, Turkey

K. Benzesik · O. Yucel · E. Acma
Metallurgical and Materials Engineering Department, Faculty of Chemical and Metallurgical Engineering, Istanbul Technical University, 34469, Maslak, Istanbul, Turkey

O. Coban
Machine and Metal Technologies Department, Gedik Vocational School, Istanbul Gedik University, 34913 Pendik, Istanbul, Turkey

S. Barela et al. (eds.), *Magnesium Technology 2023*, The Minerals, Metals & Materials Series,
https://doi.org/10.1007/978-3-031-22645-8_18

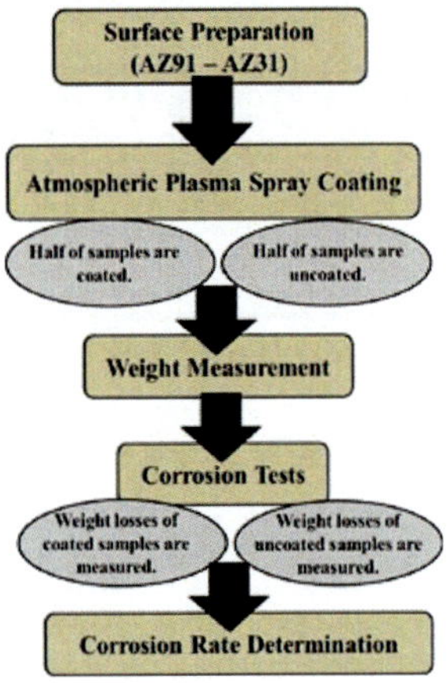

Fig. 1 Flow chart of atmospheric plasma spray coating

of Metals. Obtained weight loss values are converted to average corrosion rate according to this standard.

0.9% isotonic serum was used as artificial body fluid in corrosion experiments. According to the standard, the volume of the test solution should be in the range of 0.2–0.4 ml/mm^2. For this reason, 200 ml of test solution was used in all corrosion tests. Corrosion tests were carried out at a constant 37.5 °C [11].

The corrosion rate was calculated according to Eq. (1).

$$\text{Corrosion rate} = (K \times W)/(A \times T \times D) \tag{1}$$

K = constant
T = exposure time in hours
A = area in cm^2
W = weight loss in g
D = density in g/cm^3
K is taken as 8.76×10^4 (mm/y).

Results and Discussion

The hydroxyapatite used in experimental studies is 99% pure and spherical in shape. It contains trace amounts of heavy metals and sulfates (Heavy metals < 10 ppm, sulfates < %0.048).

SEM image at 3000 magnifications after plasma spray coating applied on AZ31 Mg alloy is given in Fig. 3. The porous structure of the coating can be seen. Although it looks spherical, the needle-like geometry of the grains is seen. This is extremely important. This provides a significant advantage in terms of adhesion.

Figure 4 shows the SEM image at 3000 magnifications after plasma spray coating applied on AZ91 Mg alloy. Although spherical shaped hydroxyapatite is preferred because the flow of powder is an important parameter when coating with plasma spray, it is extremely important to see needle-like grains on the surface after plasma. It provides an advantage in terms of adhesion of the coating.

The coating thickness of the samples was measured on stereo microscope images. The coating thicknesses for both AZ31 and AZ91 were determined as 120 µm.

One of the important parameters when coating with the electrostatic coating method is the application distance and the number of applications. In the preliminary experiments we have done, it has been observed that if the number of applications is more than 1, the coating thickness is too much and therefore the bonding is not sufficiently realized. For this reason, a single application was made while coating. Coating was done on both AZ31 and AZ91 from 20, 40 and 60 cm distances.

Samples must be heat treated to finalize the coating process. While determining the heat treatment temperature, both the melting temperature of Mg alloys and the sintering temperature of hydroxyapatite should be considered. As a result of the trials, it was determined that 450 °C and 24 h were optimum. Heat treatments were carried out at these temperatures and times.

By comparing the SEM images of the coatings, it was determined that the coatings made from 20 cm gave the best results and the coating thicknesses were measured with the help of a stereo microscope. Coating thicknesses for both alloys were found to be approximately 120 µm. However, when the electrostatic spray coating is compared with the plasma spray coating, it has been observed that the electrostatic spray lags behind the plasma spray in terms of

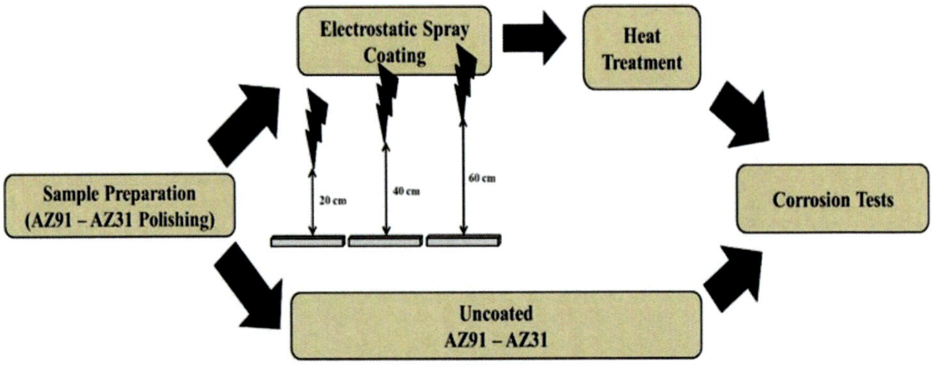

Fig. 2 Flow chart of the electrostatic coating

Fig. 3 SEM image after plasma spray coating on AZ31 Mg alloy (3000x)

homogeneity. The main reason for this is that the applied process is done by human hands.

Figure 5 shows the Sem image of AZ31 coated with hydroxyapatite by electrostatic spray method from 20 cm and Fig. 6 shows the Sem image of AZ91 coated with hydroxyapatite by electrostatic spray method from 20 cm.

In the corrosion tests, the corrosion rate of the uncoated samples was determined primarily. These results were compared with the corrosion rates of the samples coated with different methods and it was determined how much the coatings reduced the corrosion rates.

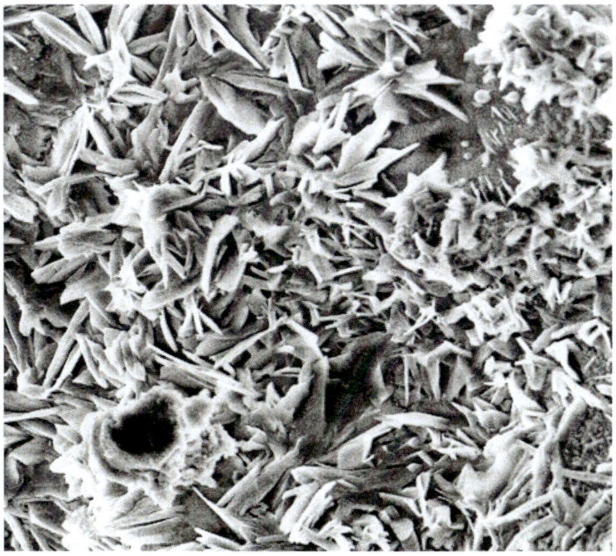

Fig. 4 SEM image after plasma spray coating on AZ91 Mg alloy (3000x)

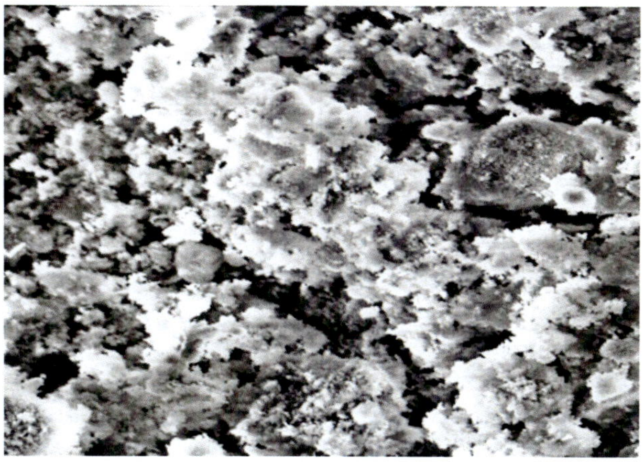

Fig. 5 SEM image after electrostatic spray coating on AZ31 Mg alloy from 20 cm (3000x)

In Fig. 7, the effects of different coating methods on the corrosion rate of these two alloys are given together.

According to Fig. 7, when the corrosion rates of different coating methods are compared in the first hour, it is seen that the corrosion rate of the samples obtained with plasma spray is lower. This shows that the coating obtained by plasma spray slows down the diffusion of chlorine to the surface compared to the electrostatic coating. In addition, the behavior of corrosion rates shows that the protective layer formed in the first 1 h in the electrostatic coating is formed in the first 24 h in the layer formed in the plasma spray coating. This difference shows that the continuity of the coatings obtained with plasma spray is higher than the electrostatic coatings, thus slowing the diffusion of chlorine more.

In addition, the fact that the coatings do not deteriorate in both methods shows that the protective layer is still intact, and no pits are formed. This is since the coating slows down

Fig. 6 SEM image after electrostatic spray coating on AZ91 Mg alloy from 20 cm (3000x)

Fig. 7 Corrosion rates of different alloys and coating methods

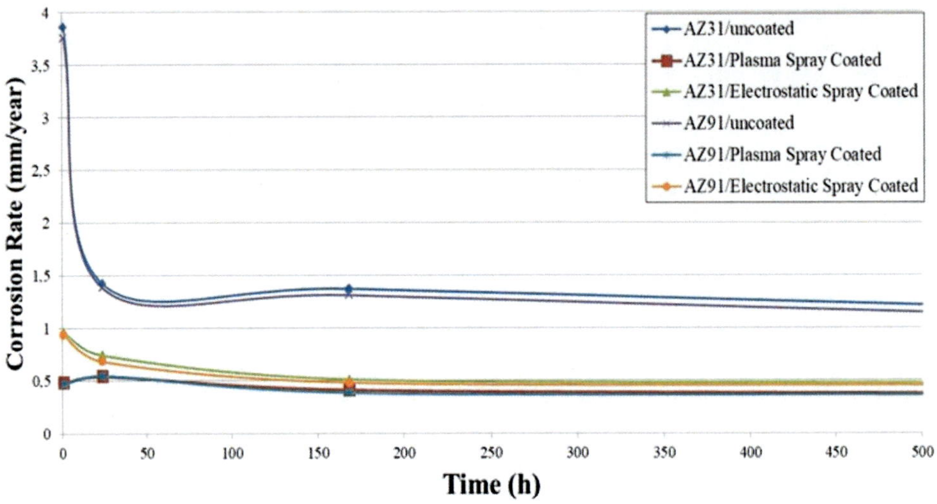

the diffusion of chlorine and keeps the chlorine concentration below a certain level.

When the coating methods are compared, it has been observed that plasma spray coating increases corrosion resistance more than electrostatic coating. The main reason for this is that the coatings made with plasma spray coating are more continuous and homogeneous.

References

1. S. Agarwal, J. Curtin, B. Duffy, and S. Jaiswal, "Biodegradable magnesium alloys for orthopaedic applications: A review on corrosion, biocompatibility and surface modifications," *Materials Science and Engineering C*. 2016 https://doi.org/10.1016/j.msec.2016.06.020
2. G. Ballerini, U. Bardi, R. Bignucolo, and G. Ceraolo, "About some corrosion mechanisms of AZ91D magnesium alloy," *Corros Sci*, vol. 47, no. 9, pp. 2173–2184, 2005, https://doi.org/10.1016/j.corsci.2004.09.018
3. A. Zieliński and S. Sobieszczyk, "Corrosion of titanium biomaterials, mechanisms, effects and modelisation," *Corrosion Reviews*, 2008, https://doi.org/10.1515/corrrev.2008.1
4. M. Esmaily *et al.*, "Fundamentals and advances in magnesium alloy corrosion," *Prog Mater Sci*, vol. 89, pp. 92–193, 2017, https://doi.org/10.1016/j.pmatsci.2017.04.011
5. A. Atrens, M. Liu, and N. I. Zainal Abidin, "Corrosion mechanism applicable to biodegradable magnesium implants," 2011. https://doi.org/10.1016/j.mseb.2010.12.017.
6. S. Virtanen, "Biodegradable Mg and Mg alloys: Corrosion and biocompatibility," 2011. https://doi.org/10.1016/j.mseb.2011.05.028.
7. Y. Xin, C. Liu, K. Huo, G. Tang, X. Tian, and P. K. Chu, "Corrosion behavior of ZrN/Zr coated biomedical AZ91 magnesium alloy," *Surf Coat Technol*, vol. 203, no. 17–18, pp. 2554–2557, 2009, https://doi.org/10.1016/j.surfcoat.2009.02.074
8. L. Wang, B. P. Zhang, and T. Shinohara, "Corrosion behavior of AZ91 magnesium alloy in dilute NaCl solutions," *Mater Des*, vol. 31, no. 2, pp. 857–863, 2010, https://doi.org/10.1016/j.matdes.2009.07.049
9. F. Witte *et al.*, "In vitro and in vivo corrosion measurements of magnesium alloys," *Biomaterials*, 2006, https://doi.org/10.1016/j.biomaterials.2005.07.037
10. Y. F. Zheng, X. N. Gu, and F. Witte, "Biodegradable metals," *Materials Science and Engineering R: Reports*, vol. 77. Elsevier Ltd, pp. 1–34, 2014. https://doi.org/10.1016/j.mser.2014.01.001
11. ASTM, "Standard guide for laboratory immersion corrosion testing of metals," *ASTM International*, 2012.

In Situ Study of the Degradation Behaviour Under Load of Mg1.8Y0.6Zn(1Ag) Using Synchrotron Tomography

D. Tolnai, B. Hindenlang, J. Bohlen, J. Pereira da Silva, J. Gu, A. Louapre, D. C. F. Wieland, and F. Wilde

Abstract

The addition of Y and Zn improves the property profile of Mg by forming a Long Period Stacking Ordered (LPSO) phase. The addition of Ag can further improve the degradation and mechanical properties. Owing to their biocompatibility and degradation under physiological conditions, these alloys are considered for temporary implants. Mg1.8Y0.6Zn and with the addition of 1 wt% of Ag were extruded and subjected to in situ tensile testing in air and submerged in simulated body fluid. At selected load intervals, synchrotron tomography was performed in order to follow the crack initiation and propagation in 3D. The results show the effect of corrosive media on the mechanical failure. Furthermore, the results are used to develop a computational model on stress corrosion cracking that could be applied in failure prediction of implants and designing alloys for structural and medical applications.

Keywords

Mg–Y–Zn alloys • In situ synchrotron tomography • Tensile test • Degradation

D. Tolnai (✉) · B. Hindenlang · J. Pereira da Silva · J. Gu · A. Louapre · D. C. F. Wieland
Institute of Metallic Biomaterials, Helmholtz-Zentrum Hereon, Max-Planck Strasse 1, 21502 Geesthacht, Germany
e-mail: domonkos.tolnai@hereon.de

J. Bohlen
Institute of Material and Process Design, Helmholtz-Zentrum Hereon, Max-Planck Strasse 1, 21502 Geesthacht, Germany

J. Pereira da Silva
Department of Mechanical Engineering, Federal University of São Carlos, Rod. Washington Luis Km 235, São Carlos, 13565-905, Brazil

F. Wilde
Institute of Materials Physics, Helmholtz-Zentrum Hereon, Max-Planck Strasse 1, 21502 Geesthacht, Germany

Introduction

The application of magnesium (Mg) in many structural applications is an ideal solution instead of heavier parts based on its excellent specific mechanical properties. Although Mg alloys have moderate strength, poor corrosion resistance, and ambient temperature formability, research efforts have been made to overcome these issues to enable the wide use of Mg alloys. One relatively new application is the production of Mg-based biodegradable medical implants, where controlled degradation is not detrimental but an advantage [1].

Mg and the human bone have similar mechanical properties; therefore, the implant can provide optimal support during the healing period [2]. The degradation of Mg-based implants under physiological conditions increases the patient comfort and reduces health care costs compared to standard implants made of, e.g. titanium as a second surgery for their removal is unnecessary. The addition of Y and Zn to Mg is an effective way to enhance the mechanical property profile of Mg. Furthermore, the favourable processing conditions give rise to the formation of the Long Periodic Stacking Ordered Structure (LPSO) that provides superior strength without deteriorating the ductility [3]. The biocompatibility of Y and Zn also allows to use WZ alloys as biodegradable implant materials [4, 5].

Numerous research has been performed on the degradation process itself, and how it is influenced by the surrounding tissue [6]. This complex process is effected by many parameters, e.g. material composition, the mechanical load on the implant prior and after the implantation, but also the ionic composition of the corrosive media and the content of organic molecules in the environment [7]. In some studies, it was observed that in the corrosive environment mechanical loads even at moderate stress levels can lead to crack formation resulting in Stress Corrosion Cracking (SCC) [8], which results in the loss of integrity and consequently the failure of the implant. There are several

S. Barela et al. (eds.), *Magnesium Technology 2023*, The Minerals, Metals & Materials Series,
https://doi.org/10.1007/978-3-031-22645-8_19

parameters influencing the SCC (e.g. environment, microstructure, strain, strain rate, etc.), whereas the strain and degradation rates play a pivotal role. SCC occurs when both the degradation and the mechanical load simultaneously contribute to the failure process; otherwise, the damage is caused purely by degradation or mechanical strain. The effect of degradation media on the crack initiation and propagation in the case of Mg alloys is not yet fully understood. Synchrotron based X-ray tomography allows to observe the crack initiation and propagation in the material subjected to simultaneous degradation- and mechanical-load conditions [9]. The aim of this study is to investigate the damage evolution in Mg1.8Y0.6Zn(1Ag) under tensile load in air and Simulated Body Fluid (SBF) with X-ray tomography. Applying SBF as a degradation medium was motivated by our goal to perform experiments under near-physiological conditions.

Experimental Methods

The alloys were produced by permanent mould indirect chill casting [10]. The pure Mg was melted in an electric resistance furnace under protective atmosphere of 2 vol% SF_6 and Ar. The alloying elements were added also as pure materials. After mixing, the melt was held at 720 °C for 10 min and, afterwards, was poured into a steel mould preheated to 660 °C. After 5 min isothermal holding, the mould was quenched into water at a rate of 10 mm s^{-1} until the top of the melt was in line with the cooling water level. Two alloys were cast with the compositions of Mg1.8Y0.6Zn in pure form and with the addition of 1 wt% of Ag, respectively. The ingots were homogenized at 400 °C for 24 h and subsequently quenched in water. The indirect extrusion was performed at 400 °C with a speed of 1 mm s^{-1} at an extrusion ratio of 1:25.

The metallographic characterization was performed using a TESCAN Vega SB-U III scanning electron microscope (SEM) with an energy dispersive X-ray spectrometer (EDXS). The micrographs were obtained in backscatter electron (BSE) mode using an accelerating voltage of 15 kV.

For the in situ synchrotron tomography experiments, cylindrical tensile specimens were machined from the extruded bar longitudinally with a gauge length of 5 mm and a gauge diameter of 2 mm with a 0.1 mm deep notch in the middle. This approach was necessary to meet the imaging conditions at the synchrotron and to predefine the region of crack formation. The in situ synchrotron radiation tomography was performed at the P05 beamline of PETRA III, DESY (Deutsches Elektronen-Synchrotron) [11, 12]. The setup is shown in Fig. 1.

During the tensile tests at different force levels, tomography was performed. 4000 radiographies were recorded

Fig. 1 Tomographic setup at the P05 beamline for tensile testing with **a** the sample submerged in SBF and **b** the fully assembled tensile machine

with a monochromatic beam at an energy of 38 keV and an acquisition time of 230 ms per radiogram at the nearest possible sample-to-detector distance of 60 mm. The specimens were deformed at room temperature at an initial strain rate of 10^{-3} s^{-1} in air and SBF, respectively. The tests were terminated at fracture. The reconstruction resulted in tomographic volumes with a $(1.8\ \mu m)^3$ voxel size. The filtering was done by iterative non-local means of denoising [13]. For 3D rendering, the Avizo Fire software was used.

Results and Discussion

The optical micrographs of the material in as extruded state are shown in Fig. 2.

Due to the simultaneous addition of Y and Zn, both alloys contain the LPSO phase. This hard phase is rich in the alloying elements and increases the strength of the material. The microstructure of the Ag modified alloy shows the presence of more intermetallic phases in the interdendritic region (Fig. 2b). The stress–strain curves recorded during the tensile tests at the synchrotron tomography measurements are shown in Fig. 3.

We note that the kinks in the force curves are signs of relaxation originating from the parallel imaging as the tensile test had to be halted for the time of the tomographic measurement. The results of the tensile test show that there is a difference in the two alloys in the tests performed without degradation media, whereas the Ag modified alloys reach a higher UTS and exhibit a higher ductility; 254 MPa and 12.26% for Mg1.8Y0.6Zn and 309 MPa and 16.23% for the Mg1.8Y0.6Zn1Ag, respectively. This can be attributed to the higher volume fraction of the LPSO phase (Fig. 2b), albeit the tests are performed in a non-standard way on an insufficient number of samples due to the restricted availability of synchrotron beam time. The two alloys show a similar behaviour in the tests performed in SBF with a UTS of 177 and 174 MPa and a ductility of 7.76% and 8.55% for Mg1.8Y0.6Zn and Mg1.8Y0.6Zn1Ag, respectively. These

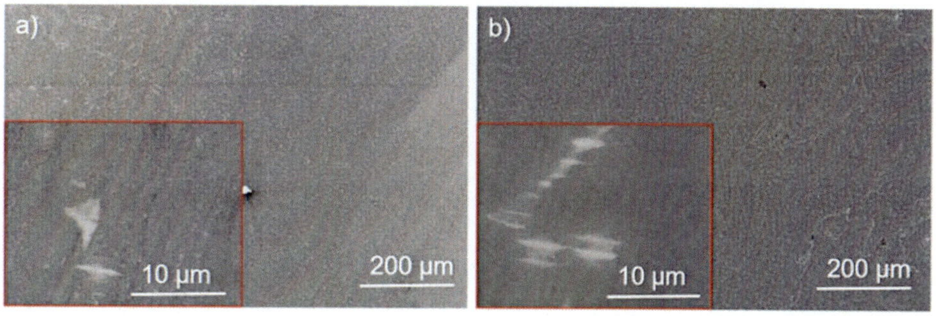

Fig. 2 BSE micrographs of the material **a** Mg1.8Y0.6Zn and **b** Mg1.8Y0.6Zn1Ag

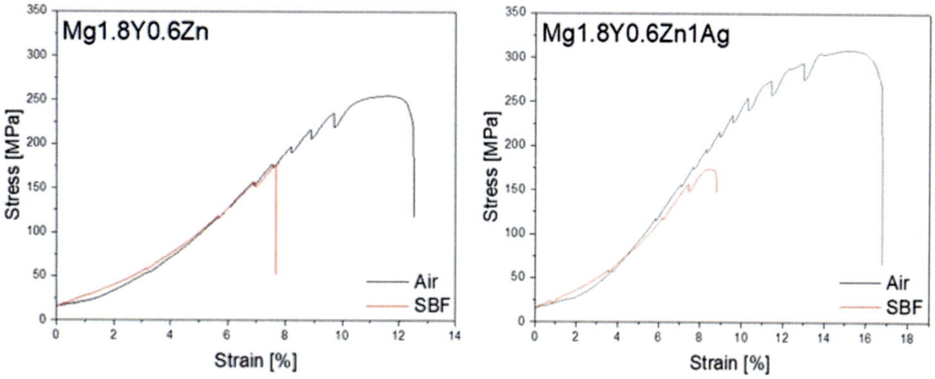

Fig. 3 Stress–strain curves obtained during the tensile tests in air and in SBF

lower values, compared to the tests performed in air, suggest a simultaneous effect of mechanical and degradation load on the samples.

Reconstructed tomographic slices acquired during the tensile testing of Mg1.8Y0.6Zn in air are shown in Fig. 4.

The tomographic results show the evolution of damage. At 1 MPa pre-load, only some surface roughness can be observed on the sample. As the load is increased, there is no significant change until about 198 MPa, when cracks start to develop. The damage is accumulated at the surface, as the cracks initiate from the notch and develop inwards the sample.

The reconstructed tomographic slices acquired during the tensile testing of Mg1.8Y0.6Zn in SBF are shown in Fig. 5.

The results of the tensile test in the SBF show a different damage evolution to the counterpart in the air. The damage accumulation starts from the beginning, owing to the degradation. This leads to an earlier failure. The cracks in this case are also originating from the surface of the sample, but even at lower stresses reach deeper into the bulk of the sample.

A 3D rendering of the interface of the pre-loaded and the last stages in air and SBF are shown in Fig. 6.

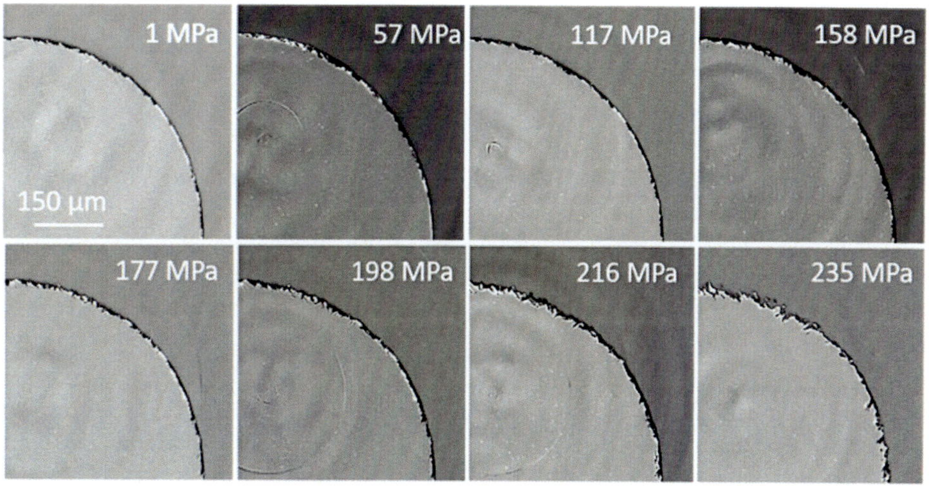

Fig. 4 Reconstructed tomographic slices acquired during the tensile testing of Mg1.8Y0.6Zn in air

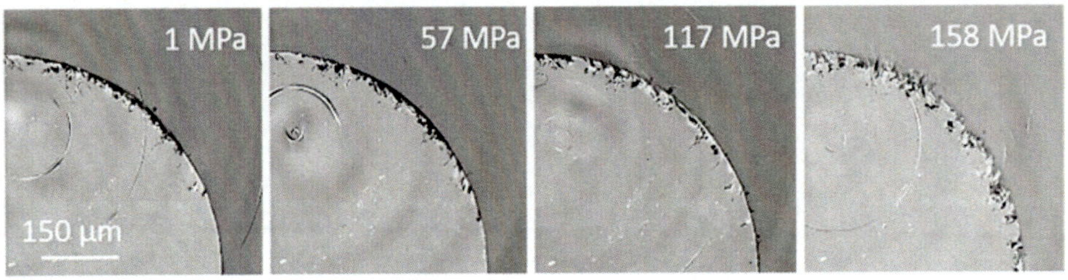

Fig. 5 Reconstructed tomographic slices acquired during the tensile testing of Mg1.8Y0.6Zn in SBF

Fig. 6 3D rendering of the tomographies from the interface of the notched area in Mg1.8Y0.6Zn at the pre-load stage and before failure

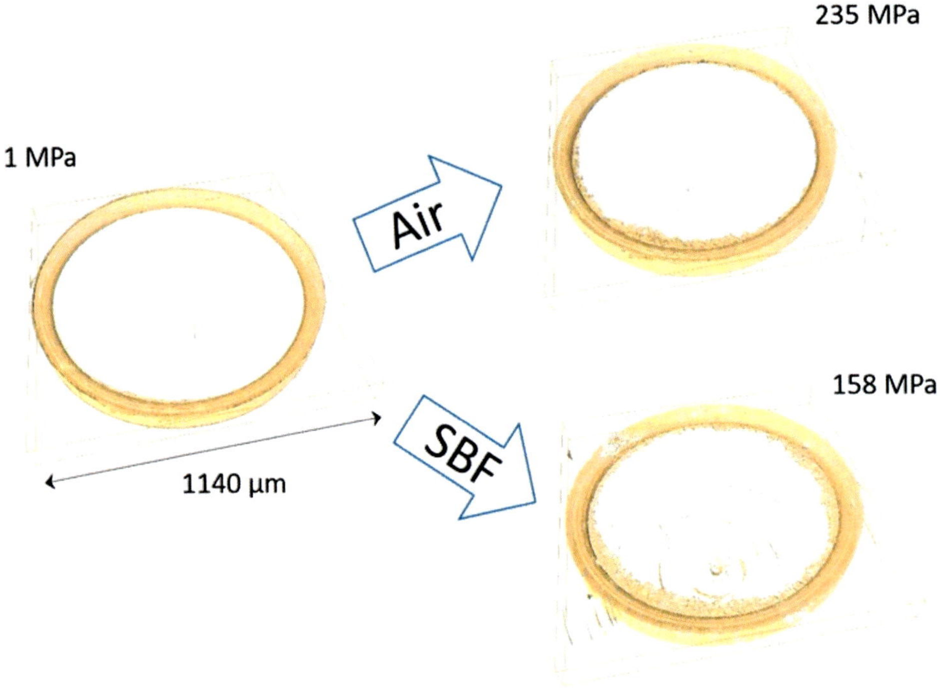

The rendering of the surfaces around the notched area before the failure shows that in the case of the test in air, the damage is localized to the surface of the sample at the notch. Also, the cracks itself are smaller than in the case of SBF at lower stress. This suggests that the material breaks in a rigid manner. On the contrary, in the case of the test in SBF the damage is more spread out in the whole perimeter of the sample, and the cracks are deeper than in the case of the sample tested in air. This suggests that the degradation has a significant effect on crack propagation as the corrosive media reaches into the crack tips.

Reconstructed tomographic slices acquired during the tensile testing of Mg1.8Y0.6Zn1Ag in air are shown in Fig. 7.

The tomographic results show a similar behaviour to the unmodified alloy. Until approximately 216 MPa, no significant damage can be observed in the sample. At 235 MPa, cracks at the perimeter of the notch start to develop and propagate until failure.

The reconstructed tomographic slices acquired during the tensile testing of Mg1.8Y0.6Zn1Ag in SBF are shown in Fig. 8.

Similar to the unmodified alloy, the damage starts already at lower stress levels. The degradation enhances the mechanical load that leads to accelerated crack propagation. At 158 MPa shortly before failure, already a significant amount of damage accumulation can be observed at the surface of the sample.

The 3D rendering of the pre-loaded and the last stages in air and SBF are shown in Fig. 9.

The 3D rendering of the notched shows a significant difference between the damage in air and in SBF. The degradation leads to a significantly larger damage accumulation before failure, even at lower load levels. This suggests, as in the case of the unmodified alloy, that the degradation media reaches into the cracks and lowers the stress necessary for the crack to propagate. The modification with silver enhances the mechanical properties in air, but

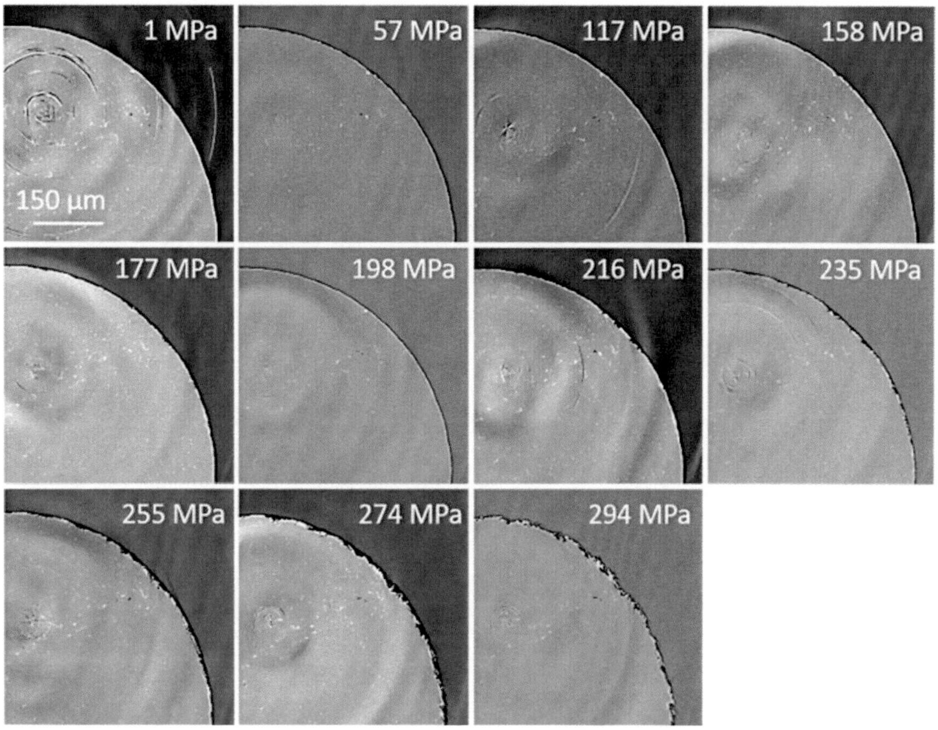

Fig. 7 Reconstructed tomographic slices acquired during the tensile testing of Mg1.8Y0.6Zn1Ag in air

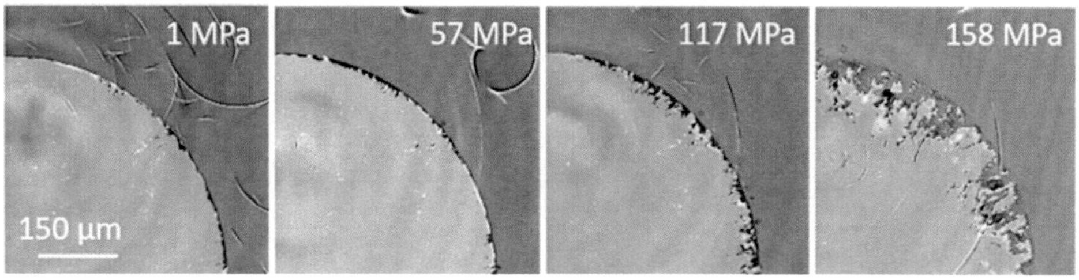

Fig. 8 Reconstructed tomographic slices acquired during the tensile testing of Mg1.8Y0.6Zn1Ag in SBF

leads to a higher susceptibility to degradation, probably due to the increased amount of particles in the Mg matrix. Although the macroscopic responses of the two alloys are very similar, the microscopic damage evolution is quite different whereas the two alloys are affected differently by degradation and the mechanical load.

Conclusions

From the analysis of the experimental results, the following conclusions can be drawn:

- The addition of 1 wt% Ag to Mg1.8Y0.6Zn increases the strength and ductility.
- The crack initiation during pure mechanical load starts at the interface, and the crack growth propagates inwards leading to the failure of the material.
- The degradation accelerates the damage evolution in the case of both alloys. The crack initiation is more likely on a surface that suffered damage from degradation. The SBF reaches into the crack tips, lowering the stress that is necessary for crack propagation and therefore leads to higher damage accumulation and an early failure of the sample.
- The two alloys have a different susceptibility to mechanical and degradation load. The unmodified Mg1.8Y0.6Zn has a lower UTS and ductility tested in air, whereas the Mg1.8Y0.6Zn1Ag is more affected by the degradation. These competing effects lead to a similar

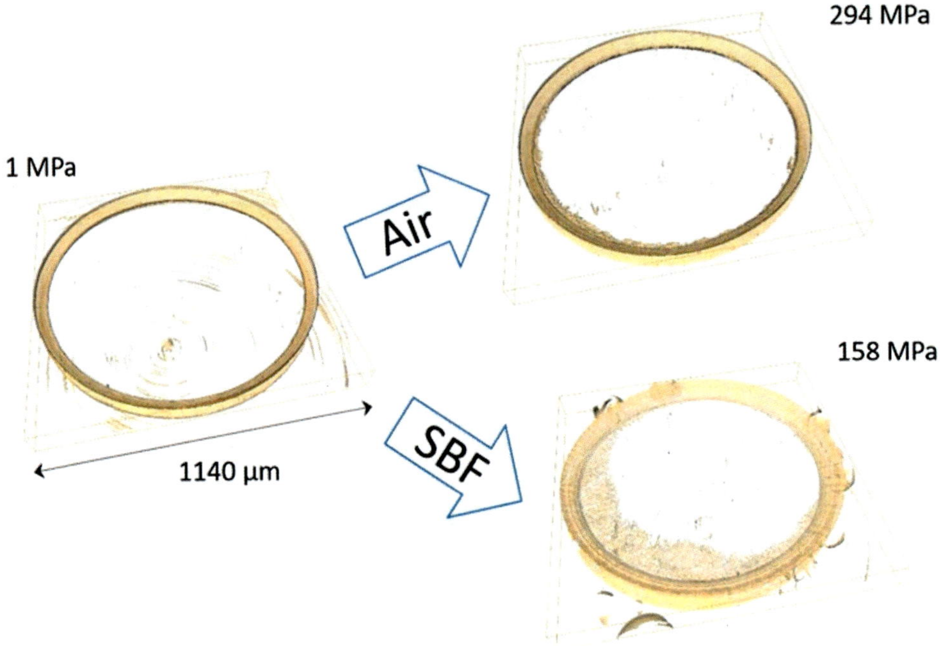

Fig. 9 3D rendering of the tomographies from the interface of the notched area in Mg1.8Y0.6Zn1Ag at the pre-load stage and before failure

mechanical failure of the two alloys originating from different microstructural processes.

Acknowledgements The authors acknowledge the Deutsches Elektronen-Synchrotron for the provision of beamline facilities in the framework of the proposal I-20220208.

References

1. H. Hermawan, D. Dubé, D. Mantovani, *Acta Biomaterialia* 6 (2010) 1693–1697.
2. Julian Moosmann et al., 'Biodegradable magnesium-based implants in bone studied by synchrotron radiation microtomography', Sep. 2017, vol. 10391. [Online]. Available: https://doi.org/10.1117/12.2275121.
3. K. Horváth et al., Magnesium Technology 2017. The Minerals, Metals & Materials Series. Springer, Cham. https://doi.org/10.1007/978-3-319-52392-7_8.
4. T. Kraus et al., Acta Biomaterialia 66 (2018) 109–117.
5. T.A. Grünewald et al., Acta Biomaterialia 31 (2016) 448–457.
6. B. Zeller-Plumhoff et al., Adv. Eng. Mater., vol. n/a, no. n/a, p. 2100197, May 2021, https://doi.org/10.1002/adem.202100197.
7. R. Willumeit-Römer et al., TMS 2018 147th Annual Meeting & Exhibition Supplemental Proceedings, Cham, 2018, pp. 275–284.
8. Y. Gao et al., Acta Biomaterialia, vol. 83, pp. 477–486, Jan. 2019, https://doi.org/10.1016/j.actbio.2018.11.019.
9. B. Zeller-Plumhoff et al., Sci. Rep. 10 (2020) 16101.
10. F.R. Elsayed et al., Materials Science Forum 690 (2011) 65–68.
11. F. Wilde et al., AIP Conference Proceedings 1741, 030035, (2016), https://doi.org/10.1063/1.4952858.
12. J. Moosmann et al., Proceedings Volume 11113, Developments in X-Ray Tomography XII; 1111318 (2019), https://doi.org/10.1117/12.2530445.
13. S. Bruns, et al., Adv. Water Res. 2017, 105, 96–107. https://doi.org/10.1016/j.advwatres.2017.04.020.

In Vitro Degradation of Magnesium Wire in Sternal-Closure-Like Conditions

Adam J. Griebel and Natalie Romick

Abstract

Even with advances in minimally invasive surgery, open heart surgery remains a mainstay of modern medicine. In open heart surgery, the sternum is bisected to allow access to the heart. After surgery, the two halves of the sternum are mechanically fixed with stainless steel wires or in some cases titanium or stainless steel cables and plates. These devices are left behind indefinitely and can in some cases cause patient discomfort or interfere with subsequent operations. An absorbable magnesium wire could allow for sufficient mechanical support while the sternum heals while avoiding the secondary complications of permanent materials. In this study, feasibility of a Mg sternal wire was assessed using a patent-pending LZ21 alloy wire and a simple sternal model mimic. Baseline mechanical and microstructural properties of 0.95 mm wire were established and compared to wire corroded in a modified Hank's fluid while under tension.

Keywords

Absorbable • Magnesium • Wire • Corrosion • Fixation

Introduction

Open heart surgery is a mainstay of modern medicine, with hundreds of thousands of procedures being performed in the United States each year [1]. In these procedures, the heart and thoracic cavity are accessed by bisecting the sternum. After the surgery, the two halves of the sternum are mechanically fixed back together. This is most commonly accomplished with annealed 316L stainless steel wires which are wrapped around the sternum in single, double, or figure-of-eight configurations and twisted together to form a simple knot [2]. In some cases, these monofilament wire-based systems can suffer from instability or be unsuitable for high-risk patients, and more complex plate and cable solutions are employed [3]. These monofilament sternal wires can also fracture and lead to secondary complications requiring further intervention [4]. A wire that provides sufficient fixation during the sternal healing period and then harmlessly absorbs into the body would mitigate this chance for secondary complications and would facilitate any necessary surgeries in the same area.

Magnesium alloys have been explored for many years as an absorbable material for medical devices, and there are several vascular and orthopedic devices on the market around the world [5–7]. Though several studies have postulated that magnesium wire could work as a sternal fixation wire, there is limited application-specific testing of such a solution. Currently, no Mg-wire-based systems have made it to market.

Fixation of the sternum with magnesium wire is a challenging feat. The wire must not degrade before the sternum is sufficiently healed, so bulk and localized corrosion must be controlled. Further, the wire will be under substantial stress during sternal healing, so resistance to stress-corrosion cracking is critical. Magnesium is inherently a lower-strength material than stainless steel, so a relatively high-strength magnesium alloy is desirable. However, substantial ductility is required to withstand the severe bending stresses sustained during the twist-knotting procedure. This combination of sufficient corrosion resistance, moderate strength, and high ductility presents a unique challenge.

LZ21 is a recently developed and patent-pending [8] alloy which may be suitable for just such an application. A small lithium addition has been found to add substantial ductility without dramatic acceleration of the corrosion rate or sacrifice of strength. Other alloys with higher lithium contents have been shown to be well tolerated, and no negative

A. J. Griebel (✉) · N. Romick
Fort Wayne Metals Research Products Corp., Fort Wayne, IN, USA
e-mail: Adam_griebel@fwmetals.com

S. Barela et al. (eds.), *Magnesium Technology 2023*, The Minerals, Metals & Materials Series,
https://doi.org/10.1007/978-3-031-22645-8_20

biological impacts from the small lithium addition are expected [9]. The alloy is strengthened with additions of nutrient metals zinc, calcium, and manganese.

The aim of this study is to explore the suitability of LZ21 wire as a sternal fixation material through a series of mechanical and stress-corrosion experiments.

Materials and Methods

LZ21 alloy material was prepared by vacuum induction melting and casting to a diameter of 50 mm. The ingot was then directly extruded to 12.7 mm diameter at 300 °C. 12.7 mm bar was reduced to a final diameter of 0.95 mm through a series of cold drawing and annealing steps, described elsewhere [9]. After drawing to the final size, the wire was annealed a final time. After annealing, wire was mounted, polished, and etched to document the microstructure.

Mechanical suitability was assessed via basic uniaxial tensile testing (127 mm gauge length, 25.4 mm/min cross-head speed, $N = 3$), twist-knot testing around a split sternum mimic, and finally tensile testing of the knotted wire by pulling the split sternum mimic halves apart.

To assess corrosion properties of the wire and how corrosion changes mechanical strength over time, wires were knotted around a Delrin sternal mimic equipped with two plastic screws threaded through on half of the mimic. After knotting, the screws were rotated 1.25 rotations after contact with the other half of the mimic was made. This was performed to ensure a consistent stress was applied to the wire. The entire wire-mimic assembly was immersed in 200 mL of a modified Hank's balanced salt solution and held in an incubator at 37 °C with a 5% CO_2 environment to buffer the solution to a pH of 7.4 ± 0.2. Samples were allowed to corrode for set durations, with three samples for each time point. Time points were 3, 7, 14, and 28 days. After degradation, two samples were tensile tested to assess residual strength, and one was imaged before sectioning to observe corrosion uniformity and depth.

Results

Representative tensile properties of annealed 0.95 mm LZ21 wire are shown in Fig. 1. All three tensile tests were very consistent, with less than 1% variation in UTS and YS and less than 5% variation in elongation. Microstructural imaging reveals a uniform, equiaxed microstructure with a fine grain size of 3 μ (Fig. 2). Twist-testing indicated the wire had sufficient ductility, with the wire able to withstand many knots without fracturing (Fig. 3). Tensile testing of the knotted wire on the sternal mimics showed wire generally but not exclusively breaking at the knot, with an average breakload of 113 N, approximately 64% of the uniaxial test.

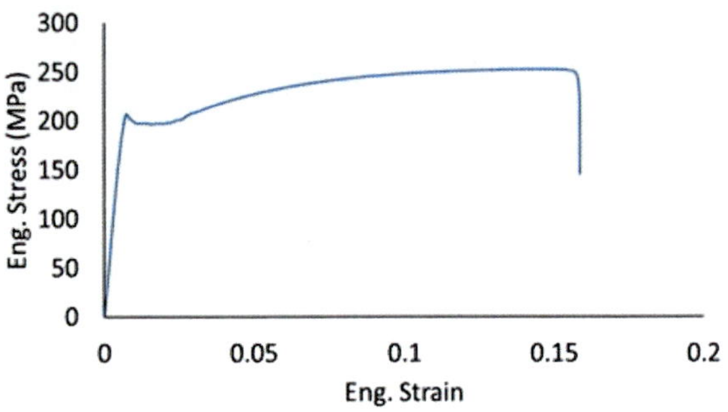

Fig. 1 Representative stress–strain curve of LZ21 wire

Representative images of the single sample from each time point after corrosion of 3, 7, 14, and 28 days are shown in Fig. 4. For both the 14- and 28-day samples, one of the three samples had fractured during the test, and this sample was selected for imaging. At 3 and 7 days, a generally uniform corrosion layer had developed across the surface of the wire. Some minor pits were forming, indicated by white surface spots. At days 14 and 28, the corrosion layer was visibly thicker with some pits which had progressed through a substantial portion of the wire's cross section (It is hypothesized that these two samples fractured during the test at a particularly large pit.). In general, corrosion was reduced on the intrados side of the wire in contact with the plastic fixture (bottom side of wires in Fig. 5) due to reduced access to Hank's solution. There were no obvious trends with corrosion and location on the fixture. Wires were then mounted and polished longitudinally to observe corrosion uniformity and penetration depth (Fig. 5). The corrosion product is readily visible on the surface of the wires. As time progresses, the corrosion layer increases in thickness and pitting becomes more pronounced. An example of a particularly large pit is shown in Fig. 5c.

Residual strength of the 2 wire samples at each of these time points is shown in Fig. 6. A general trend of loss of strength over time is clear, but the two samples at 7 and 14 days demonstrated large variability. One of the 7-day samples exhibited a breakload of 135 N, 19% higher than the uncorroded baseline of 113 N. The other 7-day sample had a breakload of 88 N. One 14-day sample also had a breakload of 88 N, but the other had a breakload of 47 N; this sample fractured at an apparent pit. The two 28-day samples were very similar, at 62 and 63 N. All samples broke near the base of the knot, except for the 14-day sample which broke at a pit away from the knot.

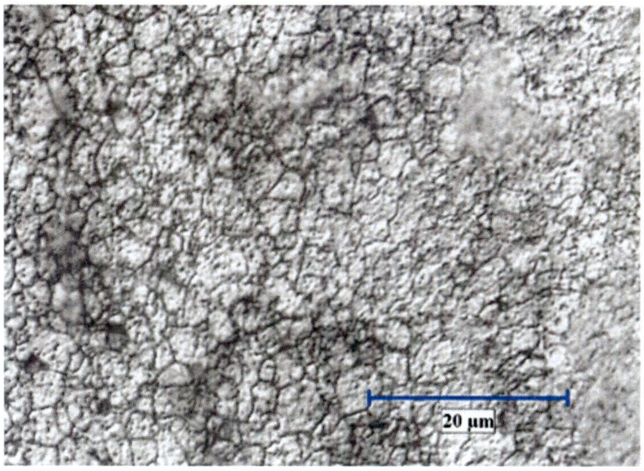

Fig. 2 Longitudinal microstructure of 0.95 mm LZ21 wire

Fig. 3 Left: wire shown knotted around a split sternal mimic, with plastic screws providing tension. Right: the sternal mimic in a tensile fixture to measure breakload

Discussion

This small study demonstrates the feasibility of LZ21-based magnesium wire as a sternal fixation material. However, it also illuminates some key challenges to address in terms of corrosion variability.

The annealed LZ21 wire had sufficient ductility to withstand the knotting procedure, which is a fundamental requirement of this sternal wire design. The strength of the wire is low compared to conventional stainless steel wires, but this can to some extent be compensated for through use of larger wires or more wires across the sternum.

2 of 3 samples had more than 50% residual strength at 28 days. Likely, this would be an adequate result as in vitro corrosion rates are generally known to be much faster (e.g. 2–10×) than in vivo rates for the same material, though this can vary widely depending on in vivo implant location and in vitro test method [10, 11]. However, the 1 of 3 samples that had already fractured at both 14 and 28 days highlights the need for further work to increase the uniformity of corrosion, at least past the critical healing period of the sternum, after which non-uniformity and variability may be tolerable. This could be achieved through enhanced wire production, absorbable coatings (either polymeric, ceramic, or both) to delay corrosion onset, or wire composite approaches [12].

Future work should include additional samples to better assess the variability in corrosion. It is possible that particular cold-working and annealing parameters may impact corrosion uniformity, and this should be assessed. Coatings,

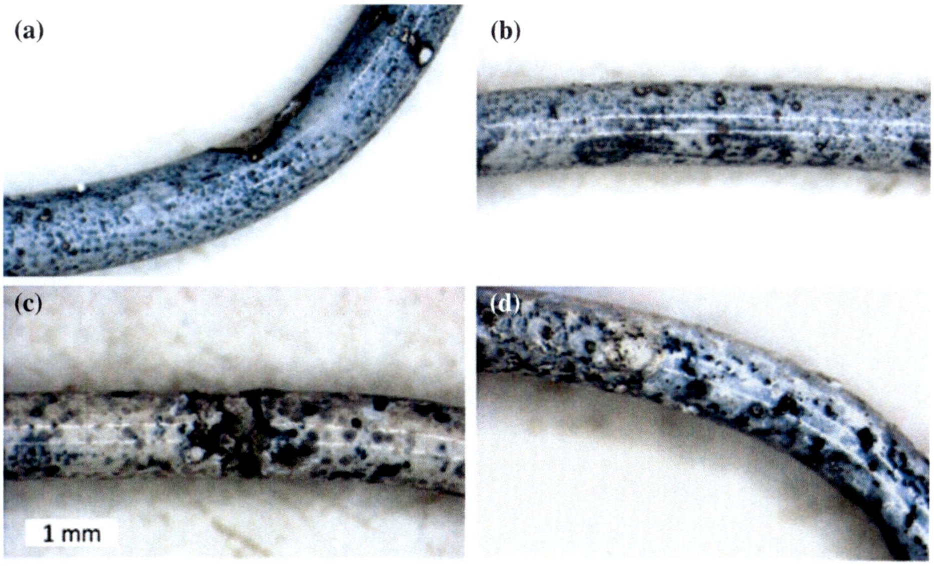

Fig. 4 Representative images of the corroded surface after 3, 7, 14, and 28 days (**a–d**)

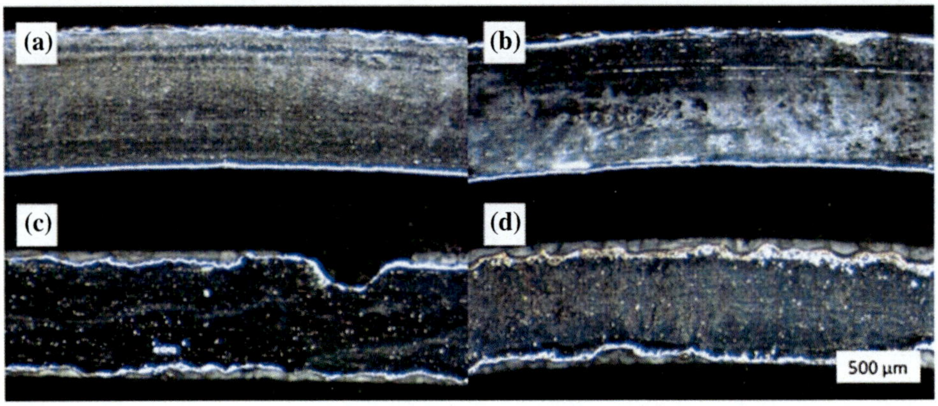

Fig. 5 Longitudinal cross-sections of degraded wire at 3, 7, 14, and 28 days (**a–d**). The top of each section is the extrados side of the wire

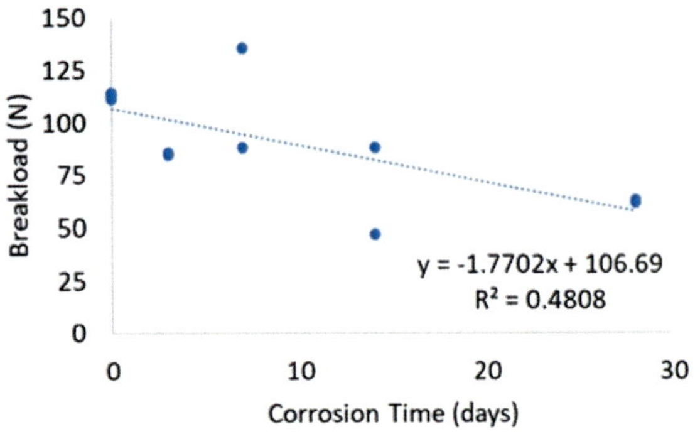

Fig. 6 Residual strength of the wire knots after 0, 3, 7, 14, and 28 days of corrosion. Note the poor fit of the trendline, highlighting the stochastic nature of corrosion

if employed, need to be able to withstand the large plastic deformation involved in twisting the knot. A worthwhile exercise would involve coating the LZ21 wire with various coatings (e.g. PLGA, PCL, PEO, MgF_2, phosphates) and observing their reaction to the knotting and their impact on corrosion rate and uniformity. Finally, an in vivo study should be conducted to confirm the relevancy of the in vitro findings.

References

1. M. Ludovic, "Fifty years of coronary artery bypass grafting," *Journal of Thoracic Disease*, vol. 10, no. 3, pp. 1960–1967, 2018.
2. A. Shafi, "Sternal closure with single compared with double or figure of 8 wires in obese patients following cardiac surgery: A systematic review and meta-analysis," *Journal of Cardiac Surgery*, vol. 36, no. 3, pp. 1072–1082, 2021.
3. A. Duzgun, "Comparison of Two Sternal Closure Techniques Based on Risk Factors: A Prospective, Observational Study," *Applied Bionics and Biomechanics*, 2021.
4. A. Amirghofran, "Ascending aorta graft pseudoaneurysm and aortobronchial fistula caused by a fractured sternal wire: a case report," *Journal of Cardiothoracic Surgery*, vol. 16, no. 348, 2021.
5. Biotronik, "BIOTRONIK Announces CE Mark for Magmaris, the First Clinically Proven Bioresorbable Magnesium Scaffold," Biotronik, 15 6 2016. [Online]. Available: https://news.biotronik.com/biotronik-announces-ce-mark-for-magmaris-the-first-clinically-proven-bioresorbable-magnesium-scaffold/. [Accessed 7 9 2021].
6. PRNewswire, "QualiMed Innovative Medizinprodukte GmbH, a Q3 Medical Devices Ltd Company, Receives European CE Mark Approval for UNITY-B™ Percutaneous Balloon Expandable Biodegradable Biliary Stent," PR Newswire, 21 6 2021. [Online]. Available: https://www.prnewswire.com/news-releases/qualimed-innovative-medizinprodukte-gmbh-a-q3-medical-devices-ltd-company-receives-european-ce-mark-approval-for-unity-b-percutaneous-balloon-expandable-biodegradable-biliary-stent-301316379.html. [Accessed 8 9 2021].
7. Biber, "Bioabsorbable metal screws in traumatology: A promising innovation," *Trauma Case Reports*, vol. 8, pp. 11–15, 2017.
8. A. Griebel and J. Schaffer, "Magnesium-based absorbable alloys". WO Patent 2020/247383 A1, 3 June 2019.
9. J. R. F. Z. N. Angrisani, "Biocompatibility and degradation of LAE442-based magnesium alloys after implantation of up to 3.5 years in a rabbit model," *Acta Biomaterialia*, vol. 44, no. 15, pp. 355–365, 2016.
10. A. Griebel and J. Schaffer, "Absorbable Filament Technologies: Wire-Drawing to Enable Next-Generation Medical Devices," *Magnesium Technology*, pp. 323–327, 2016.
11. P. Bowen, J. Drelich and J. Goldman, "A new in vitro-in vivo correlation for bioabsorbable," *Materials Science and Engineering Part C*, 2013.
12. J. Walker, S. Shadanbaz, N. Kirkland, E. Stacfe, T. Woodfield, M. Staiger and G. Dias, "Magnesium alloys: Predicting in vivo corrosion with in vitro immersion testing," *Journal of Biomedical Materials Research Part B*, vol. 100B, pp. 1142–1150, 2012.
13. J. Schaffer, "Biodegradable composite wire for medical devices". USA Patent 9561308B2, 07 02 2017.

Influence of Corrosion Extent on Residual Tensile Strength and Corrosion Fatigue Properties of an Mg−Y−Nd Alloy Characterized by μCT

B. Clausius, N. Wegner, S. Jeyavalan, H. Hartweg, F. Walther, and P. Maier

Abstract

This study focuses on the influence of inhomogeneous corrosion on the quasi-static and cyclic properties of an Mg−Y−Nd alloy. Uniform corrosion is essential for biodegradable implant materials to avoid notching effects and premature failure. Previous studies on extruded Mg−3Y−3RE have shown a pitting tendency in Ringer's solution at 37 °C. Tensile tests on pre-corroded samples with different corrosion extents, adjusted by corrosion time, allow the correlation to residual tensile strength. Corrosion fatigue tests enable determination of the influence of the pitting factor on service life and monitoring of the evolution of pits through electrochemical parameters. μCT scanning of corroded samples before and after the tests allows the identification of weak spots. Through 3D-μCT analysis, it is possible to quantify the corrosion extent and thus conclude that the amount and size of corrosion pits are critical for shorter corrosion times, whereas the residual load-bearing area is decisive for longer corrosion times.

Keywords

Mg−RE alloys • Corrosion morphology • 3D-μCT analysis • Residual tensile strength • Corrosion fatigue

Extended Abstract

This study focuses on the influence of inhomogeneous corrosion on the quasi-static and cyclic properties of an Mg−Y−Nd alloy. Uniform corrosion is essential for biodegradable implant materials to avoid notching effects and premature failure. Previous studies on extruded Mg−3Y−3RE have shown a pitting tendency in Ringer's solution at 37 °C. On this basis, tensile tests are performed on pre-corroded specimens with different corrosion extents, adjusted by corrosion time, to correlate with the residual tensile strength. Corrosion fatigue tests are used to determine the influence of a mechanical-corrosive superimposed loading and the respective pitting factors on service life. The application of electrochemical measurement techniques allows the monitoring of damage evolution. μCT scanning of corroded specimens before and after the mechanical tests enables the identification of weak spots. Through 3D-μCT analysis, it is possible to quantify the corrosion extent and thus to conclude that the amount and size of corrosion pits are critical for shorter corrosion times, whereas the residual load-bearing area is decisive for longer corrosion times.

In general, the corrosion behavior and the resulting stability of temporary implant materials have to be adapted to the healing process. The tendency of Mg alloys to pitting due to their inhomogeneous and porous corrosion layers [1, 2] can lead to stress peaks under load and thus to crack initiation and premature failure. However, for a functional phase with sufficient and predictable mechanical stability, the focus is set on homogeneous corrosion. The Mg alloy WE43 (yttrium, rare earth (RE)) is widely investigated [3–6] and is already in clinical application [7, 8]. RE elements have a passivating and grain refining effect so that a continuous network acting as a corrosion barrier is formed, and both effects reinforce each other [9, 10]. Besides the corrosion rate, the corrosion morphology and the pitting susceptibility are increasingly being investigated [11]. Previous research showed that Mg−3Y−3RE has a low resistance to pitting corrosion with a corrosion rate of 1.18 ± 0.12 mm/year and a pitting factor of 25, resulting in pits with a depth of up to 600 μm in immersion tests [1]. Furthermore, it was shown, for the systems Mg−Y−Nd and Mg10Gd, that the pitting factor decreases with increasing

B. Clausius · S. Jeyavalan · P. Maier (✉)
University of Applied Sciences Stralsund (HOST), Stralsund, Germany
e-mail: petra.maier@hochschule-stralsund.de

N. Wegner · H. Hartweg · F. Walther
Chair of Materials Test Engineering (WPT), TU Dortmund University, Dortmund, Germany

S. Barela et al. (eds.), *Magnesium Technology 2023*, The Minerals, Metals & Materials Series,
https://doi.org/10.1007/978-3-031-22645-8_21

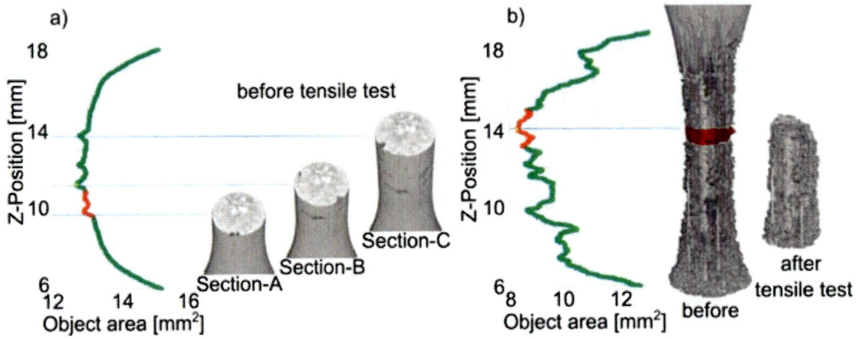

Fig. 1 Remaining cross-sectional area after corrosion of Mg−3Y−3RE tensile specimens, red curves indicate fractured area: **a** corrosion time of 24 h (cross-sections before tensile test), **b** corrosion time of 168 h (before and after tensile test). Tensile specimens: initial gauge length of 9 mm, initial diameter of 4 mm at the smallest cross-section, and a radius of 200 mm towards it

corrosion rate [1, 2]. In addition to new measurement and test methods for characterizing material behavior and methods for calculating service life, the focus is on developing application-oriented test conditions and associated measurement techniques [12].

Figure 1 shows the remaining cross-sectional area after two different corrosion times (24 and 168 h) of Mg−3Y−3RE (WE33) tensile specimens. The green curve represents the measured cross-sectional area along the corroded gauge length, whereas the red curve marks the fractured area. Figure 1a displays selected cross-sections (Section-A, -B, and -C), chosen by their corrosion pit amount and size, after a corrosion time of 24 h and before the tensile test. The specimen fails under the influence of corrosion pitting (Section-A with adjacent pits), resulting in an increased stress intensity and not under the influence of the smallest remaining cross-section (Section-B). Figure 1b, in contrast, reveals overlapping pitting corrosion after 168 h so that the specimen fails at the smallest remaining cross-section. For Mg−3Y−3RE, at short corrosion times, the local corrosive attack and, thus, the corrosion pits seem to play a dominant role.

By correlating the corrosion time and the absolute or normalized residual tensile force (Fig. 2a and b), it was found that after 24 h, the residual tensile force for Mg−3Y−3RE is about 90% of the initial value and decreases moderately for extended periods. For these tests, the variation of corrosion time only serves to achieve different corrosion morphologies. Hereafter, it is decisive to provide quantitative parameters on the corrosion extent and shape [9]. In this context, the evaluation of corrosion pits, based on ASTM standard G46 [13], will be carried out.

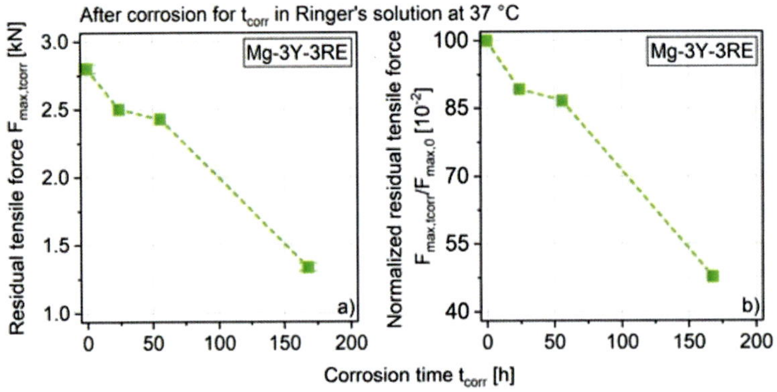

Fig. 2 Results of tensile tests after different corrosion times **a** absolute residual tensile force, **b** normalized residual tensile force

Fatigue tests in air and Ringer's solution at 37 °C are performed to determine the immediate corrosion influence on the fatigue properties. The first estimation of the fatigue strength is made by evaluating the material response during continuous load increase tests (LIT) in air. Based on the plastic strain amplitude $\varepsilon_{a,p}$, loss energy density w_L, and change in temperature ΔT, the fatigue strength is estimated between 157 and 164 MPa (Fig. 3). Based on this, the constant amplitude tests (CAT) in Ringer and as a reference in air are carried out at a maximum stress of σ_{max} = 160 MPa. Thus, the tests in air aim at a number of cycles to failure N_f of approximately the maximum number of cycles $N_{limit} = 3 \cdot 10^6$. Figure 4 shows distinct differences in the number of cycles to failure between both testing conditions. To detect the material reaction through the corrosion potential, CAT in corrosive medium is instrumented by a three-electrode system. It can be seen that the estimation by LIT is appropriate as the specimen fails below N_{limit}.

For the tests in air, cyclic hardening is evident from the plastic strain amplitude, whereas hardly any material reaction is visible close to failure. For the corrosion fatigue tests, the corrosion potential and the nominal plastic strain amplitude are evaluated analogously: Mg−3Y−3RE exhibits an almost constant course of the corrosion potential without a pronounced material reaction before failure. For the nominal plastic strain amplitude, no remarkable early material reaction, comparable to the tests in air, is recognizable. As a result of the corrosive superimposition and the significantly increased pitting factor (see [1]), a drastic decrease in the number of cycles to failure and, hence, the corrosion fatigue properties occurred. Despite the moderate decrease in residual tensile strength for corrosion times $t_{corr} \leq 24$ h (Fig. 2), the corrosion morphology under corrosive-mechanical loading appears to have a tremendous influence on the service life. It should be noted that the exposure time of approx. 2.5 h (2 h for the saturation of

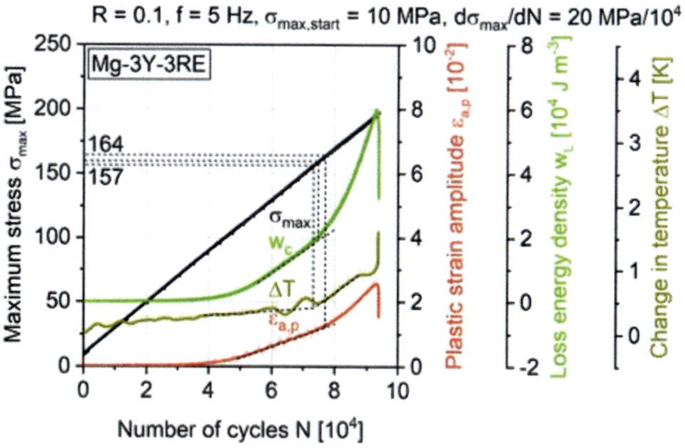

Fig. 3 Development of plastic strain amplitude, loss energy density, and change in temperature in a continuous load increase test of Mg−3Y−3RE in air

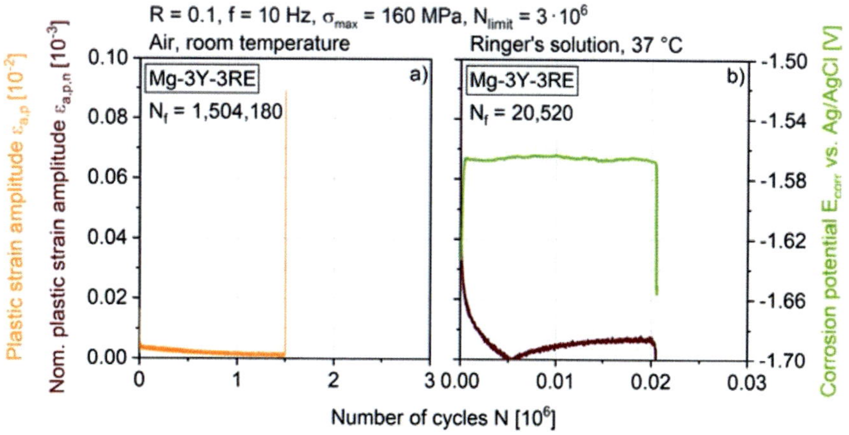

Fig. 4 Development of **a** plastic strain amplitude (in air) and **b** corrosion potential (in Ringer's solution at 37 °C) in constant amplitude tests of Mg−3Y−3RE at a maximum stress of 160 MPa

open circuit potential (OCP) and 0.5 h fatigue test time) is significantly shorter than, for example, in Fig. 1a, nevertheless, having a strong influence on the corrosion fatigue properties. 3D-μCT characterization of the failed specimens will provide data on the amount and size of the pits formed during corrosion fatigue loading.

Acknowledgments The development of the experimental setup for corrosion fatigue investigations is funded by the German Research Foundation (Deutsche Forschungsgemeinschaft, DFG) within the research project "Mechanism-based characterization of the texture influence on the corrosion and corrosion fatigue properties of zinc-based wrought magnesium alloys" (project no. 461435286).

References

1. Maier, P. et al. (2019) Mechanical and corrosion properties of two precipitation hardened Mg-Y-Nd-Gd-Dy alloys with small changes in chemical composition. JOM 71 (4) 1426-1435.
2. Maier, P. et al. (2018) Precipitation hardening on mechanical and corrosion properties of extruded Mg10Gd modified with Nd and La. Metals 8 (640).
3. Witte, F. et al. (2005) In vivo corrosion of four magnesium alloys and the associated bone response. Biomaterials 26 (17) 3557-3563.
4. Zhang, X.B. et al. (2012) Biocorrosion properties of as-extruded Mg-Nd-Zn-Zr alloy compared with commercial AZ31 and WE43 alloys. Mater. Letters 66 (1) 209-211.
5. Wegner, N. et al. (2019) Corrosion and corrosion fatigue properties of additively manufactured magnesium alloy WE43 in comparison to titanium alloy Ti-6Al-4V in physiological environment. Materials 12 (18) 2892.
6. Gu, X.N. et al. (2010) Corrosion fatigue behaviors of two biomedical Mg alloys - AZ91D and WE43 - In simulated body fluid. Acta Biomater. 6, 4605-4613.
7. Biotronik AG. Available online: www.biotronik.com/de-de/products/coronary/magmaris (accessed on 06.08.2022).
8. Seitz, J.M. et al. (2016) Magnesium-based compression screws: A novelty in the clinical use of implants. JOM 68, 1177-1182.
9. Ding, W. (2016) Opportunities and challenges for the biodegradable magnesium alloys as next-generation biomaterials. Regen. Biomater. 3 (2) 79-86.
10. Hanzi, A.C. et al. (2010) On the in vitro and in vivo degradation performance and biological response of new biodegradable Mg-Y-Zn alloys. Acta Biomater. 6 (5) 1824-33.
11. Van Gaalen, K. et al. (2021) Automated ex-situ detection of pitting corrosion and its effect on the mechanical integrity of rare earth magnesium alloy - WE43. Bioact. Mater. 8, 545-558.
12. Wegner, N. et al. (2022) Mechanical in vitro fatigue testing of implant materials and components using advanced characterization techniques. J. Biomed. Mater. Res. B 110 (4), 898-909.
13. Metals Handbook Ninth Edition (2008), Failure Analysis and Prevention, American Society for Metals, Standard Guide for Examination and Evaluation of Pitting Corrosion. ASTM G46-94 (Reapproved 2005); Materials Performance, 176.

Part IV

Deformation and Advanced Processing

Barrel Finishing of Magnesium Alloys

Nina Petersen, Björn Wiese, and Norbert Hort

Abstract

Barrel finishing is a well-established process in the industry for the targeted machining of surfaces. Especially small parts in large quantities with simple geometries can be easily machined this way. When investigating the biological behaviour of degradable magnesium implant alloys, platelets of 1 cm diameter and a thickness of slightly more than one mm are often used. Due to statistical requirements, a higher number of pieces are necessary. In addition, the geometry is comparatively simple. To ensure a certain reproducibility, all samples should also have comparatively similar surfaces. In this work, different abrasives and various process parameters are investigated to answer the question of whether barrel grinding is a viable process for producing reproducible specimens.

Keywords

Surface roughness • Barrel finish • Mg alloys

Introduction

Magnesium (Mg) has been used as an implant material in medical practice since the beginning of the nineteenth century [1], as it has a number of advantages over implant materials such as Titanium. Its mechanical properties are closer than other metals to those of bone, it leads to better bone integration and higher bone implant strength, and the mechanical properties can be optimized for specific applications by varying the Mg alloys composition and the processing [2].

By using Mg alloys as implant materials, a second operation can be avoided, as the Mg dissolves in the body´s aqueous environment to form Mg hydroxide and water. This avoids additional stress for the body, and a renewed risk of infection is eliminated [3]. Since Mg occurs as a natural trace element in the body, it has good biocompatibility and there is no risk of allergic reactions, as is the case with implants alloyed with chromium, cobalt, or nickel [4, 5].

During the fabrication and shaping of the Mg specimens, impurities such as iron and nickel from the machining tools are introduced onto and into the surface. These materials lead to galvanic corrosion, which primarily increases the initial degradation rate of the Mg [6]. A too high initial degradation rate leads to a high production of gases as a product of the degradation, and these lead to a deterioration of the healing process, especially at the beginning of the healing process [7]. Therefore, a surface treatment to remove these residues is recommended. In addition to removing contaminants, a surface treatment can reduce surface roughness, which influences the initial degradation rate [8]. According to an article by Gawlik et al. [9], implant surfaces with roughness values of Sa or Ra < 0.2 µm are more favourable for initial cell adhesion and cell viability. Higher roughness values should be avoided, as increased degradation is to be expected and, as a consequence, greater local alkalinization occurs [9].

While treatments such as acetic etching of the surface of Mg alloys have been studied [9, 10], the process of barrel finishing has not yet been studied for the use of Mg implant surface treatment. With this, it might be possible to treat large specimen quantities at once, with high uniformity and reproducibility of the results as is the case for specimens treated in the industry. In this work, Mg–1Zn–0.04Zr and

N. Petersen · B. Wiese · N. Hort (✉)
Institute of Metallic Biomaterials, Helmholtz Zentrum Hereon, 21502 Geesthacht, Germany
e-mail: norbert.hort@hereon.de

N. Petersen
Institute of Materials Physics and Technology, Hamburg University of Technology, 21073 Hamburg, Germany

N. Hort
Institute of Product and Process Innovation, Leuphana University Lüneburg, 21335 Lüneburg, Germany

S. Barela et al. (eds.), *Magnesium Technology 2023*, The Minerals, Metals & Materials Series,
https://doi.org/10.1007/978-3-031-22645-8_22

Mg–elGd alloys have been barrel finished with different abrasives and changes in parameter, to study the effect of the process on surface roughness and thickness of the specimens.

Experimental

Two Mg alloys (Table 1) in the shape of small discs of 9–10 mm diameter and approx. 1.5 mm thickness were used for barrel finishing. The Mg–1Zn–0.04Zr discs were turned from the strand. The Mg-2Gd alloy was extruded at 400 °C with a ram speed of 2.2 $\frac{\text{mm}}{\text{s}}$ [11], then turned to a diameter of 9 mm and cut from the strand with a mill. A barrel finishing machine from *effer-handel* (effer-handel, Aalen, Germany) with a round barrel with a volume of 3 l was used. The rotational speed can be varied in steps via the voltage; for most experiments, a rotational speed of 28 rpm was used.

The specimens were placed in the barrel with different abrasives from *effer-handel* (Table 2) and were finished for varying times and speeds to see the influence of the parameters on the surface and geometry. The specimens were finished with and without a compound. With polishing beads, a polishing compound from *effer-handel* with a concentration of 20 ml on 1 l of tap water was used. With abrasives, the cleaning and degreasing agent R35 (*Rösler Oberflächentechnik*, Untermerzbach, Germany) with a concentration of 10 $\frac{\text{g}}{\text{L}}$ were used. When using a compound with the abrasives, the barrel was placed on a wedge with an angle of approx. 25.5° to prevent leakage. For each abrasive, three to five specimens were taken from the barrel after 0.5, 1, 2, 5.5, and 24 h to measure the roughness and thickness over time.

After barrel finishing, the specimens were cleaned in an ultrasonic bath of ethanol for 5 min and dried under warm air. For roughness analysis, pictures of each specimen were taken with a VK-X1050 *Laserscanning-Microscope* (*Keyence Corporation*, Osaka, Japan) and analysed with *Keyence MultiFileAnalyzer* (Version 2.1.3.89). To determine the surface roughness, surface measurements were performed over an area of 3500 × 3500 μm, placing the lower right corner of the area in the center of the specimens. The filters for the evaluation of the measurement results were selected according to DIN EN ISO 25178-3 [12]. The low pass filter (S-filter) was set to 8 μm. For the high pass filter (L-filter), a value of 0.08 mm was chosen.

To measure the thickness, at least three specimens of each run were measured with a MT30B thickness gauge (*Dr. Johannes Heidenhain GmbH*, Traunreut, Germany) in combination with a *Heidenhain* 300.105.002 digital readout. Three measurements per specimen were obtained to have a statistically valuable result.

Results and Discussion

Rotational Speed

Three different rotational speeds (4.5, 28, and 76 rpm) have been used for 1000 rotations on Mg–1Zn–0.04Zr specimens in nut granule to investigate the influence on the specimen. In Fig. 1, the surface roughness as well as the thickness of the specimen are shown over the rotational speed of the barrel, in comparison to the as received specimens (AR) values.

Almost no change in surface roughness in comparison to the AR values can be seen at 4.5 and 76 rpm. This can be explained by the simple mechanics behind the barrel finishing—with a too low rotational speed, the movement of the abrasive as well as the acting forces against the specimens is too low to achieve a change. With a too high rotational speed, the tools as well as the specimen will be pressed towards the outer wall of the barrel and the movement of the tools against the specimen is too low once again. As a significant difference in surface roughness was visible at 28 rpm, this rotational speed was used for the following tests.

There is no significant change in thickness for the different rotational speeds, this is probably due to the low abrasiveness of the nut granule as well as the short processing time. To see a significant difference in thickness, a lot of material would have to be removed from the specimen; this cannot be achieved with the nut granule, which is meant for polishing specimen surfaces.

Processing Time, Abrasives, and Compounds

In Fig. 2a–d, the surface roughness of the specimens including standard deviation is shown over the processing time for the different abrasives. For a better classification of the results, the AR roughness values are plotted in each diagram at hour zero. As can be seen, a significant reduction

Table 1 Chemical composition of the alloys (wt. %), roughness, and thickness including the standard deviation of the specimens, as received

Chemical composition (wt.%)	Roughness (μm)	Thickness (mm)
Mg–1Zn–0.04Zr	1.367 ± 0.082	1.502 ± 0.011
Mg–2Gd	1.734 ± 0.232	1.534 ± 0.016

Table 2 Abrasives with measurements

Abrasive	Size (mm)
Nut granule	1.7–2.4
Polishing beads (macrolane)	Ø3.5
Plastic cones	Height 10
Corn granule	0.5–1

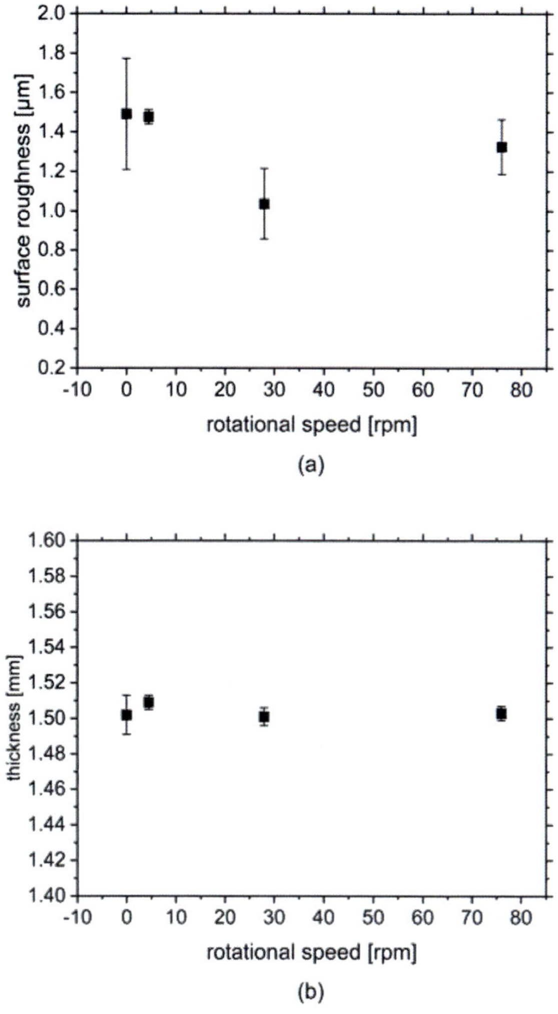

Fig. 1 **a** Surface roughness over rotational speed of the barrel and **b** thickness over rotational speed of the barrel; both for Mg–1Zn–0.04Zr after 1000 turns in nut granule with the reference value of the AR specimen at $x = 0$

of the surface roughness over time could be achieved with all of the used abrasives. Furthermore, the standard deviation could be decreased for almost all abrasives with a processing time longer than 2 h with the exception of corn granule and polishing beads with compound.

The corn granule and the polishing beads are both light weight without any sharp edges and meant for polishing specimen surfaces. Therefore, it is not surprising that the changes in surface roughness are not that strong, as the abrasives cannot remove a lot of material and the deep groves on the specimen surface remain (see next section). Especially for the polishing beads in combination with the polishing compound, the lubricating effect seems to be very high, as the roughness without the compound is significantly lower.

For nut granule, the surface roughness values after 24 h were similar to the ones after processing in plastic cones. In comparison to the corn granule, the nut granule has sharper edges and is over all harder and heavier. This explains the differences in results between corn and nuts. As both materials are of natural origin, a usage for implants would be of advantage, as no harmful chemical residues would have to be removed from the specimens' surface.

As the plastic cones were the strongest abrasives used and advertised to be specifically for flat surfaces, it was to be expected to see the strongest influence on the surface roughness. The abrasive surface was significantly bigger than that of the other abrasives. The surface roughness with and without compound is very similar after 24 h; however with the R35 compound, the low roughness values could be achieved much faster after only 1 h of barreling. This is probably due to a strong corrosion of the specimens in the R35 compound.

Nevertheless, a surface roughness below a value of Sa=0.2 µm, as was a defined goal by Gawlik et al. [4], could not be achieved with the used abrasives. Because of a high standard deviation in the AR discs surface roughness, it is hard to obtain statistically valuable results for the process, especially in weak abrasives or polishing agents such as the corn granule (Fig. 2b). A pre-sorting of the specimens by surface roughness before barrel finishing would bypass this problem, but would be a much greater investment of time, especially for high amounts of specimens, and is therefore not an economic solution.

A significant reduction of the thickness could not be achieved without the use of *Rösler* R35 compound. With this, in combination with the plastic cones, the thickness was reduced by approx. 150 µm (Fig. 2h). Presumably this is due to a corrosion of the specimen in the compound (Fig. 3).

Surface Geometry

As mentioned before, the Mg–1Zn–0.04Zr specimens are produced by turning them from the strand. Because of this, a specific surface geometry is produced, which can be seen in

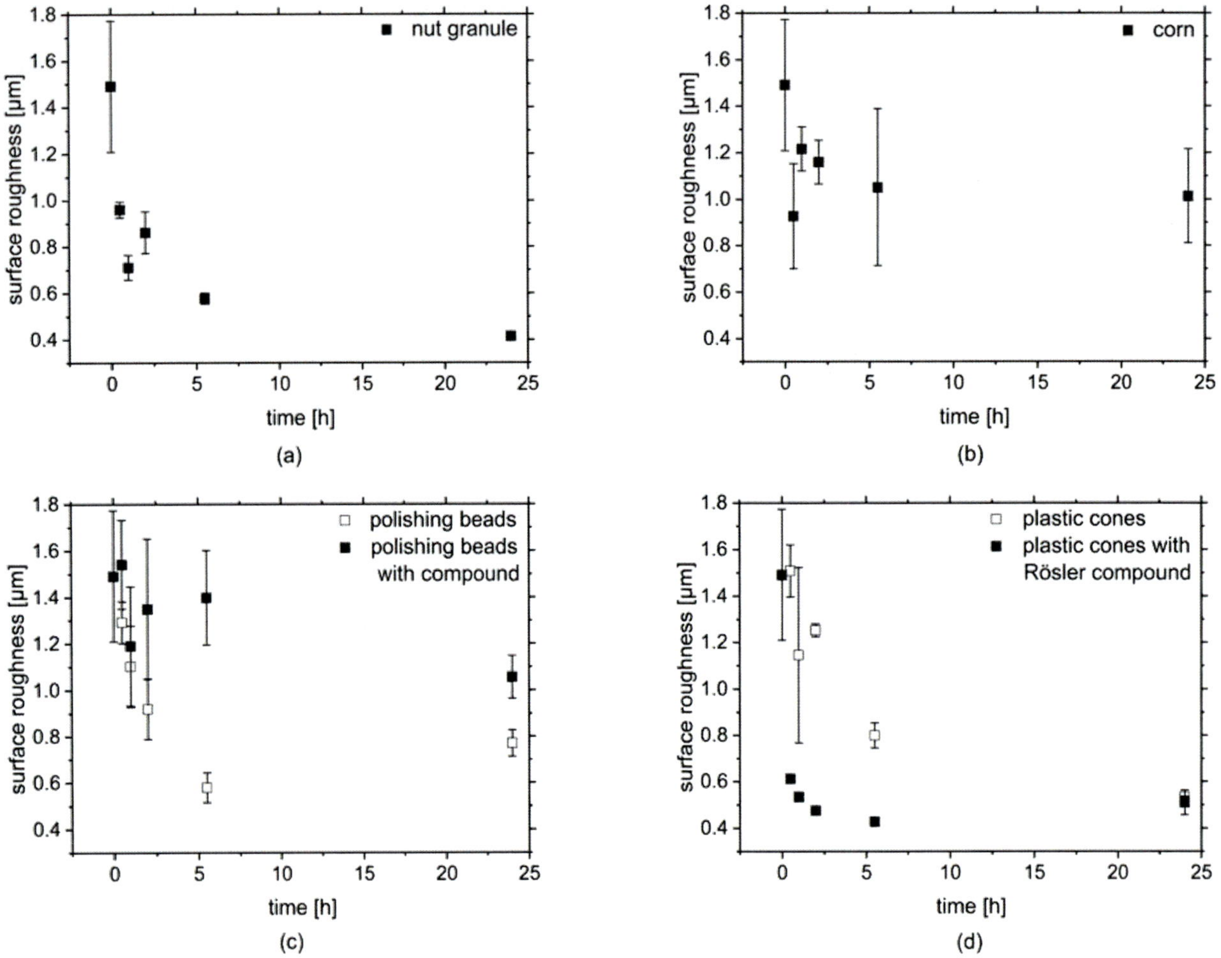

Fig. 2 Surface roughness of Mg–1Zn–0.04Zr discs over time after barrel finishing with AR values at $x = 0$ in **a** nut granule, **b** corn granule, **c** polishing beads with and without polishing compound, and **d** plastic cones with and without R35 compound

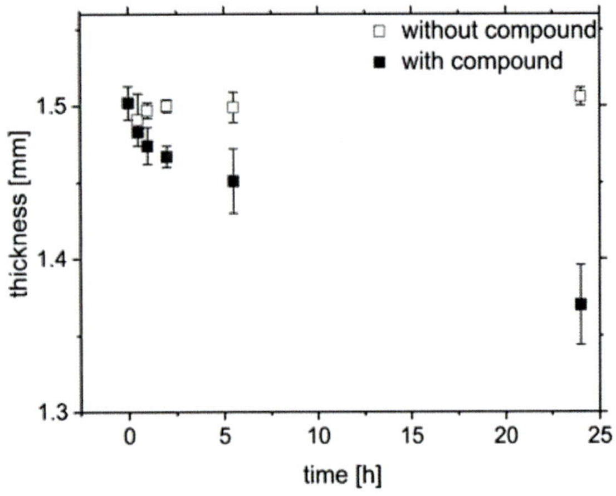

Fig. 3 Thickness of the Mg–1Zn–0.04Zr discs after barrel finishing in plastic cones with and without R35 compound with AR values at $x = 0$

Fig. 4a. The circular grooves on the surface are a big contribution to surface roughness and increase the corrosion of the specimens. A removal of the grooves is therefore a desired result.

With barrel finishing, the grooves could be reduced in all specimens, especially for a processing time of 24 h, which can also be seen in the decreasing surface roughness (Fig. 2). With the R35 compound, the grooves could be removed completely, as can be seen in Fig. 4b, after only 2 h of barrelling in plastic cones with R35 compound. As discussed before, this is probably due to the strong corrosion of the R35 compound, which removed a lot of material from the specimen surface. Due to this, signs of corrosion are visible on the surface, and because of the uncontrolled corrosion in the compound it is hard to get a defined surface. To clean the surface of the corroded material, a barrel finish in

Fig. 4 Surface of Mg–1Zn–0.4Zr specimen **a** AR and **b** after barrel finishing with plastic cones in R35 compound for 2 h

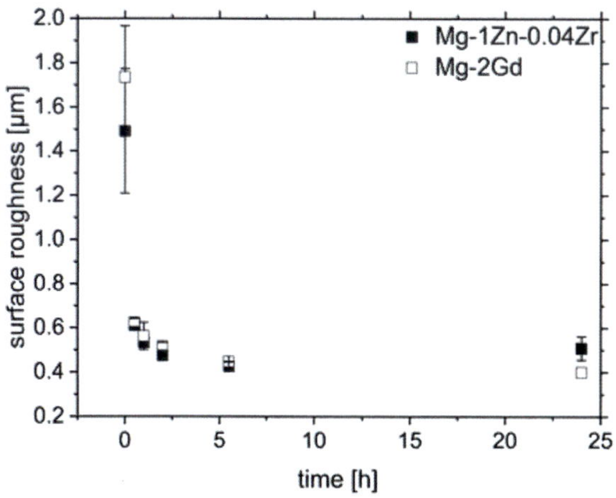

Fig. 5 Surface roughness over time after barrel finishing in plastic cones with R35 compound of Mg–1Zn–0.04Zr and Mg–2Gd with AR values at $x = 0$

two steps might be useful, for instance a barrelling with plastic cones and R35 followed by a barrelling in nut granule. This could also be helpful to reduce the surface roughness further, as the R35 will remove the grooves from the specimen surface in a short time, and the nut granule will polish the already smooth surface.

Application of the Process to Other Alloys

As many different alloys are used to produce the discs, a uniform process with similar surface roughness values for all alloys is desired. Therefore, a Mg–2Gd alloy was processed similar to the Mg–1Zn–0.04Zr specimens in plastic cones with R35 compound. In Fig. 5, the Mg–2Gd alloy (b) can be seen in comparison to Mg–1Zn–0.04Zr (a). Even though the surface roughness of the Mg-2Gd alloy used to be approx. 0.350 µm higher, after a processing time of only 0.5 h the roughness values are astonishingly similar and continue to be similar over the whole processing time. With that in mind, the choice of the Mg alloy might not be a crucial parameter for the resulting surface roughness.

Conclusion

Barrel finishing can be used to change the surface roughness and the thickness of Mg alloy discs. It is possible to obtain reproducible results, even when using the same procedure for different alloys. With the choice of different abrasives and compounds, a desired surface roughness can be produced. Nevertheless, a surface roughness under Sa=0.2 µm could not be realised with the used abrasives and compounds. With a high deviation in the AR surface roughness values, it is hard to validate the results, especially for light abrasives, yet a pre-sorting by surface roughness is too time consuming. Therefore, it is recommended to use a combination of different aggressive abrasives and polishing agents one after the other. The results so far indicate that plastic cones with R35 compound should be used to reduce coarse surface roughness, followed by a final polishing with nutshells. The thickness of the specimen did not significantly change with the used abrasives, only in combination with the R35 compound which strongly corroded the specimen surface.

References

1. J. M. Seitz, A. Lucas, and M. Kirschner, "Magnesium-Based Compression Screws: A Novelty in the Clinical Use of Implants," *JOM*, 68 (2016), 1177–1182.
2. B. Wiese, J. Harmuth, R. Willumeit-Römer, and J. Bohlen, "Property Variation of Extruded Mg–Gd Alloys by Mn Addition and Processing," *Crystals*, 12 (2022), 1036.
3. A. J. Suda, E. Heilgeist, M. Tinelli, and O. E. Bischel, "High early post-operative complication rate after elective aseptic orthopedic

implant removal of upper and lower limb," *J. Orthop. Res.*, 36 (2018), 1035–1039.
4. T. Kraus, S. Fischerauer, S. Treichler, E. Martinelli, J. Eichler, A. Myrissa, S. Zötsch, P. J. Uggowitzer, J. F. Lö_er, and A. M. Weinberg, „The influence of biodegradable magnesium implants on the growth plate," *Acta Biomaterialia*, 66 (2018), 109–117.
5. R. Walter and M. B. Kannan, "Influence of surface roughness on the corrosion behaviour of magnesium alloy," *Materials and Design*, 32 (2011), 2350-2354.
6. A. Atrens, G. L. Song, Z. Shi, A. Soltan, S. Johnston, and M. S. Dargusch, "Understanding the corrosion of mg and mg alloys," *Elsevier*, 2018.
7. D. Noviana, D. Paramitha, M. F. Ulum, and H. Hermawan, "The effect of hydrogen gas evolution of magnesium implant on the postimplantation mortality of rats," *J. Orthop. Transl.*, 5 (2016), 9–15.
8. T. L. Nguyen, A. Blanquet, M. P. Staiger, G. J. Dias, and T. B. F. Woodfield, "On the role of surface roughness in the corrosion of pure magnesium in vitro," *J. Biomed. Mater. Res. Part B*, 100 B (2012), 1310–1318.
9. M. Gawlik, B. Wiese, V. Desharnais, T. Ebel, and R. Willumeit-Römer, "The Effect of Surface Treatments on the Degradation of Biomedical Mg Alloys A Review Paper," *Materials*, 11 (2018), 2561.
10. M. Gawlik, B. Wiese, A. Welle, J. González, V. Desharnais, J. Harmuth, T. Ebel, and R. Willumeit-Römer, "Acetic Acid Etching of Mg–xGd Alloys," *Metals*, 9 (2019), 117.
11. J. Harmuth, B. Wiese, J. Bohlen, T. Ebel, and R. Willumeit-Römer, "Wide Range Mechanical Customization of Mg–Gd Alloys With Low Degradation Rates by Extrusion," *Front. Mater.*, 6 (2019), 201.
12. DIN. EN ISO 25178-3:2012-11, Geometrische Produktspezifikation (GPS)—Oberflächenbeschaffenheit: Flächenhaft—Teil 3: Spezifikationsoperatoren (ISO 25178-3:2012); Beuth Verlag GmbH: Berlin, Germany, 2012.

Effect of Minimum Quantity Lubrication on Machinability of Magnesium RZ5 Alloy: A Comparative Study

Arabinda Meher and Manas Mohan Mahapatra

Abstract

Magnesium alloys are used for structural applications in automobile and aerospace industries because of their lower density. Most magnesium-based components produced by casting or forming are of near net shape. Further machining is required to achieve dimensional accuracy and surface finish. So, magnesium machining is a topic of interest in many industries. In the present study, the machinability of magnesium RZ5 alloy (Mg-4.25 Zn-0.54 Zr-1.25 Ce) is evaluated with and without the use of cutting fluid. The cutting force and surface roughness are significantly affected by the feed rate and depth of cut; however, it is minutely affected by the cutting speed of the material. The use of cutting fluid with minimum quantity lubrication reduces the cutting force and increases the surface finish of the material. Due to the high-temperature gradient, smaller size chips were observed with lesser tool wear during machining of magnesium RZ5 alloy with minimum quantity lubrication.

Keywords

Magnesium alloy • Machining • Cutting force • Surface roughness • Minimum quantity lubrication

Extended Abstract

In the search for lightweight structural material, magnesium is found to be the most energy-efficient material because of its lighter density [1]. Magnesium alloys have potential structural applications in different components of automobile and aerospace industries because of their high strength-to-weight ratio and high stiffness-to-weight ratio [2, 3]. Machining of the as-cast or forged magnesium alloy is necessary to achieve dimensional accuracy and better surface finish for many engineering applications [4]. So, magnesium machining is a topic of interest in many industries. Earlier research studies mostly focused on analyzing the machinability of different alloys like Fe, Ti, Cu, Mg, and Al alloys. However, the study on the machinability of magnesium RZ5 alloy is not observed in the contemporary literature. Magnesium RZ5 alloy is mostly used in light applications like aerospace engine casing. Machining of such components is necessary for different industrial applications. Machining of the materials can be carried out at different operating condition and environments. Earlier studies observed that using cutting fluid enhances the machinability of the materials [5].

In the present study, an effort has been made to investigate the microstructure and machinability of magnesium RZ5 alloy (Mg-4.25 Zn-0.54 Zr-1.25 Ce) in dry turning conditions and with minimum quantity lubrication (MQL) environment. MQL condition reduces the amount of cutting fluid sprayed to a near-dry condition in the cutting zone. This condition is considered an environment-friendly cooling technique due to the lower cutting fluid consumption. The microstructure of the RZ5 alloy is analyzed using a field emission scanning electron microscope (FESEM), indicating a hexagonal structure with uniform grain distribution. Figure 1 shows the FESEM micrographs of magnesium RZ5 alloy. The elemental composition of magnesium RZ5 alloy is analyzed using energy dispersive X-ray spectroscopy (EDX) and shown in our earlier study [6], which is

A. Meher (✉)
Department of Mechanical Engineering, University Centre for Research and Development, Chandigarh University, Mohali, Punjab 140413, India
e-mail: am39@iitbbs.ac.in

A. Meher · M. Mohan Mahapatra
School of Mechanical Sciences, Indian Institute of Technology Bhubaneswar, Bhubaneswar, Odisha 752050, India

S. Barela et al. (eds.), *Magnesium Technology 2023*, The Minerals, Metals & Materials Series,
https://doi.org/10.1007/978-3-031-22645-8_23

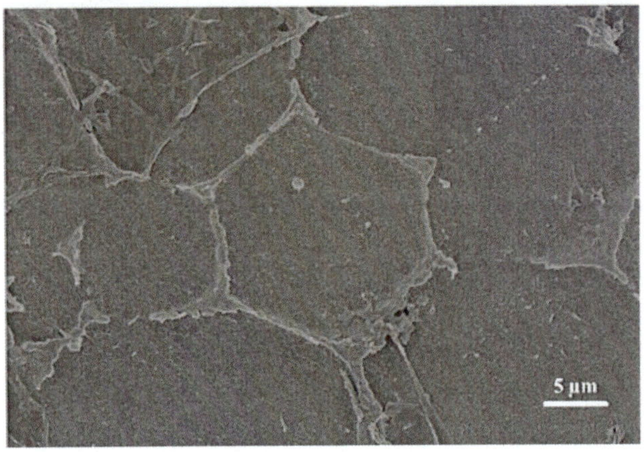

Fig. 1 FESEM micrograph of magnesium RZ5 alloy

equivalent to the actual composition of industrial-grade magnesium RZ5 alloy. The X-ray diffractometer (XRD) phase diagram showed the formation of the magnesium phase along with other alloying phases like $Mg_{17}Ce$ and $CeZn_3$. The detailed XRD phase diagram of magnesium RZ5 alloy is discussed in our earlier study [7].

The magnesium RZ5 alloy was prepared in a cylindrical shape of 50 mm in diameter and 300 mm in length to investigate the machinability of the materials. Kistler made piezoelectric dynamometer (model—9257B) was used to measure the forces acting during the machining of the materials. The dynamometer was connected to a charge amplifier, and the amplified force signal was displayed on the computer monitor using DynoWare software. The machinability of the magnesium RZ5 alloy is investigated in terms of cutting force and surface roughness with the variation of machining parameters like cutting speed (50, 100 and 150 m/min), feed rate (0.12, 0.24 and 0.36 mm/rev), and depth of cut (0.5, 0.75 and 1.0 mm) in dry turning and MQL environment. The MQL environment was developed using a mist lubricating system during machining. The lubricant was prepared with the help of a stirrer mixture using the oil (STRUB Stabillo Cut1) and water in the proportion of 1:20. The surface roughness of the machined surface of the RZ5 alloy at different machining conditions was measured using a stylus-based surface roughness tester (Mitutoyo Surftest SJ-210). The tool wear and chip morphology in dry turning and MQL condition is analyzed using an optical microscope and FESEM.

Machinability of the RZ5 alloy indicates that the cutting force during machining of RZ5 alloy significantly increased with an increase in feed rate and depth of cut; however, cutting speed has a significantly lesser effect on the cutting force of the materials. From Fig. 2a, it can be observed that the cutting force during the machining of RZ5 alloy increased with an increase in depth of cut at a constant cutting speed of 50 m/min at different feed rates in dry turning conditions and turning with MQL due to the higher materials removal rate. Lower cutting force was observed during machining using the MQL system compared to dry turning at each machining condition, which might be due to lesser friction at the cutting interface. It was observed from Fig. 2b that the surface roughness of the materials increased with an increase in feed rate and depth of cut, and it decreased with an increase in cutting speed. The feed rate during machining severely affects the surface roughness of the material as compared to cutting speeds and depth of cuts. The surface roughness of the RZ5 alloy decreased while machining in the MQL environment compared to dry turning for each operating parameter, which can be observed in Fig. 2b.

Figure 3 shows the FESEM micrograph of the machining chips in dry turning conditions and using an MQL environment. Discontinuous and smaller-size chips were generated due to the high-temperature gradient at the cutting interface due to the cooling technique introduced by the MQL system during the machining of RZ5 alloy. A build-up edge formation at the tip of the tool was observed during the machining of the RZ5 alloy at a lower cutting speed in dry turning conditions; whereas very mild tool wear was observed during the

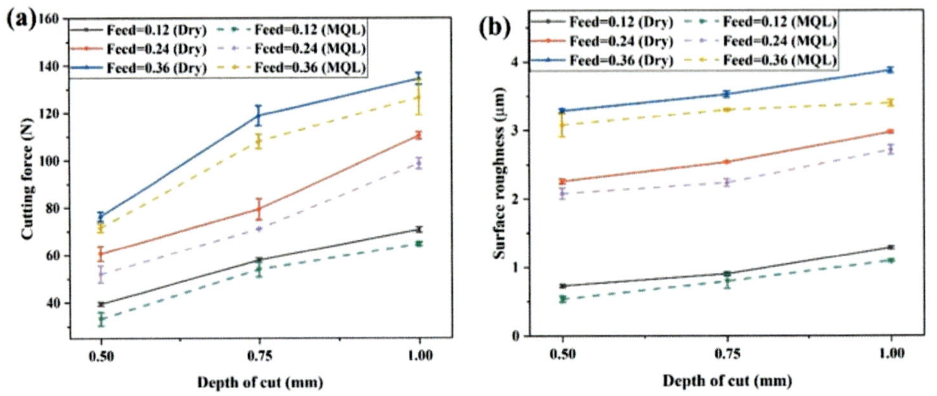

Fig. 2 Effect of depth of cut on **a** cutting force and **b** surface roughness at different feed rates at the cutting speed of 50 m/min for magnesium RZ5 alloy in dry turning and MQL conditions

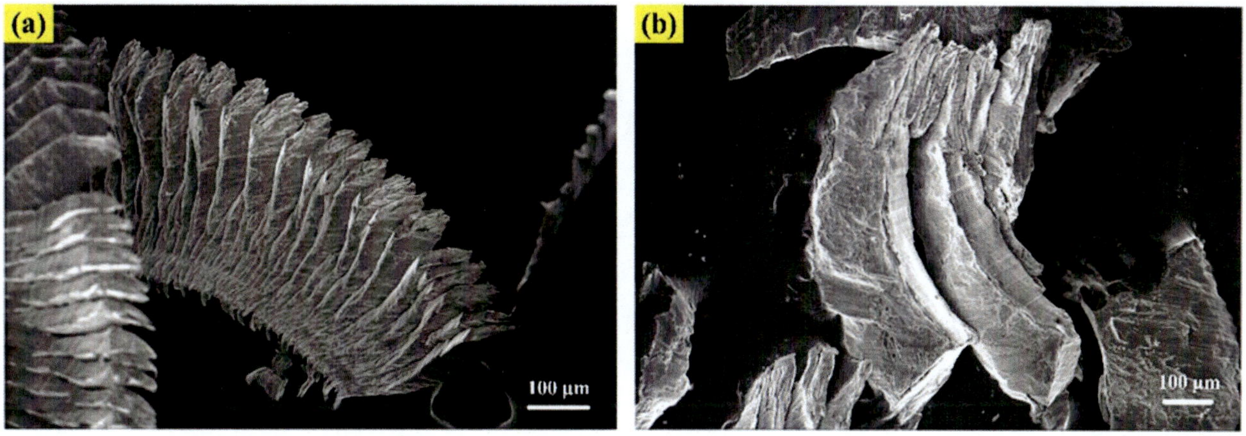

Fig. 3 FESEM micrograph of chips generated during machining of magnesium RZ5 alloy in **a** dry turning condition and **b** MQL environment

machining of RZ5 alloy using the MQL system due to the proper cooling and lubrication at the tooltip.

References

1. K.U. Kainer, Magnesium Alloys and their Applications, 2016. https://doi.org/10.1080/10667857.1998.11752784
2. A.A. Luo, Magnesium casting technology for structural applications, J. Magnes. Alloy. 1 (2013) 2–22.https://doi.org/10.1016/j.jma.2013.02.002
3. L. Ceschini, A. Dahle, M. Gupta, A.E.W. Jarfors, S. Jayalakshmi, A. Morri, F. Rotundo, S. Toschi, R.A. Singh, Aluminum and Magnesium Metal Matrix Nanocomposites, 2017.https://doi.org/10.1007/978-981-10-2681-2
4. L. Lu, S. Hu, L. Liu, Z. Yin, High speed cutting of AZ31 magnesium alloy, J. Magnes. Alloy. 4 (2016) 128–134.https://doi.org/10.1016/j.jma.2016.04.004
5. S. Khandekar, M.R. Sankar, V. Agnihotri, J. Ramkumar, Nano-cutting fluid for enhancement of metal cutting performance, Mater. Manuf. Process. 27 (2012) 963–967.https://doi.org/10.1080/10426914.2011.610078
6. A. Meher, M.M. Mahapatra, P. Samal, P.R. Vundavilli, Synthesis , Microstructure and Mechanical Properties of Magnesium Matrix Composites Fabricated by Stir Casting, Mater. Today Proc. 18 (2019) 4034–4041.https://doi.org/10.1016/j.matpr.2019.07.346
7. A. Meher, M.M. Mahapatra, P. Samal, P.R. Vundavilli, Abrasive Wear Behaviour of TiB_2 Reinforced In Situ Synthesized Magnesium RZ5 Alloy based Metal Matrix Composites, Met. Mater. Int. 27 (2020) 3652–3665.https://doi.org/10.1007/s12540-020-00746-1

Influence of Preforging in Extrusion as Well as in Equal Channel Angular Pressing in EXtrusion (ECAPEX) on the Properties of Magnesium Rods

René Nitschke and Sören Mueller

Abstract

Magnesium predominantly exhibits a hexagonal lattice structure. The anisotropic distribution of activatable slip systems often leads to crystallographic preferential orientations during forming, whereby the anisotropic crystallographic properties show up macroscopically as anisotropic mechanical properties in components. Efforts to improve materials and optimize them should not only consist of mitigating or even preventing such anisotropic component properties. There is great potential to exploit them constructively. For this purpose, a 6 × 3 process matrix was designed using the magnesium alloy AZ31 as an example to investigate the influence of different forging as well as forming routes on the properties of extruded round bars. Six different initial states of extrusion billets and three different extrusion dies were used, with one die designed for Equal Channel Angular Pressing in EXtrusion (ECAPEX). Forming behavior as well as microstructure and mechanical properties were investigated. Selected examples are used to present the results and evaluate the various influencing factors.

Keywords

Magnesium • Extrusion • ECAPEX

Introduction

Magnesium materials and their alloys can be effectively formed into near-net-shape profiles or intermediates using the extrusion process, but usually exhibit distinct preferential orientations of crystallographic orientation as has been reported many times in the literature cf. [1–3]. In the case of alloy AZ31, for example, it is described that the basal plane in the forming zone is oriented perpendicular to the highest compressive stress encountered. In general, this causes a preference for the perpendicular orientation of the c-axis to the extrusion direction and consequently an increase in the pole density distribution of the {0001} basal planes tilted by 90° with respect to the extrusion direction (ED). If round bars are extruded from circular billets, a rotationally symmetric forming zone is formed immediately upstream of the extrusion channel, so that a uniform rotationally symmetric preferred orientation of the basal planes parallel to the extrusion axis can also be detected. Furthermore, depending on the extrusion parameters temperature, extrusion ratio, and extrusion speed, other textures can be formed in the course of recrystallization and grain growth which, for example, result in an increased probability of a perpendicular orientation of the basal planes to the extrusion direction, so that the c-axes are thus oriented parallel to ED. Corresponding dependencies can also be formulated for the pyramidal and prismatic slip systems. However, the forming behavior at room temperature is preferentially determined by the basal slip systems as well as the twinning, so that further interpretation can be dispensed with at this point.

The mechanical properties and the forming behavior of metallic materials at room temperature are primarily determined by the crystal lattice, i.e. the activatability of slip systems and, in the case of magnesium, also by the activation of twinning. Since alloy AZ31 does not exhibit a phase change and therefore the crystal structure and thus the slip systems cannot be influenced, the remaining variables that can be changed are the grain size and the texture. The activations of both the slip systems and the twinning systems can be described in dependencies on the grain size with the aid of the Hall–Petch relationship according to Eq. (1) [4, 5]:

$$\sigma_a = \tau_c + k \cdot d^{-1/2}. \quad (1)$$

R. Nitschke (✉) · S. Mueller
Extrusion Research & Development Center, TU Berlin,
Gustav-Meyer-Allee 25, 13355 Berlin, Germany
e-mail: rene.nitschke@strangpressen.berlin

S. Barela et al. (eds.), *Magnesium Technology 2023*, The Minerals, Metals & Materials Series,
https://doi.org/10.1007/978-3-031-22645-8_24

Here, σ_a describes the yield stress, τ_c the critical shear stress (CRSS) of the respective system, k the respective Hall–Petch constant, and d the grain size. Since the Hall–Petch constant for twinning is larger than that for dislocation sliding, twinning is preferentially activated with increasing grain size, or dislocation sliding is the dominant deformation mechanism with decreasing grain size cf. [6–11]. Furthermore, Schmidt's shear stress law can be used to determine the resulting shear stresses as a function of spatial orientation. Table 1 shows values given in the literature for the critical shear stresses of the sliding and twin systems of the AZ31 alloy e.g. [12]. Both grain size and texture can be influenced in the extrusion of AZ31 magnesium alloy by selecting suitable process parameters.

A lower temperature, for example, generally results in smaller grain sizes [13]. The strain path in the forming zone can also influence the local texture [14]. In this project, extrusion with normal dies was compared with ECAPEX.

The use of materials for structural applications generally consists of assembling components consisting of materials with mechanical properties that are as isotropic as possible into load-adapted, i.e. anisotropic, assemblies. This often involves joining different materials, such as plastics and metals in the automotive sector. During the subsequent recycling process, the challenge arises of separating these different materials from one another by type. However, due to their anisotropic material properties, materials such as magnesium offer the possibility of using anisotropic properties to take load adaptation aspects into account as early as the component development stage, thereby avoiding the mixing of different material groups under certain circumstances. Possible applications and manufacturing routes have been addressed, for example, in [15]. By using the AZ31 and ME21 alloys together at 50% by volume each for an extruded hybrid flat section, a 25% improvement in energy absorption under bending load was demonstrated by exploiting the different microstructure during extrusion of the two alloys.

Materials and Methods

Billet Pretreatment

In order to produce starting material that was as homogeneous as possible, cast material of the alloy AZ31 was first homogenized for 15 h at a temperature of 420 °C, then extruded on the 8 MN extrusion press also available at ERDC at an extrusion ratio of R_1 = 14 to form a round bar with a diameter of 30 mm. From this, extrusion billets with a length of 50 mm and an outer diameter of 28.5 mm were machined. These billets were further treated on five different routes, so that finally six configurations were available for extrusion tests. Pretreatment parameters of the extrusion billets are shown in Table 2.

The forging was performed rotationally symmetrically once each in a total of 8 radial directions A to H according to Fig. 1, so that an angle of 22.5° lay between the forging directions.

The billets were forged to a resulting height of 27 mm. At room temperature, all eight operations were performed in sequence. Forging at elevated temperatures was performed in two passes A to D (red) and C to H (black). To prevent excessive cooling of the billets, the forging pad used was also heated to the process temperature in each case.

Die Design

Furthermore, the influence of the die design should be included in the investigations. For this purpose, a flat as well as a stepped die face was to be used and, in addition, a die with integrated ECAP process. In order to provide as high as possible of the available process forces for the forming, the tests should be carried out in indirect extrusion mode.

In this process, the extruded profile must pass through the hollow die in the direction of extrusion. Thus, in order to maintain the direction of extrusion, double shearing by the same angle in each case is necessary. The first task was therefore to come up with a die design for the complicated ECAPEX die. Within the framework of flow simulations using the FEM software DEFORM, a favorable shear angle of 135° was determined. From this and from the other boundary conditions, a possible geometry for the die could be derived. The aim was to deflect the profile of the ECAPEX by at least its own width in order to avoid "flowing around the corner" and to support the required shear. For the existing container diameter of 30 mm, the specifications could be implemented with a rod diameter of 8 mm. This corresponds to an extrusion ratio R_2 = 14:1, so the extruded rods, together with the first extrusion to produce

Table 1 CRSS—values (MPa) given in the literature for the critical shear stresses of the sliding and twin systems of the AZ31 alloy e.g. [12]

⟨a⟩-slip			{10–12} ⟨10-11⟩	{10–11} ⟨1012⟩
Basal	Prismatic	Pyramidal	ttw	ctw
10–30	50–90	60–100	15–30	76–153

Table 2 Pretreatment parameters of the billets (A = as extruded, W = heat treated, n = not forged, c = cold forged, h = hot forged)

Pretreatment	Billet designation					
	A_n	A_c	A_h	W_n	W_c	W_h
525 °C/3 h				X	X	X
Forged at RT		X			X	
Forged at 250 °C			X			X

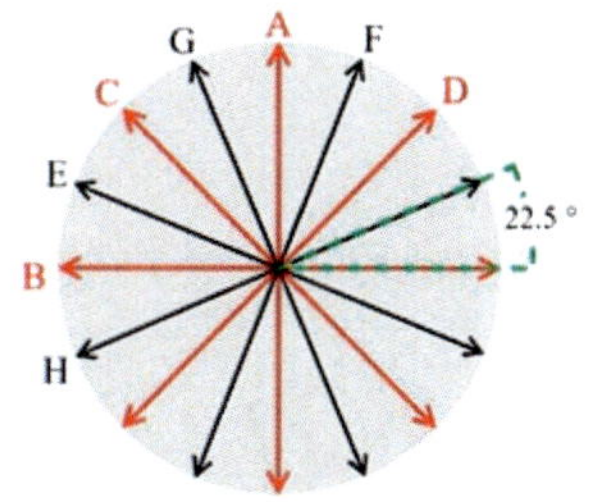

Fig. 1 Forging route for the pretreatment of the extrusion billets

Fig. 3 ECAPEX die used

the starting material, have a total extrusion ratio of $R_{total} = 196{:}1$.

A three-part die and a corresponding hollow punch were designed. The die parts were aligned to each other by three alignment pins and could be separated from each other for the removal of the extrusion residue. For the conventional extrusion tests, two different die faces were designed. First, a conical die with an opening angle of $2\alpha = 150°$ was designed. Based on the opening angle, a die with a stepped face was designed. Figure 2 shows the die geometries schematically, and Fig. 3 shows a representation the ECAPEX die used.

Extrusion was first carried out with the conventional dies C and G at process temperature of 250 °C. This means that all tools involved as well as the billet material are heated and had the same temperature. During the first extrusion test with the ECAPEX tool, it was found that the tool load would become too high at 250 °C, so the process temperature was raised to 300 °C for the ECAPEX tool. In order to be able to compare the results with respect to the extrusion behavior as well as the microstructure and the mechanical properties, at least indirectly, selected extrusion tests were additionally carried out with normal tools at 300 °C. The extrusion tests shown in Table 3 were carried out. This table also serves to assign a specific nomenclature to the tests.

Characterization of Microstructure and Texture

To characterize the microstructure and texture, samples were cut perpendicular to the extrusion direction (ED). The microstructure was examined under an optical imaging microscope (OIM) in the ED plane. Texture was analyzed using an X-ray diffractometer (XRD), also in the ED plane. Each sample was embedded, ground, and then etched. The specimens for microstructure analysis were also chemically polished with a CP2 agent consisting of 100 ml ethanol, 12 ml hydrogen chloride, and 8 ml nitric acid, and then polished with an etching solution consisting of 70 ml ethanol, 30 ml distilled water, 15 ml acetic acid, and 4.2 g picric acid.

Characterization of Mechanical Properties

The mechanical properties in the tensile and compression test in the ED direction were determined with an MTS 810

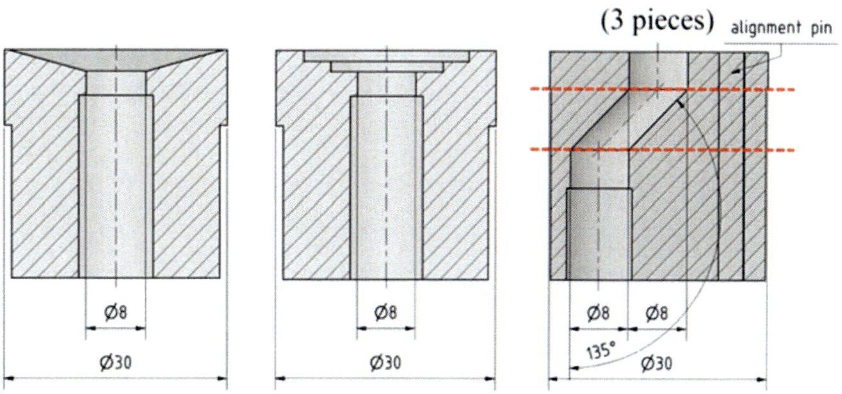

Fig. 2 Schematic representation of the extrusion dies used from left to right (conical die = C, gradueted die = G, ECAPEX die = E)

Table 3 Overview of the extrusion tests carried out

Temperature (°C)	Die type	Billet designation					
		A_n	A_c	A_h	W_n	W_c	W_h
250	C	X	X	X	X	X	X
	G	X	X	X	X	X	X
300	E	X	X	X	X	X	X
	C	X			X	X	

universal testing machine in accordance with DIN EN ISO 6892-1 and DIN 50106. The tests were carried out with tensile specimens with an initial length L_c of 40 mm and a diameter of 5 mm and compression specimens with a length of 8 mm and also a diameter of 5 mm. The strain rate chosen for all tests was $2.5 \times 10^{-4}\ s^{-1}$.

Results

Extrusion Behavior

For the evaluation of the extrusion behavior, two force values can basically be read from each force curve: on the one hand, the initial peak force that frequently occurs with magnesium materials, the so-called F_{max} (blue) and additionally a force value in the stationary range, i.e. approximately in the middle of the extrusion trail, F_{stat} (red). In the context of the investigations presented here, the force requirement in the stationary range is used as a comparative value. In Fig. 4, this is exemplified by the billet configuration W_n, extruded with the G die at 250 °C.

First of all, the forging process reduced the average extrusion force requirement by around 15% from approx. 320 to 275 kN (Fig. 5). A significant difference in the force requirement between the two dies can only be seen for the non-forged billets, with G showing the higher values.

For the forged billets, the extrusion force requirement only tends to be higher when using the G dies. No significant difference in the force requirement between the W and A billet configurations is discernible, only a tendency towards slightly higher values for the W configuration.

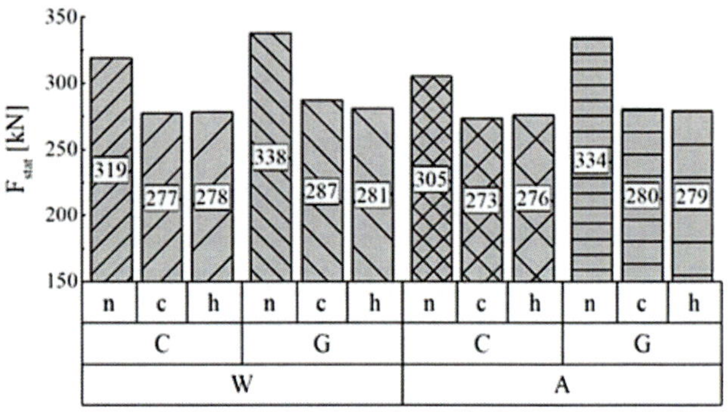

Fig. 5 Steady-state force requirement when using the C and G dies (250 °C)

The highest determined value for the stationary extrusion force was obtained when the billet configuration W_n with the G die with F_{stat} = 338 kN and the lowest with the billet A_c with the C die with F_{stat} = 273 kN. The force requirement decreases when the temperature is increased. Interestingly, the heat treatment at 525 °C does not influence the relative reduction of the extrusion force requirement for the non-forged billets, which is 28% for both configurations W_n and A_n (Fig. 6). The W_c example shows that the extrusion force reducing effect of forging, which was clearly noticeable at 250 °C, no longer occurs at a process temperature of 300 °C. A reduction in extrusion force requirement due to the temperature increase is only about 20% at this point. The relative average reduction in extrusion force due to the temperature increase is thus about 25% for the comparative extrusions.

It had to be ensured that the E die could be used for all six planned extrusion tests. By increasing the process temperature to T = 300 °C, the maximum pressure could be reduced to below 800 MPa, so that damage to the die was very unlikely.

The extrusion force requirement in the steady-state process range increased by an average of about 150 kN due to the two additional shear forming operations compared with the extrusion of the non-forged billets with the C and G dies at 250 °C (Figs. 5 and 6).

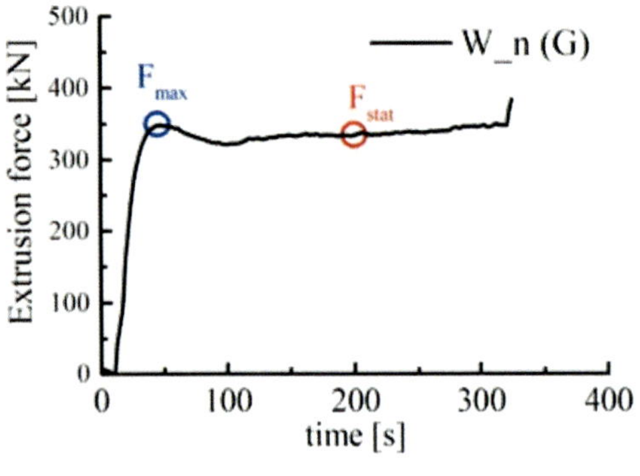

Fig. 4 Representation of the force development using the example W_n, extruded with the G die

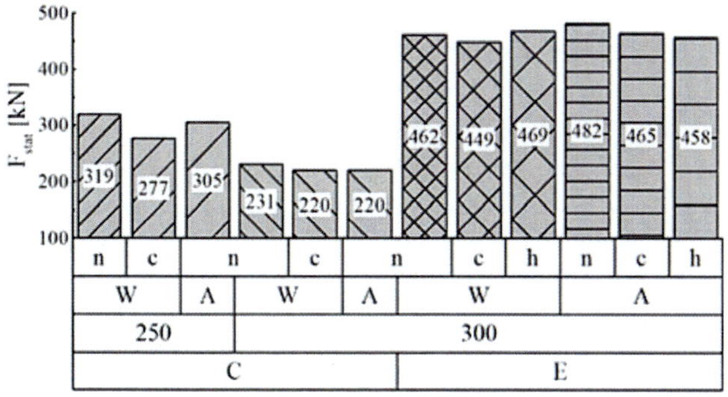

Fig. 6 Steady-state force requirement when using the C and E dies (250 °C/300 °C)

In the tests with the non-forged and cold-forged A-billets, each had a higher force requirement than the comparable W-billets. In the hot forged billets, however, the W-billet showed a higher force requirement.

However, all six tests together show a standard deviation of only $S = 11$ kN in F_{stat} which corresponds to a coefficient of variation of $S/\dot{x} = 2.4\%$, so that these force differences must be understood rather as a tendency. Therefore, it can be concluded that no significant force differences in the steady-state process range can be detected between the individual billet configurations when using the E dies.

Microstructure and Texture

Grain Size

The OIM showed a completely recrystallized, predominantly homogeneous microstructure for all samples. Selective grain growth was observed only in isolated cases. In a similar way as for the extrusion forces, the grain sizes determined in the cross-section are also plotted.

Figure 7 first shows the values for the round bars extruded at 250 °C. The values range from 4.2 to 6.3 µm, with the initially heat-treated non-forged billet W_n showing the highest value and the non-heat-treated cold-forged billet A_c the lowest. The non-forged billets do not show a clear trend compared to the forged ones. However, the cold forged billets tend to have a slightly smaller grain size compared to the hot forged billets. Overall, the level of grain sizes is very low, and no significant effects can be identified due to the closely spaced values.

Figure 8 shows the grain sizes extruded at 250 and 300 °C with the C and E dies. As expected, the use of the C die shows a dependence of the grain size on the temperature, whereby the grain size also increases with increasing process temperature. An increase in grain size can also be observed if the extrusion billets were first subjected to heat treatment. With heat treatment, the grain sizes increase into the range of 10 µm and higher. Without upstream heat treatment, the grain sizes remain at a low level between 6 and 8 µm. However, no significant dependence on the forging route can be detected. Nevertheless, it should be noted that without upstream heat treatment the hot forged route shows the highest grain size while with upstream heat treatment the same route shows the lowest grain size.

However, the two values determined, 8.1 µm for the A_h route and 9.9 µm for the W_h route, are quite close to each other, so that it can certainly be assumed that hot forging produces a comparable starting situation for the ECAPEX process, which in turn produces comparable results in the area of grain size.

Finally, in this section the influence of the ECAPEX process on the grain size compared to the conical die will be viewed (Fig. 8). It can be seen that when processing the untreated billets in the as extruded condition, both the conical die and the ECAPEX die were able to achieve the smallest grain sizes. Compared to the other routes in this plot, the billet A_n additionally sheared in the E die has a lower value of 5.9 µm than the billet extruded with the C die. In contrast, the round bars produced with the E die from the billet W_n (11.6 µm) and W_c (12.5 µm) have the highest determined grain sizes in these investigations.

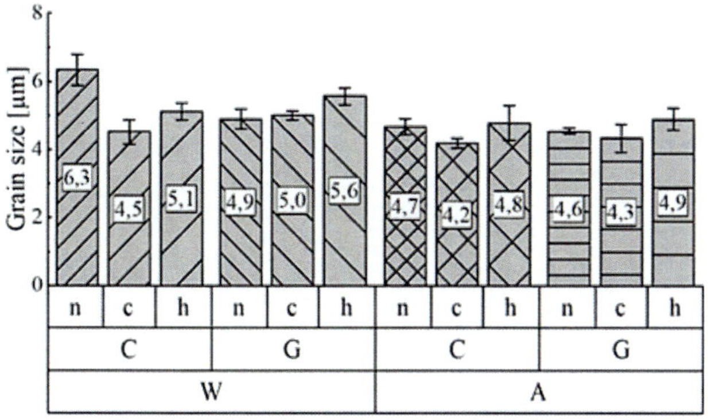

Fig. 7 Grain sizes when using the C and G dies (250 °C)

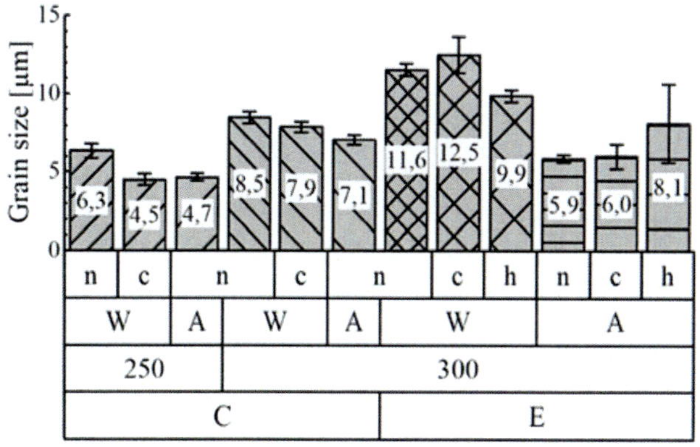

Fig. 8 Grain sizes when using the C and E dies (250 °C/300 °C)

Texture

The textural considerations refer only to the orientation of the {0002} basal planes, which most likely influence the activation of the dislocation slip and the twinning at room temperature. An exemplary representation of the pole density distributions is shown in Fig. 9. Representative for the C and G this is shown as the {0002} pole figure of the A_n rods extruded at 250 °C. For the E, the A_n rods are extruded at 300 °C. For the E die at 300 °C, the exemplary representation is given for the W_c billet. The corresponding maximum intensities of the XRD measurements are shown in Figs. 10 and 11.

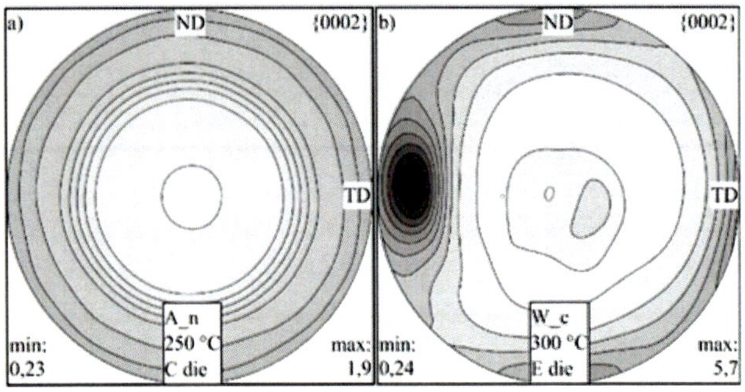

Fig. 9 Exemplary representation of {0002}-pole density distributions

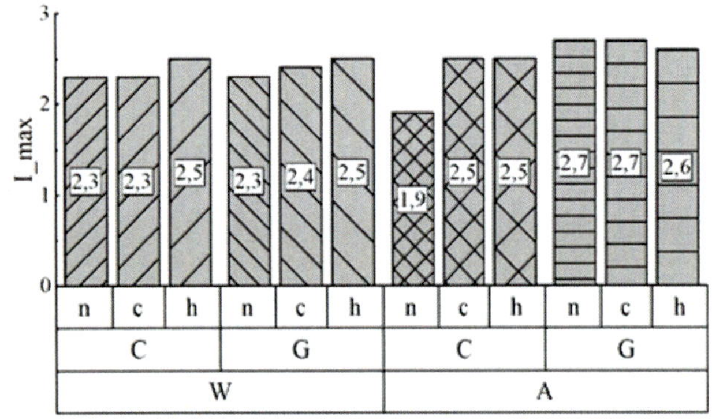

Fig. 10 {0002}-Maximum intensities of the XRD measurements (C and G dies, 250 °C)

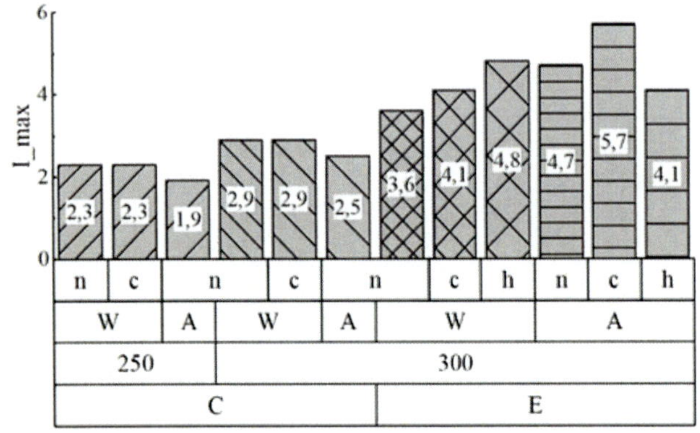

Fig. 11 {0002}-Maximum intensities of the XRD measurements (C and E dies, 250 °C and 300 °C)

Rotationally symmetric extrusion results in a rotationally symmetric pole density distribution of the basal planes. The distribution shown in Fig. 9a is similar in all rods extruded with the C and G dies. In this case, the highest intensity values are found tilted by 90° to the extrusion direction with a decreasing tendency towards the extrusion direction.

The determined maximum intensities (Fig. 10) do not show any significant dependencies for the W billets. Only for the W_h billets slightly increased values are shown for both dies. For the A-billets, the low intensity of the non-forged billets is noticeable when they are extruded with the C die.

In contrast to the use of the G dies (2.7), the intensity is only 1.9 and thus represents the lowest value determined. Apart from this outlier, all measurements show values between 2.3 and 2.7 and are thus relatively close to each other. Using the C dies at higher process temperature results in a significant increase of the maximum intensities. Interestingly, the A_n billet also shows the lowest value at 300 °C with 2.5 (Fig. 11).

The shear plane in the 135° ECAPEX die has an angle of 67.5° to the extrusion direction in the standard case. It is therefore not surprising that the highest maximum intensities of all ECAPEX strands are found in this region, asymmetrically on one side. Slight increases in intensity are found in each case rotated by 90° around ED. Compared to the strands which were rotationally symmetrically transformed, the ECAPEX rods show consistently higher maximum intensities. The highest overall value is achieved by the A_c rod.

Mechanical Properties

The mechanical properties are usually closely linked to the microstructure. In particular, the grain size and, under certain circumstances, the texture have an influence on the achievable properties. Figures 12 and 13 show the characteristic TYS values determined.

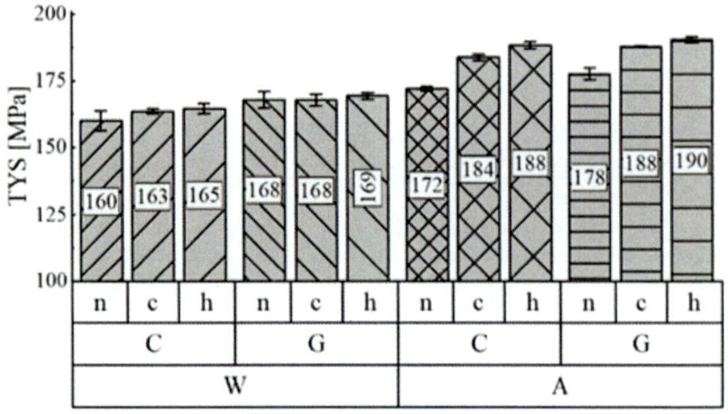

Fig. 12 TYS values for C and G dies (250 °C)

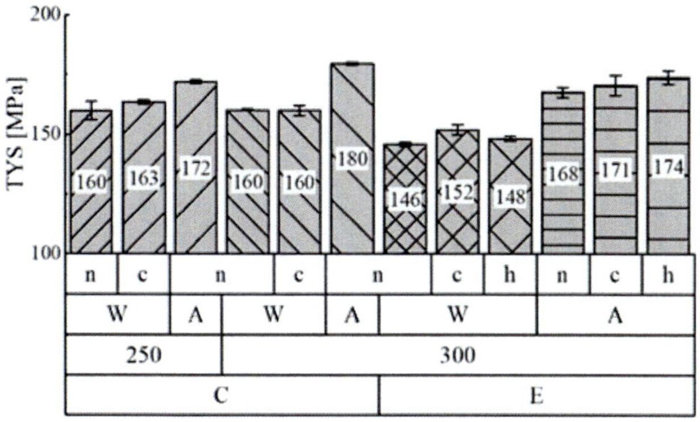

Fig. 13 TYS values for C and E dies (250 °C/300 °C)

The highest yield strength with TYS = 190 MPa was found in the configuration A_h (G/250 °C) and the lowest with TYS = 146 MPa in the configuration W_n (E/300 °C). At 250 °C, the A billets consistently exhibit the higher TYS values compared to the corresponding W billets. In addition, the higher values are always achieved with the G dies, even if only slightly. It can also be seen that, in a comparison of the pretreatment, the hot forged billets always exhibit the highest TYS values and the non-forged billets the lowest.

Using the C dies at 300 °C results in only slight changes in TYS compared to 250 °C. Here, the slight increase of 8 MPa to TYS = 180 MPa for the A_n (C/300 °C) configuration represents the largest change.

Using the E dies, the A billets show significantly higher values of TYS ≈ 170 MPa compared to TYS ≈ 149 MPa for the W billets. The consistently low level of the values for the W billets is particularly striking in the plot in Fig. 13.

Conclusion

Extrusion with additional shear forming results in an increased extrusion force requirement, as can be clearly seen in Fig. 6. Here, the force requirement for the untreated billets A_n increased by about 120% and for the heat-treated billets by about 100%. From this it can be estimated that if the ECAPEX extrusions were carried out at 250 °C, extrusion forces of up to about 750 kN could be expected, which would correspond to a tool stress of approx. 1200 MPa. Due to the partially low wall thicknesses, this would have resulted in a failure of the die.

An influence of billet pretreatment on the extrusion forces was only clearly discernible at the lower process temperature of 250 °C. In this case, it was shown that preforging can reduce the extrusion force requirement by up to 17%. In the area of grain size, no clear correlations to pretreatment could be identified. The textures determined are in line with expectations. The higher maximum intensities determined in the area of the ECAPEX routes cannot be attributed any excessive significance insofar as they do not allow any conclusions to be drawn about the absolute orientation distribution of the basal planes. Likewise, in the respective preferred orientations, the *c*-axes have comparable angular positions with respect to ED, so that a texture effect can hardly be represented.

It is therefore not surprising that the TYS values determined in the course of this work are most likely to correlate with the grain sizes. Between Figs. 7 and 12, an inverse correlation can at least be guessed at, since the absolute values are very close to each other in each case. This becomes clearer when looking at Figs. 8 and 13, where there are significant differences both in the grain sizes and in the TYS values, which clearly show an inverse correlation.

A further evaluation of the experiments in this field will include a closer look at the SDE as well as the interpretation of corresponding simulation calculations.

Funding We are grateful for the financial support by the Deutsche Forschungsgemeinschaft (DFG, German Research Foundation)—383412845.

References

1. Gall, S., Müller, S., Reimers, W.: Microstructure and mechanical properties of magnesium AZ31 sheets produced by extrusion, Int. J. Mater. Form 6 (2013) 187–197.
2. Gall, S., Huppmann, M., Mayer, H. M., Müller, S., Reimers, W.: Hot working behavior of AZ31 and ME21 magnesium alloys, J Mater Sci 48 (2013) 473–480.
3. Gensch, F.: Grundlegende Charakterisierung des Strangpressprozesses pressnahtbehafteter Magnesium-Hohlprofile - Experimentelle und numerische Betrachtungen (in german), Ph. D. thesis TU Berlin, 2018.
4. Hall, E. O.: The Deformation and Ageing of Mild Steel: III Discussion of Results, Proc. Phys. Soc. B 64 (1951) 747–753.
5. Petch, N. J.: The Cleavage Strength of Polycrystals, J. Iron Steel Inst. 173 (1953) 25–28.
6. Armstrong, R. W., Worthington, P. J.: A constitutive relation for the deformation twinning in body centered cubic metals, In: Metallurgical Effects at High Strain Rates, Plenum Press, New York, 1973, 401–414.
7. Barnett, M. R., Keshavarz, Z., Beer, A. G., Atwelll, D.: Influence of grain size on the compressive deformation of wrought Mg-3Al-1Zn, Acta Mater. 52 (2004) 5093–5103.
8. Barnett, M. R.: A rationale for the strong dependence of mechanical twinning on grain size, Scr. Mater 59 (2008) 696–698.
9. Bohlen, J., Dobron, P., Swiostek, J., Letzig, D., Chemilk, F., Lukac, P., Kainer, K. U.: On the influence of the grain size and solute content on the AE response of magnesium alloys tested in tension and compression, Mater. Sci. Eng. A 462 (2007) 302–306.
10. Meyer, M. A., Vöhringer V., Lubarda, A.: The onset of twinning in metals: a constitutive description, Acta Mater. 49 (2001) 4025–4039.
11. Wang, Y., Choo, H.: Influence of texture on Hall-Petch relationships in an Mg alloy, Acta Mater. 81 (2014) 83–97.
12. Agnew, S. R., Brown, D. W., Tomé, C. N.: Validating a polycrystal model for the elastoplastic response of magnesium

alloy AZ31 using in situ neutron diffraction, Acta Mater. 54 (2006) 4841–4852.

13. Müller, S.: Weiterentwicklung des Strangpressens von AZ Magnesiumlegierungen im Hinblick auf eine Optimierung der Mikrostruktur, des Gefüges und der mechanischen Eigenschaften, (in german), Ph.D. thesis TU Berlin (2007)
14. Gensch, F.: Investigation of the sheared layer in porthole die extrusion of magnesium alloys, 11th International Conference on Magnesium Alloys and Their Applications | Old Windsor, UK (2018).
15. Nitschke, R.: Untersuchungen zum Strangpressen von Magnesium-Hybridprofilen, (in german), Ph.D. thesis TU Berlin (2021).

Microstructure and Properties of Wrought Mg–Gd–Y–Zn–Zr (VW94) Alloy

Joshua Caris, Janet Meier, Vincent Hammond, and Alan Luo

Abstract

Long Period Stacking Order forming Mg-Rare Earth-Zn alloys have attracted much interest due to high strength in the cast and wrought conditions. These alloys may find niche applications where high strength and thermal stability are desired for lightweight components. This presentation will detail the recent results of a production run Mg–Gd–Y–Zn–Zr alloy, cast and extruded at scale with 380 MPa tensile strength. Aging studies post extrusion and post forging resulted additional strengthening. Microstructure as well as extrusion and forging conditions will be linked to equilibrium phase diagrams.

Keywords

Magnesium • Forging • Extrusion • Rare earth

Introduction

Mg alloys are one of the lightest structural materials, which have great potential for weight-savings and CO_2 emission reduction. Mg alloys additionally have high specific strength and stiffness, superior damping performance, good biocompatibility, large hydrogen storage capacity, and high theoretical specific capacity for battery applications, etc. Hence, magnesium and its alloys have been applied in the field of aerospace, automotive, and 3C (computers, communications, and consumer electronics). In addition, the application of Mg and Mg alloys in biomedical and energy sectors has attracted increasing attention. Excellent reviews of recent trends in magnesium alloy research have been written [1–3]. In specific, reviews of magnesium extrusion alloys [4] and forging alloys [5] have also been collected.

Despite the emerging and critical significance of Mg alloys as a class of structural engineering materials, Mg extrusion alloys have had only minimal usage and industrial penetration to date, which is similar to other wrought Mg products (including sheet and forgings) [3–5]. The first issue is the properties of the extrudate, with Mg extrusions having lower strength than Al extrusions, poor formability, and tension–compression yield asymmetry. The second issue is commercial viability, with some Mg alloys being expensive (depending on alloying elements) and with only few Mg alloys capable of being extruded at high enough speeds (relative to aluminum alloys) to be viable. The third issue is the generally poor corrosion resistance of Mg extrusion alloys, when compared to Al wrought alloys.

Through recent federally sponsored projects, Terves, LLC has investigated high strength magnesium alloys as well as alloys with improved formability and extrudability, in collaboration with The Ohio State University. The present work summarizes findings from a recent study of a Mg–Gd–Y–Zn–Zr alloy (VW94) produced at commercial scale with properties evaluated in the extruded as well as forged conditions.

Experimental Procedure

The Mg–Gd–Y–Zn–Zr alloy was permanent mold cast under protective SF_6/CO_2/Dry Air cover gas into an extrusion billet 230 mm (diameter) × 760 mm (length). The billet was solutionized at 510 °C/12 h with a forced air cool. The billet was then lathe turned and trimmed to dimensions 222 mm (diameter) × 749 mm (length) and extruded at 400 °C with an extrusion ratio of 13 and ram speed of 0.42 mm/s. The

J. Caris (✉)
Terves, LLC, 24112 Rockwell Dr., Euclid, OH 44107, USA
e-mail: jcaris@tervesinc.com

J. Meier · A. Luo
Department of Materials Science and Engineering, The Ohio State University, Columbus, OH 43210, USA

V. Hammond
DEVCOM Army Research Laboratory, Aberdeen Proving Ground, Harford County, MD, USA

S. Barela et al. (eds.), *Magnesium Technology 2023*, The Minerals, Metals & Materials Series,
https://doi.org/10.1007/978-3-031-22645-8_25

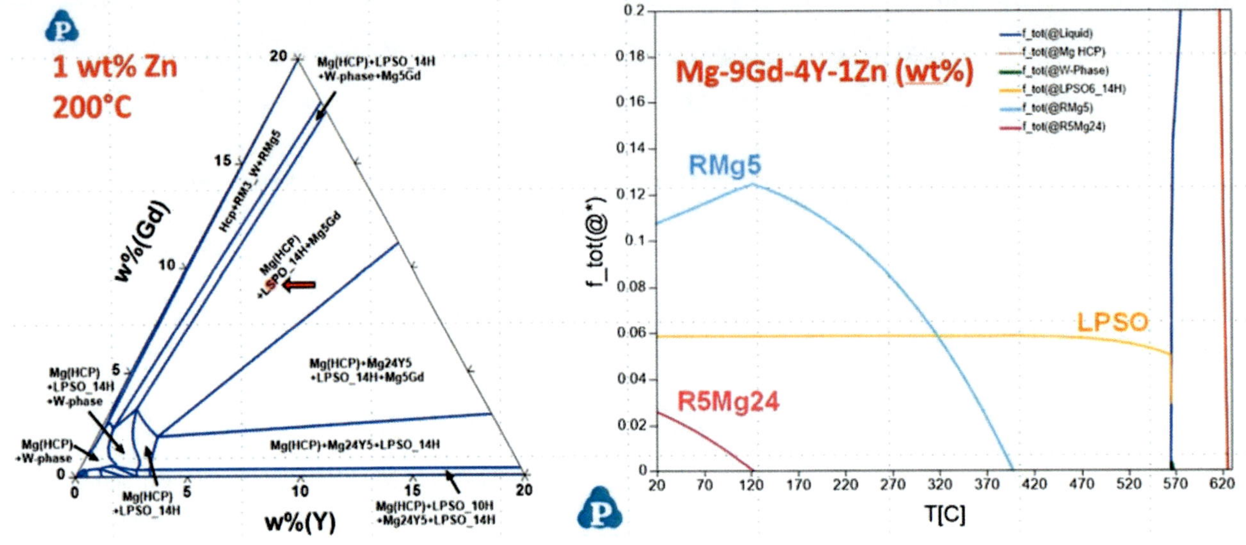

Fig. 1 Ternary isothermal section in the Mg-Gd-Y-1Zn (wt%) system at 200 °C as provided by PANDAT (left) [8]. The composition Mg-8Gd-4Y-1Zn is highlighted with a red dot in the three phase field α-Mg + LPSO + Mg5Gd. The equilibrium cooling diagram for Mg–9Gd–4Y–1Zn (wt%) is presented as well (right)

resulting extrusion diameter was 64 mm. The extrusion exit temperature was measured at 416 °C with a contact thermocouple. Lathe turnings from the extruded material were tested for chemistry via ICP-OES with the resulting composition Mg-8.20Gd-3.40Y-1.04Zn-0.24 Zr (wt%).

Portions of the extrusion were turned to forging billets 51 mm (diameter) × 64 mm (length). Upset forging was performed by compressing the circumferential surface with a screw driven press. Initial forging trials (Billets #1, #2, #3, and #4) were performed in a single step (#1–300, #2–350, #3–400, and #4–450 °C) with a platen speed of 2.54 mm/s and platen temperature of 200 °C. No lubricant was used for any of the forgings. Billet #7 was forged in two steps at 425 °C with a target overall 75% reduction in thickness. The platen speed was 1.27 mm/s and the platen temperature was 288 °C. Billets #9 and #10 were forged in three steps at 425 °C with a target overall 75% reduction in thickness. The platen speed was 0.42 mm/s and the platen temperature 288 °C.

Portions of the as extruded material were subjected to an aging study at temperatures 200, 225, 250 °C in open atmosphere furnace with a water quench. Aging was monitored by Rockwell B hardness testing performed with the load parallel to the extrusion direction. As extruded and peak aged material were tested under quasistatic tensile conditions according to ASTM E8 with the tensile direction parallel to the extrusion direction. Portions of the as forged material were also tested in tension with the load parallel to the original extrusion direction. An aging study was conducted on the as forged material as well with the loading direction parallel to the original extrusion direction.

Optical microscopy was completed on the as extruded material in the direction transverse and parallel, longitudinal, to the extrusion direction. In addition, optical microscopy was completed on as forged Billet #7 and Billet #9 in the direction transverse and parallel, longitudinal, to the original extrusion direction. Forged billets #7, #9, and #10 were imaged under X-ray for defect analysis.

Results

CALPHAD Results

The CALculation of PHAse Diagrams (CALPHAD) approach [6, 7] based on computational thermodynamics, has matured over the past few decades to calculate phase diagrams and predict phase equilibrium for complex multi-component systems [8]. This section summarizes calculated phase diagrams, using Pandat [9] software and PanMagnesium 2021 database for the Mg–Gd–Y–Zn system. A ternary isothermal (200 °C—selected due to this temperature being common for age hardening) section in the Mg–Gd–Y–1Zn (wt%) system is presented in Fig. 1(left) [8]. A composition (Mg-9Gd-4Y-1Zn (wt%)) with similar phase evolution to the experimental alloy is indicated by a red arrow and dot in a three phase field (Mg(HCP) + LPSO_14H + Mg5Gd). An equilibrium cooling diagram for Mg-9Gd-4Y-1Zn (wt%), presenting the mol fraction phases present as a function of temperature, is shown on the right in Fig. 1.

Extrusion Results

Optical microscopy at 50x and 500x is presented in Fig. 2 for the as extruded material in the transverse (a, c) and longitudinal (b, d) orientations. Elongated phases are evident in the longitudinal image (Fig. 2b). In the 500x images (c, d), the yellow boxes highlight regions of dynamic recrystallization with equiaxed grain structure.

An aging study was conducted on portions of the as extruded material with aging temperatures of 200, 225, and 250 °C selected. Hardness, Rockwell B (HRB), was performed with the loading direction parallel to the extrusion direction and is presented in Fig. 3. The as extruded material has HRB ranging from 47 to 53. Significant hardening, from HRB 50 to HRB 70 is attained by aging at 200 °C. Significant aging time (300 h) is necessary to achieve peak hardness with aging at 200 °C. Peak hardness from aging at 225 and 250 °C is achieved after 40–50 h aging, but the maximum hardness decreases with increasing temperature above 200 °C.

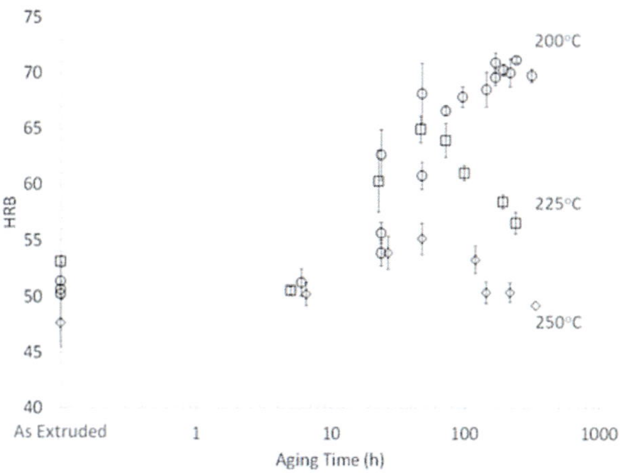

Fig. 3 Hardness, Rockwell B, results for as extruded material aged 200, 225, and 250 °C for times up to 350 h. Individual data points are an average of (6) indents with a standard deviation indicated by vertical bars. Peak aging is observed after 300 h at 200 °C

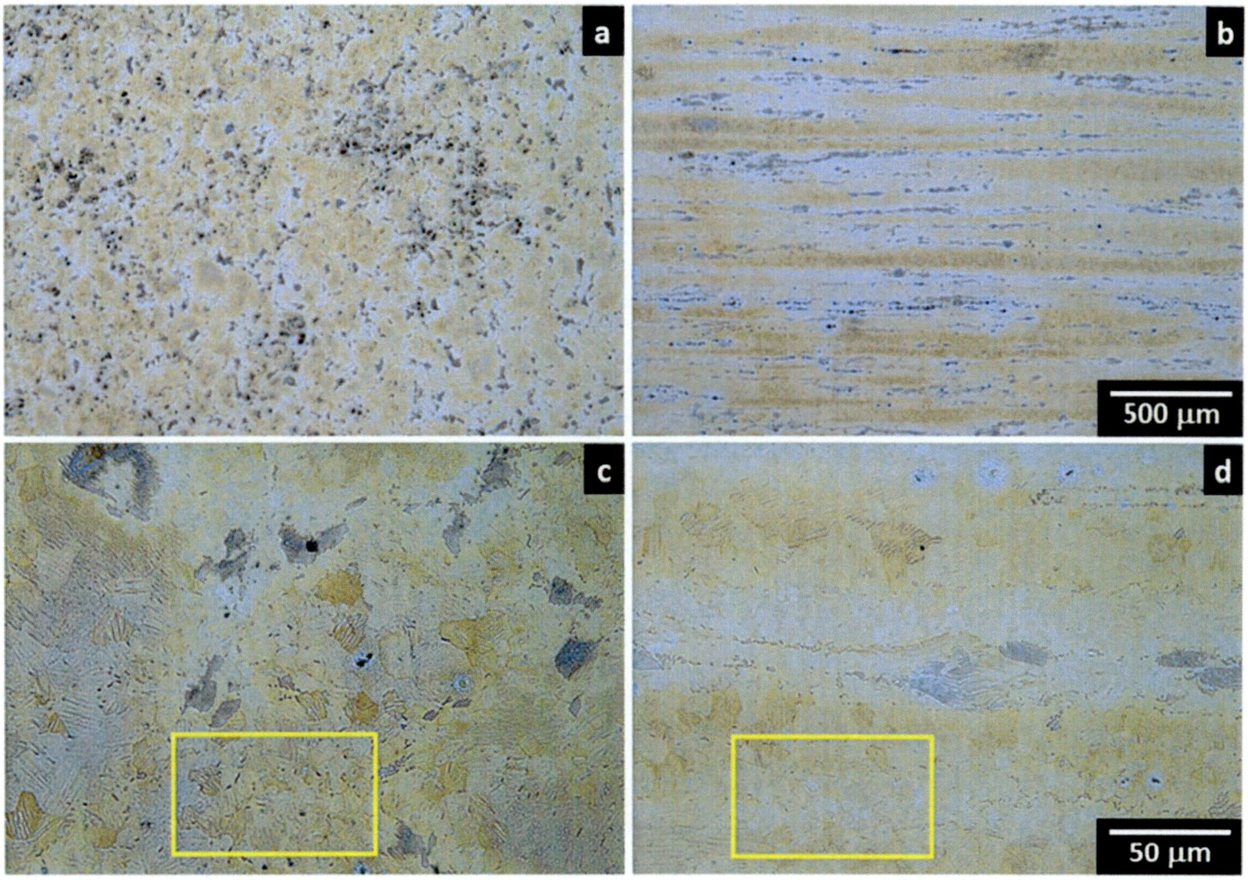

Fig. 2 Optical microscopy 50x (**a**, **b**) and 500x (**c**, **d**) for as extruded material. Images from the direction transverse to the extrusion direction are presented (**a**, **c**) and parallel, longitudinal on the (**b**, **d**). Equiaxed grains are highlighted in the yellow box shown in both transverse and longitudinal orientations

Table 1 Room temperature, quasistatic tensile properties of the as extruded as well as aged (*200 °C/48 h with water quench) material

Condition	$YS_{0.2\%}$ (MPa)	UTS (MPa)	e_f (%)
As extruded	268 ± 13	304 ± 18	5.8 ± 1.0
Aged*	392 ± 9	399 ± 7	1.5 ± 1.0

The tensile direction is parallel to the extrusion direction. Data presented is an average of three tests with one standard deviation

Room temperature tensile testing was conducted on as extruded as well as aged (200 °C/48 h with a water quench) extruded material. The tensile loading is parallel to the extrusion direction. The results are presented in Table 1 as an average and standard deviation of three tests. Aging results in a significant increase in both 0.2% offset yield strength and ultimate tensile strength as well as a decrease in strain to failure.

Forging Results

Four of the 51 mm (diameter) × 64 mm (length) billets were upset forged on the circumferential surface with a platen temperature of 200 °C and speed 2.54 mm/s corresponding to a compressive strain rate of 0.05 s^{-1}. Images of the cylindrical forging billets as well as the first four forgings are presented in Fig. 4. The billet extruded at 300 °C exhibited minimal reduction in thickness prior to brittle failure. The 350 °C sample was forged to thickness of 38 mm (26% reduction in thickness) before brittle failure. The final thickness of the billets pressed at 400 and 450 °C was 19 mm which corresponded to a 63% reduction in thickness. While the sample pressed at 400 °C did exhibit brittle failure, the sample at 450 °C exhibited only edge cracking on the two surfaces corresponding to the original flat faces of the billet.

Billets #7, #9, and #10 were all forged in multiple steps at 425 °C to a final thickness of 12.7 mm, corresponding to a total 75% reduction in thickness. Billet #7 was forged at 1.27 mm/s (strain rate of 0.025 s^{-1}) in two steps with a platen temperature of 288 °C. Billets #9 and 10 were forged at 0.42 mm/s (strain rate of 0.0083 s^{-1}) in three steps with an approximate 37% reduction per step. Images of the three forgings are presented in Fig. 5. All three forgings exhibit superficial cracking on the surfaces corresponding to the original flat faces of the cylindrical billet.

Optical microscopy for the as forged Sample #9 in the direction transverse and parallel, longitudinal, to the original extrusion direction is presented in Fig. 6. Continued refinement of phases is observed with structures elongated further in the longitudinal and slightly in the transverse.

A brief aging study was performed on as forged #7 and #9 to compare the mechanical properties resulting from the two processing routes. Rockwell B testing, with loading parallel to the original extrusion direction, is presented in Fig. 7.

Table 2 summarizes mechanical properties for the as forged and aged samples #7, #9, and #10. The loading direction of the Rockwell B hardness tests is transverse to the original extrusion direction. As observed with the aging study on Sample #9, there is a slight increase in average hardness on aging Sample #10. The tensile tests were performed with the loading direction parallel to the original extrusion direction.

Discussion

With an extrusion ratio of 13 and ram speed of 0.42 mm/s, the resulting extrusion speed (v_e) for the current researched alloy is 0.33 m/min, Table 3. Generally, this would be considered a slow extrusion speed for a magnesium alloy [10]. With a billet temperature of 400 °C and a slow extrusion speed, it would be expected, and is confirmed, that the resulting extrusion temperature would raise only slightly

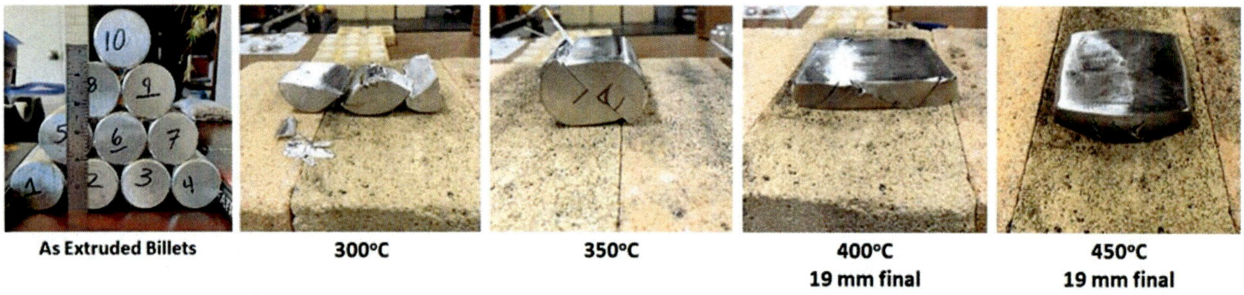

Fig. 4 Images of machined "as extruded" forging billets (left) and results of initial forging test matrix (forging temperature indicated) with platens heated to 200 °C and a platen speed of 2.54 mm/s

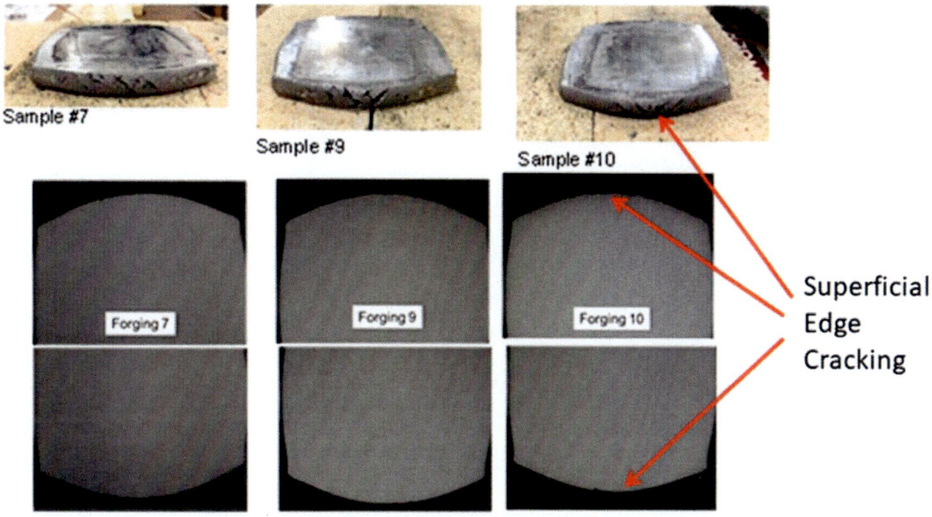

Fig. 5 Forging sample (#7, images and corresponding X-ray images with the viewing direction parallel to the compressive forging direction. The X-ray images indicate that the cracking observed to form during forging on the original flat faces of the forging billets is superficial

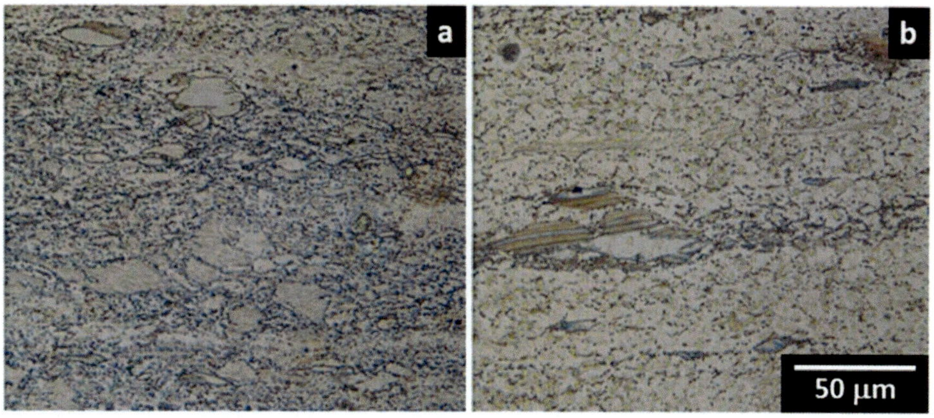

Fig. 6 Optical microscopy (500x) of forging #9 near the center of the sample. The image transverse to the original extrusion direction is shown (**a**) and parallel, longitudinal to the extrusion direction (**b**)

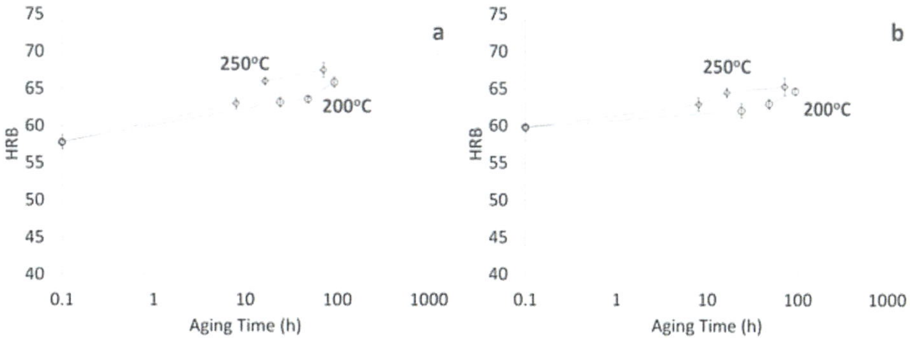

Fig. 7 Hardness, Rockwell B, results for as forged material #7 (**a**) and #9 (**b**) aged 200 and 250 °C for times up to 100 h. Individual data points are an average of (6) indents with a standard deviation indicated by vertical bars

to 416 °C. The billet and extrusion temperature are significantly lower than the Pandat predicted solidus of 565 °C, Fig. 1—equilibrium cooling diagram. Thus, an increase in billet temperature and/or speed could be possible while maintaining the viable processing window [10].

Multiple authors have investigated direct extrusion of alloys within the Mg–Gd–Y–Zn–Zr with compositions similar to the present work (Summarized in [4, 11]). Many of the early references lack extrusion details such as ram speed, while later references focus on either direct or indirect extrusion. A comparison with the current direct extruded material is from [11–13]. Yield and tensile strength, Table 4, in the extrusion direction of the present extruded material are comparable to the literature alloys;

Table 2 Hardness, Rockwell B, and room temperature tensile properties (in the original extrusion direction) for forged samples #7, #9, and #10

Forging	Condition	HRB	Tensile, in extrusion direction		
			$YS_{0.2\%}$ (MPa)	UTS (MPa)	e_f (%)
#7	As forged	59 ± 1	313 ± 1	373 ± 2	12.3 ± 0.6
#9	As forged	59 ± 2	313 ± 5	374 ± 2	12 ± 2
#10	As forged Aged*	60 ± 2 66 ± 1	– 319 ± 6	– 379 ± 6	– 2.3 ± 0.8

The hardness is an average and standard deviation of 18 measurements and the tensile properties average of 3 measurements and a standard deviation. Results are presented for forging #10 in the aged (*250 °C/72.5 h with a water quench) condition

Table 3 Details regarding extrusion conditions for the present Mg-8.20Gd-3.40Y-1.04Zn-0.24Zr extrusion as well as direct extrusions referenced from literature

Reference	Alloy	Homogenization	D_e (mm)	n	T (°C)	ER	v_e (m/min)
Present	Mg-8.20Gd-3.40Y-1.04Zn-0.24Zr	510 °C/12 h/FAC	64	1	400	13.0	0.33
[11]	Mg-9.5Gd-4Y-2.2Zn- 0.5Zr	520 °C/24 h/WQ	15	1	450	16.0	0.96
[12]	Mg-5.59Gd-3.78Y-0.54Zn-0.49Zr	320 °C/2 h + 500 °C/10 h/AC	16.7 12.5	1 1	400 400	9.0 16.0	1.62 2.88
[13]	Mg–8.2Gd–3.8Y–1Zn–0.4Zr	510 °C/12 h/WQ	60	3	400	8.0	0.72

The billet homogenization is detailed with forced air cool (FAC), air cool (AC) and water quench (WQ) from the homogenization temperature indicated. The diameter of the extrusion (D_e), number (n) of ports on the extrusion die, billet temperature (T), extrusion ratio (ER), and resulting extrusion speed (v_e) are also presented

Table 4 Details regarding the tensile properties, in the extrusion direction, of the as fabricated (F) and peak age (T5) after extrusion condition for the present alloy and extrusions referenced (Table 3) from literature

Reference	Alloy	Condition	Age	$YS_{0.2\%}$ (MPa)	UTS (MPa)	e_f (%)
Present	Mg-8.20Gd-3.40Y-1.04Zn-0.24Zr	F T5	– 200 °C/48 h/WQ	268 392	305 400	4.6 1.5
[11]	Mg-9.5Gd-4Y-2.2Zn- 0.5Zr	F T5	– 200 °C/9 h	325 358	378 417	7.8 5.7
[12]	Mg-5.59Gd-3.78Y-0.54Zn-0.49Zr	F (ER 9.0) F (ER 16.0)	– –	239 229	324 318	18.1 19.5
[13]	Mg–8.2Gd–3.8Y–1Zn–0.4Zr	F T5	– 200 °C/48 h/WQ	303 395	378 470	17 8

however, the elongation to failure is lowest of the alloys compared.

Similar to the extruded material detailed in the reference articles, the present extrusion exhibits multiple phases including dynamically recrystallized phases (highlighted with a yellow box) interspersed between elongated lamellar structures, Fig. 2. Reference [13] identified the elongated lamellar structure as containing 14H LPSO which would be anticipated from the phases present at 400–450 °C (corresponding to the extrusion conditions) on the equilibrium cooling diagram, Fig. 1.

The improvement of extrusion strength with aging is observed by multiple authors. According to Chi [13], this is mainly attributed to the following three factors. Firstly, the peak-aged alloy contains a high number density of β′ precipitates (a metastable phase in the RMg5—Fig. 1—sequence [14]) which have been generally reported to play a key role in strengthening the magnesium alloys through hindering the motion of basal dislocation. Second, the remaining LPSO phases after ageing treatment can also strengthen the alloy by a mechanism of fiber-like reinforcement. Thirdly, the basal fiber texture becomes stronger after ageing treatment. When the tensile axis is parallel to the extrusion direction, the critical resolved shear stress (CRSS) for the basal slip of the grains with a basal fiber texture is zero. So, the strengthening of the basal fiber texture can further suppress the activation of basal slip, hence resulting in the improvement of yield strength.

With aging temperature above 200 °C, the time to peak age for the extrusion decreases and the magnitude of the peak age hardness decreases, Fig. 3. This could be due to a decrease in the mol fraction of RMg5 with increasing temperature above 200 °C, Fig. 1, as well as an increase in diffusion kinetics resulting in accelerated coarsening of strengthening particles.

Upset forging of the present extruded material is most viable with a temperature window 400–450 °C and platen speed 2.54 mm/s, Figs. 4 and 5. Although die lubricant is suggested for magnesium alloys [5], it was not used for the present study. The as forged alloy exhibits further refinement to the microstructure, Fig. 6b, relative to the as extruded, Fig. 2d, with grains elongated in both the longitudinal and slightly in the transverse direction. The LPSO structure is also further broken. Tensile properties are similar, Table 2, regardless of whether the extrusion was forged #7 at 2 passes 1.27 mm/s or #9 three passes 0.42 mm/s. Aging at 200 and 250 °C after forging results in a slight increase in hardness, Fig. 7, for both Sample #7 and #9. Aging at 250 °C/72.5 h with a water quench also results in a slight increase to both yield and tensile strength (Sample #10) in Table 2. However, there is a significant drop in forging ductility.

Authors have only recently begun investigating the forging of Mg–Gd–Y–Zn–Zr alloys from the As Cast/homogenized [15, 16], previously forged [17], and extruded [18, 19] conditions, Table 5. A comparison of tensile properties between the as extruded forging stock and the resulting forging is also presented, Table 6. Similar yield and ultimate tensile properties are exhibited for the present alloy and that presented in by Tong in [18]. The [18] cast ingot was homogenized at 510 °C for12 h, and then extruded at 400 °C, with the extrusion ratio of 12:1 and ram speed of 10 mm/s (extrusion speed of 7.2 m/min). Data for Tong are collected after one forging step. It is interesting to see an increase in elongation to failure between the as extruded and forged material both for the present alloy as well as Tong. This may be due to further refinement of the coarse LPSO phases during forging.

In the works of Han, Xu and Shan [15, 16] a cast Mg–Gd–Y–Zn–Zr alloy was forged. The in-depth study of the microstructure showed a grain refinement of the α-Mg grains and precipitation of the LPSO phase (14H type) during forging. These LPSO phases are known to inhibit dislocation glide and grain growth. Moreover, the formation of β' phase precipitates during the ageing process was found. The high strength of the alloy is therefore attributed to these precipitated LPSO and b' phases. The over ageing visible in this alloy is thought to stem from precipitate free zones, grain growth and coarsening of secondary phases [5].

Conclusions

1. Extrusion and subsequent forging of the Mg-8.20Gd-3.40Y-1.04Zn-0.24Zr (VW94) alloy was demonstrated. Mechanical behavior consisting of Rockwell B hardness and tensile properties were sampled in the as fabricated and aged conditions.
2. Aging at 200 °C improved extrusion strength at the expense of ductility.
3. Forging improved both the yield and tensile strength of the as extruded material and increased ductility, attributed to a refined microstructure.

Table 5 Details regarding composition and forging conditions (the forging stock material, the temperature, *T*, for billet forging and the platen speed, *v*) for the present alloy as well as similar alloys in literature

Reference	Alloy (wt%)	Forge stock	*T* (°C)	*v* (mm/s)	Notes
Present	Mg-8.20Gd-3.40Y-1.04Zn-0.24Zr	As extruded	425	1.72	Multi step
[15, 16]	Mg-9.12Gd-2.96Y-0.68Zn-0.57Zr	Hom. 510 °C/10 h/WQ	470	–	–
[17]	Mg-9.12Gd-2.96Y-0.68Zn-0.57Zr	As forged	407	1.00	Bracket design
[18]	Mg–8.2Gd–3.8Y–1.0Zn–0.4Zr	As extruded	420	2.00	Multi directional
[19]	Mg-10Gd-4Y-1.5Zn-0.5Zr	As extruded	350	3.33	Pack forging

Table 6 Tensile properties of the as extruded forging stock and resulting forging for the present alloy (Tables 1 and 2) as well as details from literature

Reference	Condition	$YS_{0.2\%}$ (MPa)	UTS (MPa)	e_f (%)
Present	As extruded	268	304	5.8
	As forged	313	373	12.3
[18]	As extruded	276	347	10.7
	As forged	313	382	17.4
[19]	As extruded	369	410	19
	As forged	401	469	6.4

Acknowledgements This material is based upon work supported by the Army Contracting Command—Adelphi, MD under Contract No W911Q-X20-C-0001. The views and conclusions contained in this document are those of the authors and should not be interpreted as representing the official policies, either expressed or implied, of the Army Research Laboratory or the US Government. The US Government is authorized to reproduce and distribute reprints for Government purposes notwithstanding any copyright notation herein.

References

1. Song, J., Chen, J., Xiong, X., Peng, X., Chen, D. and Pan, F., 2022. Research advances of magnesium and magnesium alloys worldwide in 2021. Journal of Magnesium and Alloys, 10(4), pp.863–898.
2. Yang, Y., Xiong, X., Chen, J., Peng, X., Chen, D. and Pan, F., 2021. Research advances in magnesium and magnesium alloys worldwide in 2020. Journal of Magnesium and Alloys, 9(3), pp.705–747.
3. Abbott, T.B., 2015. Magnesium: industrial and research developments over the last 15 years. Corrosion, 71(2), pp.120–127.
4. Zeng, Z., Stanford, N., Davies, C.H.J., Nie, J.F. and Birbilis, N., 2019. Magnesium extrusion alloys: a review of developments and prospects. International Materials Reviews, 64(1), pp.27–62.
5. Papenberg, N.P., Gneiger, S., Weissensteiner, I., Uggowitzer, P. J. and Pogatscher, S., 2020. Mg-Alloys for forging applications—A review. Materials, 13(4), p.985.
6. Chang, Y.A., Chen, S., Zhang, F., Yan, X., Xie, F., Schmid-Fetzer, R. and Oates, W.A., 2004. Phase diagram calculation: past, present and future. Progress in Materials Science, 49(3–4), pp.313–345.
7. Luo, A.A., 2015. Material design and development: From classical thermodynamics to CALPHAD and ICME approaches. Calphad, 50, pp.6–22.
8. Meier, J.M., Caris, J. and Luo, A.A., 2022. Towards high strength cast Mg-RE based alloys: Phase diagrams and strengthening mechanisms. Journal of Magnesium and Alloys.
9. CompuTherm, Pandat Software: CALPHAD-based materials design, (2021). https://computherm.com/software.
10. Atwell, D.L. and Barnett, M.R., 2007. Extrusion limits of magnesium alloys. Metallurgical and Materials Transactions A, 38(12), pp.3032–3041.
11. Zheng, J., Chen, Z., Yan, Z., Zhang, Z., Wang, Q. and Xue, Y., 2022. Preparation of ultra-high strength Mg–Gd–Y–Zn–Zr alloy by pre-ageing treatment prior to extrusion. Journal of Alloys and Compounds, 894, p.162490.
12. Jin, X., Xu, W., Yang, Z., Yuan, C., Shan, D., Teng, B. and Jin, B. C., 2020. Analysis of abnormal texture formation and strengthening mechanism in an extruded Mg–Gd–Y–Zn–Zr alloy. Journal of Materials Science & Technology, 45, pp.133–145.
13. Chi, Y.Q., Zheng, M.Y., Xu, C., Du, Y.Z., Qiao, X.G., Wu, K., Liu, X.D., Wang, G.J. and Lv, X.Y., 2013. Effect of ageing treatment on the microstructure, texture and mechanical properties of extruded Mg–8.2 Gd–3.8 Y–1Zn–0.4 Zr (wt%) alloy. Materials Science and Engineering: A, 565, pp.112–117.
14. Nie, J.F., 2012. Precipitation and hardening in magnesium alloys. Metallurgical and Materials Transactions A, 43(11), pp.3891–3939.
15. Han, X.Z., Xu, W.C. and Shan, D.B., 2011. Effect of precipitates on microstructures and properties of forged Mg-10Gd-2Y-0.5 Zn-0.3 Zr alloy during ageing process. Journal of alloys and compounds, 509(35), pp.8625–8631.
16. Xu, W.C., Han, X.Z. and Shan, D.B., 2013. Precipitates formed in the as-forged Mg–Zn–RE alloy during ageing process at 250 °C. Materials characterization, 75, pp.176–183.
17. Shan, D., Xu, W., Han, X. and Huang, X., 2012. Study on isothermal precision forging process of rare earth intensifying magnesium alloy. Materials Science and Engineering: B, 177(19), pp.1698–1702.
18. Tong, L.B., Chu, J.H., Sun, W.T., Xu, C., Zou, D.N., Wang, K.S., Kamado, S. and Zheng, M.Y., 2021. Achieving an ultra-high strength and moderate ductility in Mg–Gd–Y–Zn–Zr alloy via a decreased-temperature multi-directional forging. Materials Characterization, 171, p.110804.
19. Wei, X., Jin, L., Liu, C., Wang, F., Dong, S. and Dong, J., 2021. Effect of pack-forging on microstructure and properties of Mg–Gd–Y–Zn–Zr alloy. Materials Science and Engineering: A, 802, p.140674.

Recent Advances in PRISMS-Plasticity Software for Simulation of Deformation in Mg Alloys

Mohammadreza Yaghoobi, Duncan A. Greeley, Zhe Chen, Tracy Berman, John E. Allison, and Veera Sundararaghavan

Abstract

An open-source parallel 3-D crystal plasticity finite element (CPFE) software package, PRISMS-Plasticity, is presented here as part of an overarching PRISMS Center integrated framework. A new rate-dependent twinning-detwinning model is incorporated into the framework based on an integration point sensitive scheme to model Mg alloys. The model includes both kinematic and isotropic hardening in order to handle cyclic response of structural metals. The model is validated versus high energy diffraction microscopy (HEDM) results of Mg alloys during cyclic loadings. PRISMS-Plasticity TM is another feature which has been developed as a new open-source rapid texture evolution analysis pipeline based on the Taylor model, which is integrated into the open-source crystal plasticity software, PRISMS-Plasticity. The developed framework is used to capture the effects of alloying on texture development in Mg–Zn–Ca alloys. Finally, the PRISMS-Plasticity software has been integrated with the PRISMS-PF phase-field framework to model twinning within Mg alloys.

Keywords

Magnesium • ICME • Modeling • Simulation

M. Yaghoobi (✉) · D. A. Greeley · Z. Chen · T. Berman · J. E. Allison
Department of Materials Science and Engineering, University of Michigan, Ann Arbor, MI, USA
e-mail: yaghoobi@umich.edu

V. Sundararaghavan
Department of Aerospace Engineering, University of Michigan, Ann Arbor, MI, USA

Extended Abstract

Crystal plasticity finite element (CPFE) is one of the common methods to model the mechanical response of Mg and its alloys [1]. PRISMS-Plasticity [2], which is a highly scalable CPFE software, has been developed as one of the important components of PRISMS Center at University of Michigan to investigate the mechanical response of Mg alloys. In addition to the CPFE framework itself, several pre-processing and post-processing pipelines have been developed in a way that the software can be effectively integrated with other computational tools and experiments. Three major integration pipelines include DREAM.3D to generate initial microstructures [3], PRISMS-Fatigue to conduct simulation-based fatigue analysis [4–7], and Materials Commons as the PRISMS Center information repository. Several applications have been investigated using the PRISMS-Plasticity software. A new rate-independent crystal plasticity model is implemented within the software which can capture both twinning and detwinning deformations [8]. Various Mg alloys have been calibrated and modeled using PRISMS-Plasticity including ZK60A [8], Mg–Nd [9, 10], WE43 [11–13], and Mg_4Al [14, 15].

PRISMS-Plasticity software version 1.4.0 has been recently released which includes new features such as new boundary conditions (BCs) like torsion BCs, Neumann BCs, and nodal BCs, element deletion capability to model pores and surface roughness, and user-defined visualization. Additionally, a new buffer layer feature is added to mitigate the effect of CPFE boundary conditions for the simulation of high energy diffraction microscopy experiment. The latest feature added to PRISMS-Plasticity software is PRISMS-Plasticity TM [16], which is a rapid texture evolution analysis pipeline. In support of a growing user community a You-Tube training channel has been established, virtual training sessions are held regularly and a new Virtual Machine (VM) for Windows users has been established.

S. Barela et al. (eds.), *Magnesium Technology 2023*, The Minerals, Metals & Materials Series,
https://doi.org/10.1007/978-3-031-22645-8_26

As an example of the application for the newly released PRISMS-Plasticity TM pipeline, the texture evolution of Mg-Y during rolling is investigated. Unalloyed Mg shows a very strong basal texture along the normal direction (ND) direction after cold rolling [17–19]. The addition of rare earth (RE) elements to Mg can weaken the basal texture by rotation from ND towards RD [20]. To investigate the effect of the rare-earth element Y on Mg alloy texture evolution, the texture evolution of Mg-3 wt% Y (Mg-3Y) alloy is investigated during the rolling process. The sample initially has a random texture consisting of 4096 grains of equal grain sizes. The rate-dependent crystal plasticity formulation is used here with the perfect-plasticity assumption [4]. Here, the response of Mg-3Y is simulated using four slip modes of Basal $\langle a\rangle(\{0001\}\langle 11\bar{2}0\rangle)$, Prismatic $\langle a\rangle(\{10\bar{1}0\}\langle 11\bar{2}0\rangle)$, Pyramidal $\langle a\rangle(\{\bar{1}\bar{1}22\}\langle\bar{1}\bar{1}23\rangle)$, and Pyramidal $\langle c+a\rangle$ $(\{\bar{1}\bar{1}22\}\langle\bar{1}\bar{1}23\rangle)$, along with one extension twin mode $(\{10\bar{1}2\}\langle\bar{1}011\rangle)$. The elastic constants of pure Mg at room temperature are used here which are $C_{11} = 59,400\,\mathrm{MPa}$, $C_{12} = 25,610\,\mathrm{MPa}$, $C_{13} = 21,440\,\mathrm{MPa}$, $C_{33} = 61,600\,\mathrm{MPa}$, and $C_{44} = 16,400\,\mathrm{MPa}$. The initial slip resistances of Mg-3Y reported by Wang et al. [21] are used here for different deformation systems, as summarized in Table 1. Following Yaghoobi et al. [2], the PTR scheme is used to model reorientation due to extension twinning with the parameters of $A = 0.7$ and $B = 0$. In order to mimic the rolling process along y axis (with z axis being the short

Table 1 The initial slip resistances of different deformation modes in Mg-3Y (Wang et al. [21])

Mode	Basal	Prismatic	Pyramidal <*a*>	Pyramidal <*c* + *a*>	Twinning
s_0^{α} (MPa)	12	38	36	60	40

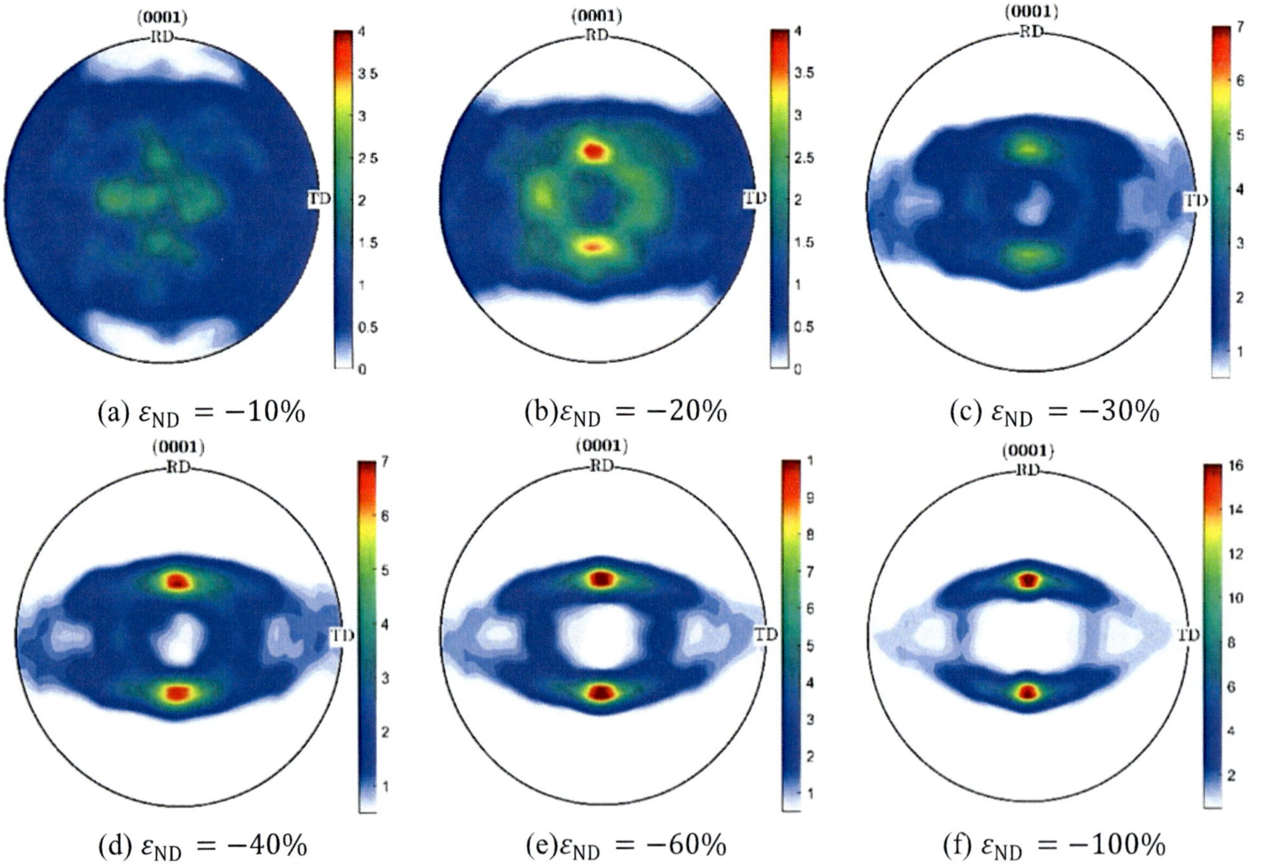

Fig. 1 The PRISMS-TM simulated evolution of basal (0001) pole figures in Mg-3Y alloy sample with initial random texture during rolling at different true strain along ND: **a** $\varepsilon_{ND} = -10\%$, **b** $\varepsilon_{ND} = -20\%$, **c** $\varepsilon_{ND} = -30\%$, **d** $\varepsilon_{ND} = -40\%$, **e** $\varepsilon_{ND} = -60\%$, and **f** $\varepsilon_{ND} = -100\%$

transverse direction), the velocity gradient tensor of $\mathbf{L}_{\text{rolling}} = \begin{bmatrix} 0 & 0 & 0 \\ 0 & 0.01 & 0 \\ 0 & 0 & -0.01 \end{bmatrix} s^{-1}$ is incorporated. The rolling simulation is conducted up to a true strain of 100% in compression in the sheet normal direction (ND). Figure 1 shows the evolution of basal (0001) pole figures at different true strains along ND. Unlike unalloyed Mg which forms a very sharp basal texture along ND during rolling, Fig. 1 shows that the addition of Y weakens the basal texture by rotation from ND towards RD [20].

Acknowledgements This work was supported by the U.S. Department of Energy, Office of Basic Energy Sciences, Division of Materials Sciences and Engineering under Award#DE-SC0008637 as part of the Center for Predictive Integrated Structural Materials Science (PRISMS Center) at University of Michigan. We also acknowledge the financial cost-share support of University of Michigan College of Engineering. This work used the Extreme Science and Engineering Discovery Environment (XSEDE), which is supported by National Science Foundation grant number ACI-1548562, through the allocation TG-MSS160003.

References

1. M. Yaghoobi, G.Z. Voyiadjis, V. Sundararaghavan, Crystal Plasticity Simulation of Magnesium and Its Alloys: A Review of Recent Advances, Crystals 11(4) (2021) 435.
2. M. Yaghoobi, S. Ganesan, S. Sundar, A. Lakshmanan, S. Rudraraju, J.E. Allison, V. Sundararaghavan, PRISMS-Plasticity: An open-source crystal plasticity finite element software, Computational Materials Science 169 (2019) 109078.
3. M.A. Groeber, M.A. Jackson, DREAM. 3D: a digital representation environment for the analysis of microstructure in 3D, Integrating materials and manufacturing innovation 3(1) (2014) 56–72.
4. M. Yaghoobi, K.S. Stopka, A. Lakshmanan, V. Sundararaghavan, J.E. Allison, D.L. McDowell, PRISMS-Fatigue computational framework for fatigue analysis in polycrystalline metals and alloys, npj Computational Materials 7(1) (2021) 1–12.
5. K.S. Stopka, M. Yaghoobi, J.E. Allison, D.L. McDowell, Effects of Boundary Conditions on Microstructure-Sensitive Fatigue Crystal Plasticity Analysis, Integrating Materials and Manufacturing Innovation (2021) 1–20.
6. K.S. Stopka, M. Yaghoobi, J.E. Allison, D.L. McDowell, Simulated effects of sample size and grain neighborhood on the modeling of extreme value fatigue response, Acta Materialia 224 (2022) 117524.
7. K.S. Stopka, M. Yaghoobi, J.E. Allison, D.L. McDowell, Microstructure-Sensitive Modeling of Surface Roughness and Notch Effects on Extreme Value Fatigue Response, International Journal of Fatigue (2022) 107295.
8. M. Yaghoobi, J.E. Allison, V. Sundararaghavan, Multiscale modeling of twinning and detwinning behavior of HCP polycrystals, International Journal of Plasticity (2020) 102653.
9. D. Greeley, M. Yaghoobi, D. Pagan, V. Sundararaghavan, J. Allison, Using synchrotron radiation to improve understanding of deformation of polycrystalline metals by measuring, modelling and publishing 4D information, IOP Conference Series: Materials Science and Engineering 580 (2019) 012017.
10. D. Greeley, M. Yaghoobi, D. Pagan, V. Sundararaghavan, J. Allison, Characterization of cyclic twin evolution in a Mg–Nd alloy using high energy X-ray diffraction microscopy, Acta Materialia, In Revision (2022).
11. S. Ganesan, M. Yaghoobi, A. Githens, Z. Chen, S. Daly, J.E. Allison, V. Sundararaghavan, The effects of heat treatment on the response of WE43 Mg alloy: crystal plasticity finite element simulation and SEM-DIC experiment, International Journal of Plasticity 137 (2021) 102917.
12. Z. Chen, M. Yaghoobi, V. Sundararaghavan, J. Allison, S. Daly, The Effects of Microstructure on Deformation Twinning in Mg WE43, Available at SSRN 4128924 (2021).
13. M. Yaghoobi, Z. Chen, V. Sundararaghavan, S. Daly, J.E. Allison, Crystal Plasticity Finite Element Modeling of Extension Twinning in WE43 Mg Alloys: Calibration and Validation, Integrating Materials and Manufacturing Innovation 10(3) (2021) 488–507.
14. M.T. Andani, A. Lakshmanan, V. Sundararaghavan, J. Allison, A. Misra, Quantitative study of the effect of grain boundary parameters on the slip system level Hall-Petch slope for basal slip system in Mg-4Al, Acta Materialia 200 (2020) 148–161.
15. M. Yaghoobi, Z. Chen, A.D. Murphy-Leonard, V. Sundararaghavan, S. Daly, J.E. Allison, Deformation twinning and detwinning in extruded Mg-4Al: In-situ experiment and crystal plasticity simulation, International Journal of Plasticity 155 (2022) 103345.
16. M. Yaghoobi, J.E. Allison, V. Sundararaghavan, PRISMS-Plasticity TM: An open-source rapid texture evolution analysis pipeline, Integrating Materials and Manufacturing Innovation (2022).
17. A.D. Rollett, S.I. Wright, Typical textures in metals, Texture and Anisotropy: Preferred Orientations in Polycrystals and their Effect on Materials Properties. Cambridge University Press, Edinburgh Building, Cambridge, CB 2 2 RU, UK, 1998. (1998) 178–238.
18. M.R. Barnett, M.D. Nave, C.J. Bettles, Deformation microstructures and textures of some cold rolled Mg alloys, Materials Science and Engineering: A 386(1–2) (2004) 205–211.
19. T.D. Berman, J.E. Allison, Coupling Thermomechanical Processing and Alloy Design to Improve Textures in Mg–Zn–Ca Sheet Alloys, JOM 73(5) (2021) 1450–1459.
20. J. Wu, L. Jin, J. Dong, F. Wang, S. Dong, The texture and its optimization in magnesium alloy, Journal of Materials Science & Technology 42 (2020) 175–189.
21. L. Wang, Z. Huang, H. Wang, A. Maldar, S. Yi, J.-S. Park, P. Kenesei, E. Lilleodden, X. Zeng, Study of slip activity in a Mg-Y alloy by in situ high energy X-ray diffraction microscopy and elastic viscoplastic self-consistent modeling, Acta Materialia 155 (2018) 138–152.

Solid-Phase Processing of Mg–Al–Mn–Ca for High Strength and Ductility

David Garcia, Hrishikesh Das, Kumar Sadayappan, Peter Newcombe, Darrell Herling, Glenn J. Grant, and Mageshwari Komarasamy

Abstract

While rare-earth Mg alloys have remarkable properties for high strength applications, lower cost alternatives are necessary for the widespread industry use of Mg. Ca added Mg alloys have shown promise as an alternative to rare-earth alloys. Ca-based precipitates can reduce basal texture, reduce casting porosity, and increase mechanical strength of cast components. However, the accumulation of Ca-based precipitates along inter-dendritic regions can severely limit ductility. Here, we apply two solid-phase processing techniques, friction stir processing and shear assisted processing and extrusion, to produce wrought microstructure sheet and extruded tubes from a cast Mg–Al–Mn–Ca alloy. Ductility of the alloy is enhanced by densification under the applied thermomechanical processing conditions, grain refinement, and refinement of (Al, Mg)–Ca-based precipitates. Solid-phase processing provides a low cost opportunity to improve the properties of cast Mg alloys and improve service life.

Keywords

Friction stir processing • Shear assisted processing and extrusion • Non-rare earth magnesium alloys

D. Garcia (✉) · H. Das · D. Herling · G. J. Grant · M. Komarasamy
Pacific Northwest National Laboratory, Richland, WA 99354, USA
e-mail: david.garcia@pnnl.gov

K. Sadayappan · P. Newcombe
CANMET—Materials Technology Laboratory, Hamilton, ON L8P 0A5, Canada

Introduction

Lightweighting has been on the forefront of research due to its importance in fuel efficiency and global energy conservation initiatives. Mg alloys in particular are a promising candidate due to their high strength to weight ratio. Pure Mg has a density of 1.7 g/cm^3 which is approximately one-fifth that of iron and two-thirds that of aluminum. The primary limitation for current Mg alloys lies in the ability to fabricate defect free components for low cost. Traditional casting methods result in porous structures with local microstructure heterogeneities [1]. This typically results in low ductility and poor fatigue life [2, 3]. Higher cost casting alternatives, such as vacuum assisted high pressure die casting, exist to reduce the frequency and size of defects, but are not economically scalable [4]. In addition to processing techniques, alloy development has been a focus to improve strength, ductility, and castability by reducing texturing and defect formation. Rare-earth Mg alloys have shown remarkable properties for high strength and mission critical applications by reducing basal texture in cast components [5], but lower cost alternatives are necessary for widespread industry adoption. Ca added Mg alloys offer a promising alternative to rare-earth alloys with low environmental impact [6]. Ca-based precipitates can reduce basal texture, reduce casting porosity, and increase mechanical strength [7, 8]. However, the accumulation of Ca-based precipitates along inter-dendritic regions can severely limit ductility. Figure 1 shows the yield strength and elongation of several Mg-based alloys, highlighting the property response of Mg–Ca alloys after thermo-mechanical processing.

Solid-phase processing techniques, such as extrusion, that lead to wrought microstructures also have the ancillary benefit of breaking down these secondary phase particles in Mg–Ca alloys [18]. Here, we apply two solid-phase processing techniques, friction stir processing (FSP) and shear assisted processing and extrusion (ShAPE), to produce locally modified sheets and extruded tubes from a cast Mg–Al–Mn–Ca

S. Barela et al. (eds.), *Magnesium Technology 2023*, The Minerals, Metals & Materials Series,
https://doi.org/10.1007/978-3-031-22645-8_27

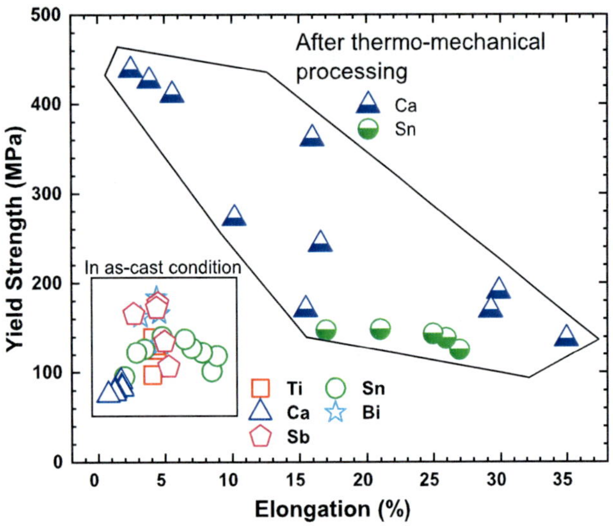

Fig. 1 An Ashby plot of yield strength and elongation for several non-rare earth Mg alloys [8–18]

alloy. The effect of the processing parameters on the properties and microstructure are observed.

Experimental

Friction Stir Processing

Friction stir processing (FSP) is a solid-state thermomechanical processing technique that enables local microstructure modification [19]. A non-consumable rotating tool is brought into contact with the surface of a workpiece. The tool rotation and applied normal force lead to frictional heating and local softening of the workpiece surface [20]. Continuous plastic deformation takes places as the softened material yields under the shear imparted by the rotational motion of the tool. The elevated temperature and imparted strain lead to dynamic recrystallization of the material which yields a refined grain structure [21].

FSP has been widely applied to magnesium alloys to improve strength, ductility, and fatigue life [22]. FSP leads to improved mechanical performance due to several microstructural modifications. Cavaliere et al. [23] demonstrated that FSP eliminates casting defects in AZ91 which slows crack propagation during fatigue. The fracture surface in the FSP condition demonstrated no evidence of crack growth along existing defects such as voids or oxides. Additionally, it is possible to selectively fabricate the texture of magnesium components as either pyramidal or basal texture depending on the processing conditions as shown in FSP of AZ91 [24]. Furthermore, there was substantial dissolution and fragmentation of the β-$Mg_{17}Al_{12}$ after FSP. Similar observations were captured by Nene et al. [14] in AXM541 where dispersion strengthening was observed due to refinement of Al–Ca particles during FSP. Elimination of casting defects, homogenization of secondary phase particles, and grain refinement have all been observed for FSP of Mg alloys.

Shear Assisted Processing and Extrusion

Shear assisted processing and extrusion (ShAPE) is a similar thermomechanical processing technique that is used to fabricate tubes, rods, or disks [25]. In ShAPE, a non-consumable rotating tool or die is brought into contact with a stationary billet of material. Similar to FSP, the rotational motion leads to frictional heating of the billet. As the temperature rises, the billet surface begins to yield and undergo continuous plastic deformation. Features on the surface of the die guide material flow toward an extrusion orifice at the center of the die surface. The thermomechanical processing conditions at the die-billet interface are similar to those undergone during FSP, so the final result yields a dynamically recrystallized microstructure with refined grains [26]. Whalen et al. [27] fabricated Mg tubes with refined secondary phase particles and basal textures not aligned with the extrusion direction. This work aims to leverage the benefits observed in FSP and ShAPE for Mg alloys and apply them to Mg–Al–Mn–Ca to produce a high strength, high ductility, and low cost Mg alloy.

Conclusion

There is substantial evidence in literature that Mg–Ca alloys have a strong response to thermomechanical processing. As such, Mg–Ca alloys are a promising alternative to rare-earth Mg alloys which have a low cost and low environmental impact. However, there remain questions regarding the microstructure evolution during thermomechanical processing and the process-microstructure relationship. This work aims to establish a thorough understanding of the process-microstructure relationship during FSP and ShAPE of Mg–Al–Mn–Ca.

References

1. Das S. (2003) Magnesium for automotive applications: Primary production cost assessment, JOM 55, 22–26. https://doi.org/10.1007/s11837-003-0204-x
2. Horstemeyer M.F., Yang N., Gall K., McDowell D.L., Fan J., Gullett P.M. (2004) High cycle fatigue of a die cast AZ91E-T4 magnesium alloy, Acta Materialia 52, 1327–1336. https://doi.org/10.1016/j.actamat.2003.11.018

3. Mayer H., Papakyriacou M., Zettl B., Stanzl-Tschegg S.E. (2003) Influence of porosity on the fatigue limit of die cast magnesium and aluminium alloys, International Journal of Fatigue 25, 245–256. https://doi.org/10.1016/S0142-1123(02)00054-3
4. Niu X.P., Hu B.H., Pinwill I., Li H. (2000) Vacuum assisted high pressure die casting of aluminium alloys, Journal of Materials Processing Technology 105, 119–127. https://doi.org/10.1016/S0924-0136(00)00545-8
5. Hantzsche K., Bohlen J., Wendt J., Kainer K.U., Yi S.B., Letzig D. (2010) Effect of rare earth additions on microstructure and texture development of magnesium alloy sheets, Scripta Materialia 63, 725–730. https://doi.org/10.1016/j.scriptamat.2009.12.033
6. García Gutiérrez I., Elduque D., Pina C., Tobajas R., Javierre C. (2020) Influence of the Composition on the Environmental Impact of a Casting Magnesium Alloy, Sustainability. https://doi.org/10.3390/su122410494
7. Kondori B., Mahmudi R. (2010) Effect of Ca additions on the microstructure, thermal stability and mechanical properties of a cast AM60 magnesium alloy, Materials Science and Engineering: A 527, 2014–2021. https://doi.org/10.1016/j.msea.2009.11.043
8. Wu G., Fan Y., Gao H., Zhai C., Zhu Y.P. (2005) The effect of Ca and rare earth elements on the microstructure, mechanical properties and corrosion behavior of AZ91D, Materials Science and Engineering: A 408, 255–263. https://doi.org/10.1016/j.msea.2005.08.011
9. Ai X., Quan G. (2012) Effect of Ti on the Mechanical Properties and Corrosion of Cast AZ91 Magnesium Alloy, The Open Materials Science Journal 6, 6–13. https://doi.org/10.2174/1874088X01206010006
10. Candan S., Unal M., Koc E., Turen Y., Candan E. (2011) Effects of titanium addition on mechanical and corrosion behaviours of AZ91 magnesium alloy, Journal of Alloys and Compounds 509, 1958–1963. https://doi.org/10.1016/j.jallcom.2010.10.100
11. Chen J., Chen Z., Yan H., Zhang F., Liao K. (2008) Effects of Sn addition on microstructure and mechanical properties of Mg–Zn–Al alloys, Journal of Alloys and Compounds 461, 209–215. https://doi.org/10.1016/j.jallcom.2007.07.066
12. Guangyin Y., Yangshan S., Wenjiang D. (2001) Effects of bismuth and antimony additions on the microstructure and mechanical properties of AZ91 magnesium alloy, Materials Science and Engineering: A 308, 38–44. https://doi.org/10.1016/S0921-5093(00)02043-8
13. Huang Q., Liu Y., Tong M., Pan H., Yang C., Luo T., Yang Y. (2020) Enhancing tensile strength of Mg–Al–Ca wrought alloys by increasing Ca concentration, Vacuum 177, 109356. https://doi.org/10.1016/j.vacuum.2020.109356
14. Nene S.S., Zellner S., Mondal B., Komarasamy M., Mishra R.S., Brennan R.E., Cho K.C. (2018) Friction stir processing of newly-designed Mg-5Al-3.5Ca-1Mn (AXM541) alloy: Microstructure evolution and mechanical properties, Materials Science and Engineering: A 729, 294–299. https://doi.org/10.1016/j.msea.2018.05.073
15. Pan H., Ren Y., Fu H., Zhao H., Wang L., Meng X., Qin G. (2016) Recent developments in rare-earth free wrought magnesium alloys having high strength: A review, Journal of Alloys and Compounds 663, 321–331. https://doi.org/10.1016/j.jallcom.2015.12.057
16. Turen Y. (2013) Effect of Sn addition on microstructure, mechanical and casting properties of AZ91 alloy, Materials & Design 49, 1009–1015. https://doi.org/10.1016/j.matdes.2013.02.037
17. Wang B., Chen X., Pan F., Mao J. (2017) Effects of Sn addition on microstructure and mechanical properties of Mg-Zn-Al alloys, Progress in Natural Science: Materials International 27, 695–702. https://doi.org/10.1016/j.pnsc.2017.11.002
18. Xu S.W., Oh-ishi K., Kamado S., Uchida F., Homma T., Hono K. (2011) High-strength extruded Mg–Al–Ca–Mn alloy, Scripta Materialia 65, 269–272. https://doi.org/10.1016/j.scriptamat.2011.04.026
19. Ma Z.Y. (2008) Friction Stir Processing Technology: A Review, Metallurgical and Materials Transactions A 39, 642–658. https://doi.org/10.1007/s11661-007-9459-0
20. Mishra R.S., Ma Z.Y., Charit I. (2003) Friction stir processing: a novel technique for fabrication of surface composite, Materials Science and Engineering: A 341, 307–310. https://doi.org/10.1016/S0921-5093(02)00199-5
21. Huang Y., Wang Y., Meng X., Wan L., Cao J., Zhou L., Feng J. (2017) Dynamic recrystallization and mechanical properties of friction stir processed Mg-Zn-Y-Zr alloys, Journal of Materials Processing Technology 249, 331–338. https://doi.org/10.1016/j.jmatprotec.2017.06.021
22. Wang W., Han P., Peng P., Zhang T., Liu Q., Yuan S., Huang L., Yu H., Qiao K., Wang K. (2020) Friction Stir Processing of Magnesium Alloys: A Review, Acta Metallurgica Sinica (English Letters) 33, 43–57. https://doi.org/10.1007/s40195-019-00971-7
23. Cavaliere P., De Marco P.P. (2007) Fatigue behaviour of friction stir processed AZ91 magnesium alloy produced by high pressure die casting, Materials Characterization 58, 226–232. https://doi.org/10.1016/j.matchar.2006.04.025
24. Yousefpour F., Jamaati R., Aval H.J. (2021) Effect of traverse and rotational speeds on microstructure, texture, and mechanical properties of friction stir processed AZ91 alloy, Materials Characterization 178, 111235. https://doi.org/10.1016/j.matchar.2021.111235
25. Overman N.R., Whalen S.A., Bowden M.E., Olszta M.J., Kruska K., Clark T., Stevens E.L., Darsell J.T., Joshi V.V., Jiang X., Mattlin K.F., Mathaudhu S.N. (2017) Homogenization and texture development in rapidly solidified AZ91E consolidated by Shear Assisted Processing and Extrusion (ShAPE), Materials Science and Engineering: A 701, 56–68. https://doi.org/10.1016/j.msea.2017.06.062
26. Whalen S., Overman N., Joshi V., Varga T., Graff D., Lavender C. (2019) Magnesium alloy ZK60 tubing made by Shear Assisted Processing and Extrusion (ShAPE), Materials Science and Engineering: A 755, 278–288. https://doi.org/10.1016/j.msea.2019.04.013
27. Whalen S., Joshi V., Overman N., Caldwell D., Lavender C., Skszek T. (2017) Scaled-Up Fabrication of Thin-Walled ZK60 Tubing Using Shear Assisted Processing and Extrusion (ShAPE), in: K.N. Solanki, D. Orlov, A. Singh, N.R. Neelameggham (Eds.) Magnesium Technology 2017 Springer International Publishing Cham pp. 315–321

The Effects of Temperature and Strain Rate on the Tensile Behaviour of Die-Cast Magnesium Alloy AE44

Trevor Abbott, Hua Qian Ang, Suming Zhu, and Mark Easton

Abstract

The tensile properties and work hardening behaviour of die-cast magnesium alloy AE44 were investigated at a wide range of temperatures from 77 to 473 K and strain rates from 10^{-6} to 10^{-1} s^{-1}. AE44 was also subjected to T5 ageing to understand the effect of heat treatment on the work hardening behaviour. Both the as-cast and T5-aged AE44 showed a continuous decrease in strength and hardening rate with increasing temperature. A positive strain-rate sensitivity was observed over the entire temperature range. Voce hardening law was used to examine the work hardening behaviour, and it appears that the athermal hardening stage previously reported for magnesium alloys does not exist in AE44 for the studied temperatures and strain rates.

Keywords

Magnesium alloys • Mechanical properties • Work hardening • Athermal hardening • Dislocation slip

Introduction

Magnesium (Mg) alloys are highly skewed towards being manufactured by pressure die-casting processes. Despite the efforts of the research community over the last 20 years, the consumption of Mg alloys by high-pressure die-casting, and the closely related thixomolding processes, remains an order of magnitude above those of all other processes combined.

Mg die-cast parts have unique microstructural characteristics that set them apart from parts manufactured by other processes. Die-castings typically have a smaller grain size (approximately 10 μm) and a random texture [1]. Studies have shown that the activity of basal slip increases markedly as the grain size decreases [2, 3], and basal slip is shown to operate in grains with orientations that a simple Schmidt factor analysis would preclude [3]. The behaviour of twinning deformation is also strongly influenced by grain size at this scale, with interactions of twins with grain boundaries and neighbouring grains becoming significant at grain sizes below about 10–20 μm [4].

The Mg-aluminium (Al) system forms the basis of die-cast alloys with Al improving castability and mechanical properties. Small quantities of manganese (Mn) are present to control iron levels and limit its detrimental effects on corrosion performance. The most common alloys are AZ91 (Mg-9Al-1Zn-0.2Mn) and AM60 (Mg-6Al-0.3Mn, all compositions in weight percent, wt%). High levels of Al in Mg solid solution have a significant impact on high temperature creep resistance and it also modifies the response to strain rate and temperature through dynamic strain ageing [5, 6]. More specialised alloys have been developed for high temperature applications and these commonly contain elements such as Cerium (Ce) and Lanthanum (La) that have a high affinity for Al and, therefore, reduce the level of Al in Mg solid solution [7]. Although more complex in composition, the deformation behaviour of these alloys is simpler to interpret with less complicating effects from dynamic strain ageing.

One of the most commonly used creep resistant Mg die-casting alloys is AE44 (Mg-4Al-4RE-0.3Mn). Our previous study showed that AE44 exhibited significant strain-rate sensitivity at room temperature, which is typical of Mg alloys with low solute levels [5, 8–10]. The present study extends the study of deformation behaviour of die-cast

T. Abbott
Magontec Limited, Sydney, NSW 2000, Australia

T. Abbott · H. Q. Ang (✉) · S. Zhu · M. Easton
School of Engineering, RMIT University, Bundoora, VIC 3083, Australia
e-mail: huaqian.ang@rmit.edu.au

T. Abbott · S. Zhu
Department of Materials Science and Engineering, Monash University, VIC, 3800, Australia

S. Barela et al. (eds.), *Magnesium Technology 2023*, The Minerals, Metals & Materials Series,
https://doi.org/10.1007/978-3-031-22645-8_28

AE44 to include the effects of temperature, strain rate, and heat treatment. The work hardening behaviour is also compared with Voce hardening law to better understand the deformation stages in AE44.

Experimental Methods

The AE44 alloy used in this work has the following composition (wt%): Al 3.86, Mn 0.38, Ce 2.61, La 1.66, and balance Mg. All samples were cast with a 250 tonne cold chamber machine to produce cylindrical tensile bars of 5.65 mm diameter and 34.5 mm gauge length. AE44 alloy was tested in both the as-cast condition (designated AE44-F) and after a heat treatment of 32 h at 200 °C (designated AE44-T5). T5 ageing leads to remarkable improvements in strength with no substantial reduction in ductility for AE alloys [11, 12].

Tensile tests were carried out using a screw-driven Instron machine at a series of temperatures from 77 to 473 K and strain rates, 10^{-1} to 10^{-6} s^{-1}. Temperatures above ambient were achieved using a furnace mounted within the Instron machine. Temperatures below ambient were achieved using a custom-made combination grip plus cryogenic liquid holder shown in Fig. 1. A 60 mm diameter rod of Al alloy 2011 was machined to create a cup to hold cryogenic liquids, such as ethanol plus dry ice and liquid nitrogen. In the bottom of the cup, an M12 threaded blind hole was machined to provide a means of attaching the threaded tensile specimens. On the outside surface beneath this hole, another M12 hole was machined to provide a connection to a 20 mm diameter, 304 stainless steel alloy rod. The use of opposing blind holes eliminated the possibility of leakage of cryogenic fluid.

Due to the complexity of equipment set up, the use of an extensometer was not practical, instead a model was used to deduce extension from the crosshead displacement. Tests at ambient temperature, with an extensometer and with the cup inverted, were first used to obtain strain and crosshead displacement data. By assuming that all deformation outside of the gauge length is elastic, an accurate tensile strain-crosshead displacement model was established to obtain the true stress–strain data. True stress–strain data were then processed to determine the work hardening rate or gradient (Θ) using curve fit function from Python libraries numpy, scipy and sklearn.

Scanning electron microscopy (SEM) was used for microstructural characterisation. Electron-backscatter diffraction (EBSD) data was collected in a FEI Nova NanoSEM at 20 kV using a step size of 0.5-μm. The sample surfaces for EBSD and SEM analyses were prepared using standard mechanical polishing procedures and were finished with 0.06 μm colloidal silica suspension.

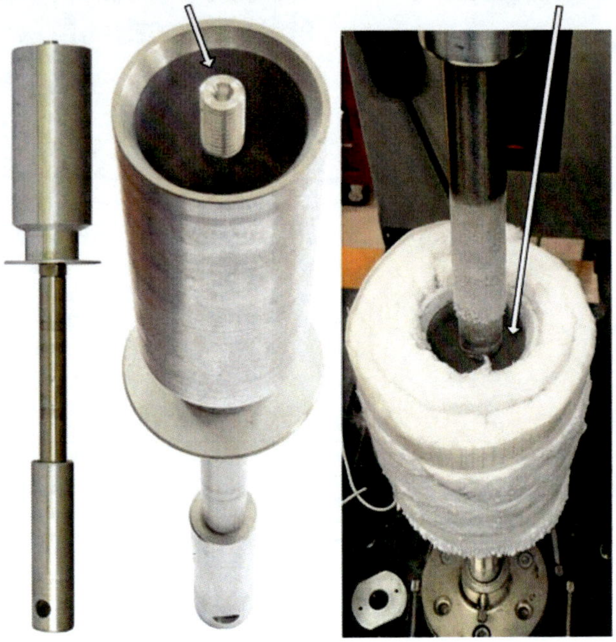

Fig. 1 Apparatus for testing at cryogenic temperatures. During testing, the cup was filled with cryogenic liquid covering the tensile specimen, and wool blanket was wrapped around the cup to provide insulation

Results and Discussion

Alloy Microstructures

The microstructures and grain size of the die-cast AE44 are shown in Fig. 2. Mg solid solution forms the primary phase which is outlined by a eutectic mixture consisting of Mg and Al–Ce–La based intermetallic phase. Several intermetallic phases have been reported in this system and the details of these are described extensively elsewhere [13]. Blocky Mg–Al based intermetallic phases are also observed. Note that there are no obvious changes in shape or size of the intermetallic phases in AE44 after T5 ageing. However, TEM examinations revealed the formation of spherical nanoscale precipitates (less than 10 nm) after ageing as previously reported [14, 15]. The AE44 used in the present study has a uniform grain size of about 8 μm and a random texture.

Effect of Temperature on Tensile Behaviour of AE44

The tensile curves for AE44-F and AE44-T5 covering temperatures from 77 to 473 K are shown in Fig. 3. Both as-cast and T5-aged conditions show a continuous variation in strength and hardening rate with temperature, and a positive strain-rate sensitivity is observed over the entire

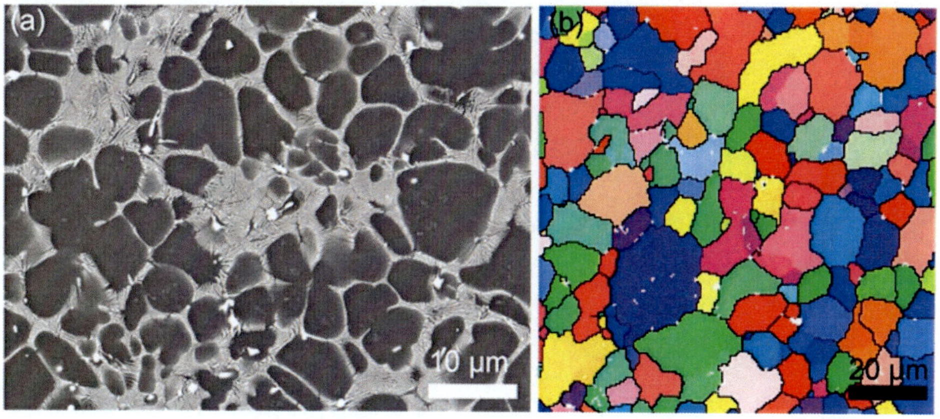

Fig. 2 (**a**) SEM backscattered electron image and (**b**) EBSD image of the studied alloy AE44

temperature range. The presence of a knee, that is, a region where the slope rapidly decreases then increases slightly, is restricted to a range of temperatures. The knee is absent at 200 K and below, and at 473 K. The knee is strongest in the AE44-T5, 373 K specimens.

The effects of temperature, heat treatment and strain rate on 0.2% proof stress, σ_y are shown in Fig. 4. Details of 0.2% proof stress measurement method can be found in previous studies [16]. Proof stress decreases with increasing temperatures for both heat treatment conditions and strain rates. In the case of AE44-F, the decrease is close to linear, with a slight increase in steepness as the temperature increases. For the AE44-T5 condition, the rate of decrease is significantly higher above room temperature. Both AE44-F and T5 conditions appear to extrapolate towards a similar value at 0 K of about 220 MPa. At high temperatures, there also appears to be a convergence towards similar values. The highest strengthening increment from heat treatment appears to occur near room temperature (295 K). At all temperatures, for both alloy conditions, lower strain rate results in lower proof stress.

The hardening rates for AE44-F and T5 are shown in Fig. 5a, b, respectively. In AE44-F, the hardening rate (Θ) varies linearly with stress indicating Voce hardening [17]. The Voce hardening law is given by Eq. 1. In AE44-T5, the 295 K specimen shows a deviation from Voce hardening, that is, the region after the knee, is curved. This is the temperature where the maximum heat treatment strengthening effect occurred as shown by the separation of the F and T5 curves in Fig. 4. The Voce hardening parameters, Θ_0 and σ_v extrapolated from Fig. 5a and the shear modulus, μ of Mg, (17 GPa [18]) were plotted against temperature for AE44-F in Fig. 5c.

$$\Theta = \Theta_0\left(1 - \frac{\sigma}{\sigma_v}\right) \quad (1)$$

Effect of Strain Rate on Tensile Behaviour of AE44

The tensile curves vary with strain rate in a similar manner to temperature but over a narrower range, as shown in Fig. 6a,

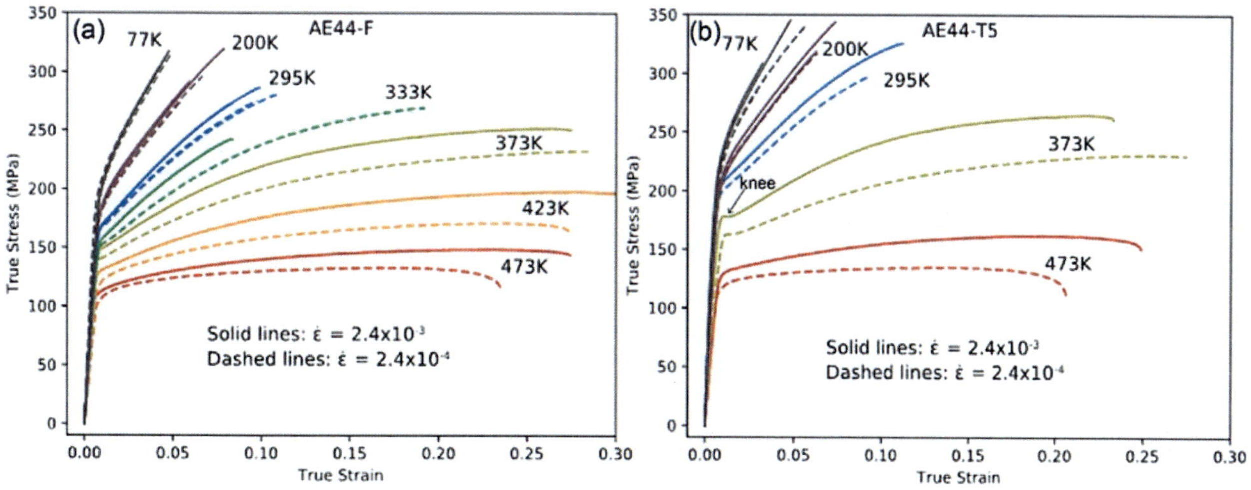

Fig. 3 Tensile curves for (**a**) AE44-F and (**b**) AE44-T5 at two strain rates and multiple temperatures

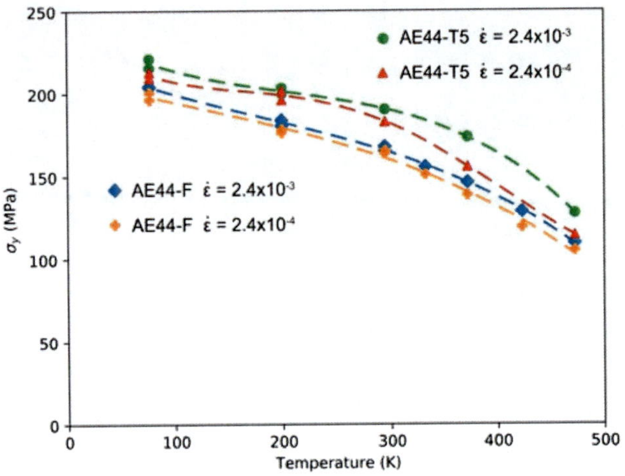

Fig. 4 Variation of 0.2% proof stress against temperature for AE44-F and AE44-T5 at two strain rates

b. The effect of strain rate on 0.2% proof stress is shown in Fig. 6c. Proof stress increases with strain rates for both AE44-F and AE44-T5, but less markedly for AE44-T5. The extrapolated linear fits intersect at 220 MPa and a strain rate of 10^7. This value of 220 MPa coincides with that for 0 K, shown in Fig. 4, and the strain rate of 10^7 is consistent with values commonly used in empirical strain-rate equations [19]. The hardening rate variations with stress are shown in Fig. 7. The best correlations to Voce hardening laws are for AE44-F at intermediate strain rates.

Athermal Hardening Discussion

The work hardening curves presented above serve as a good starting point to understand the deformation stages in hexagonal close-packed (HCP) metals, which is still under

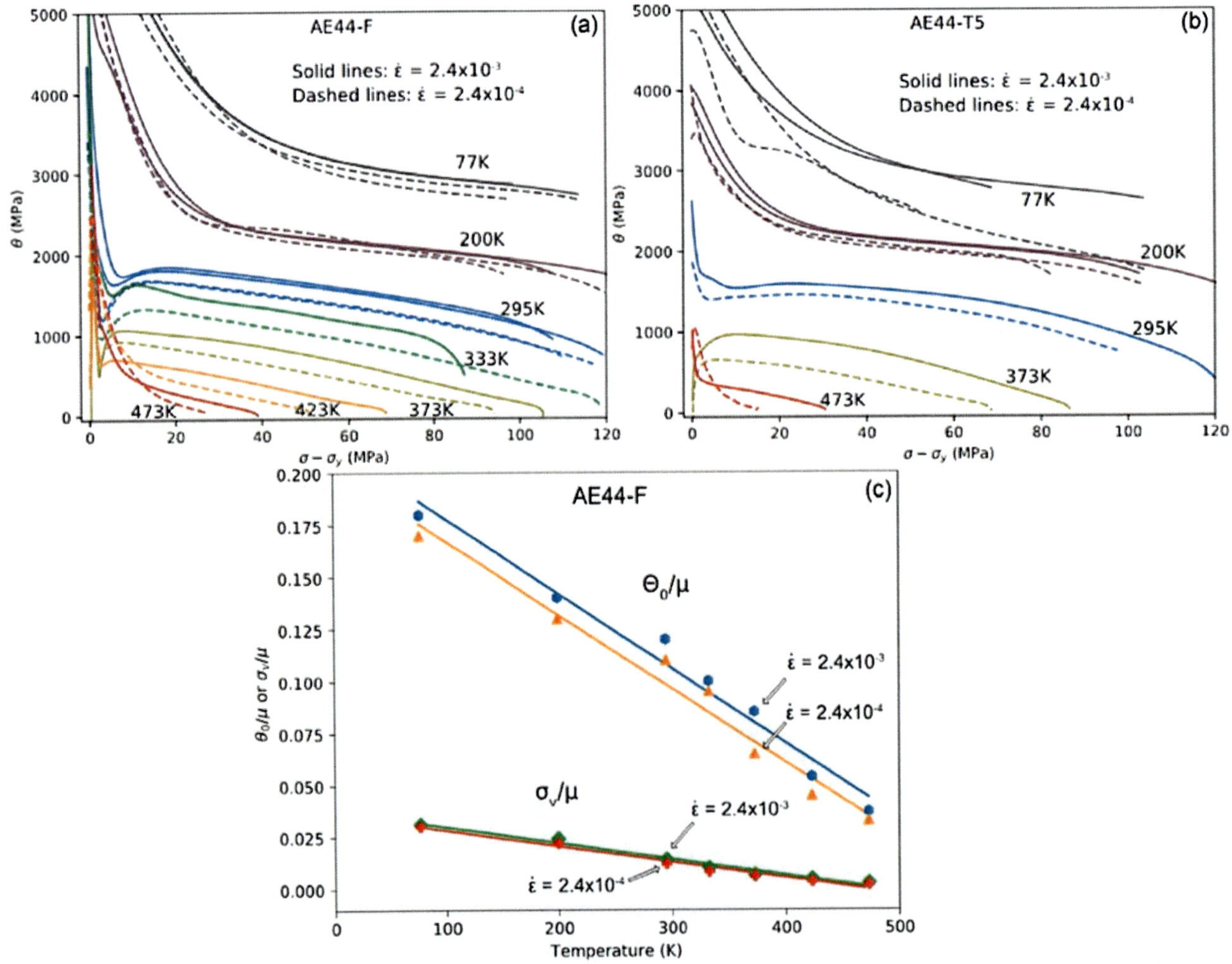

Fig. 5 Work hardening rate ($\Theta = \frac{d\sigma}{d\varepsilon}$) versus stress (relative to proof stress) for (**a**) AE44-F and (**b**) AE44-T5. Voce hardening parameters, extrapolated from (**a**), divided by shear modulus as a function of temperature for AE44-F in (**c**)

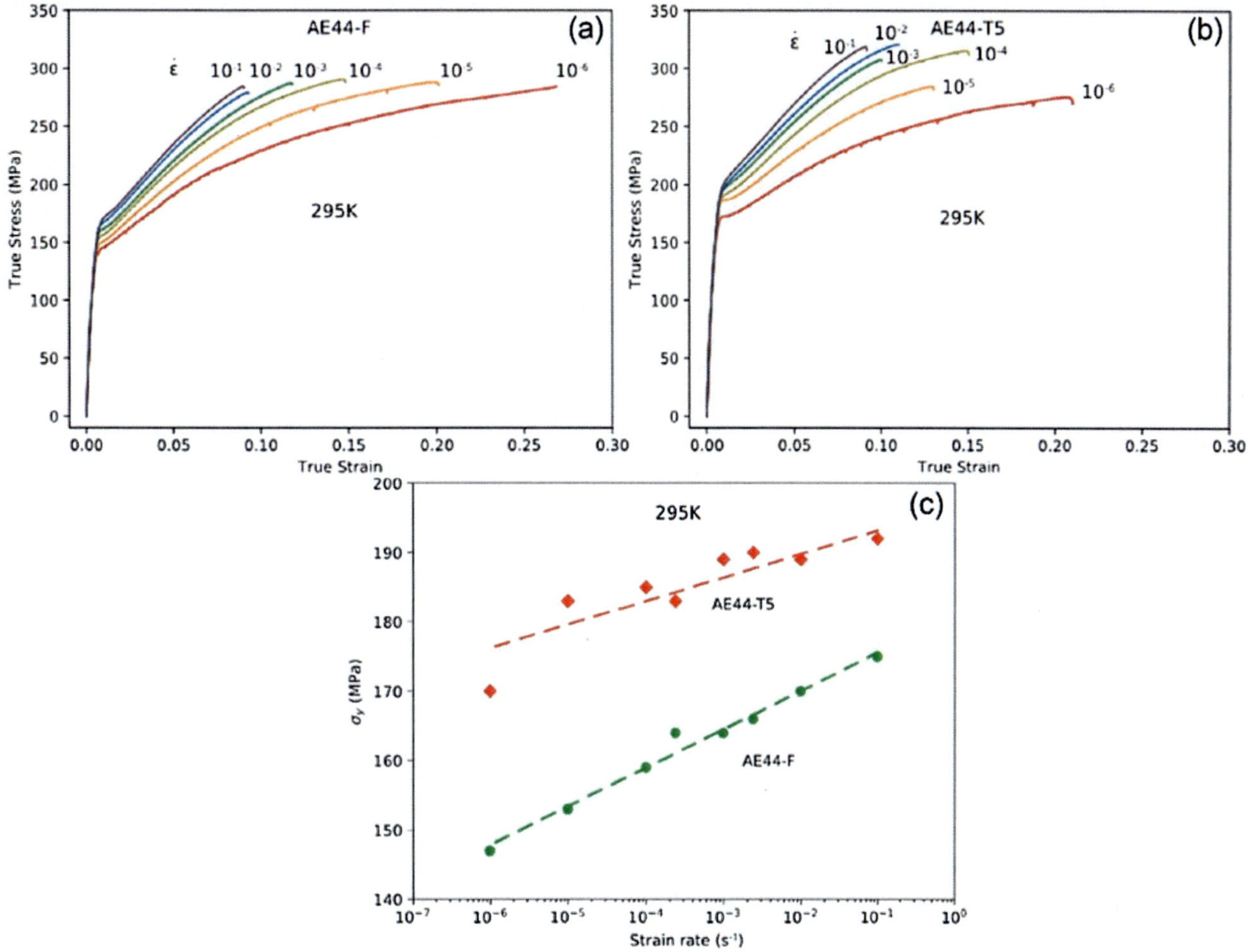

Fig. 6 Tensile curves for (**a**) AE44-F and (**b**) AE44-T5 for multiple strain rates at 295 K, and their corresponding 0.2% proof stress values in (**c**). Data replotted from Ref. [5]

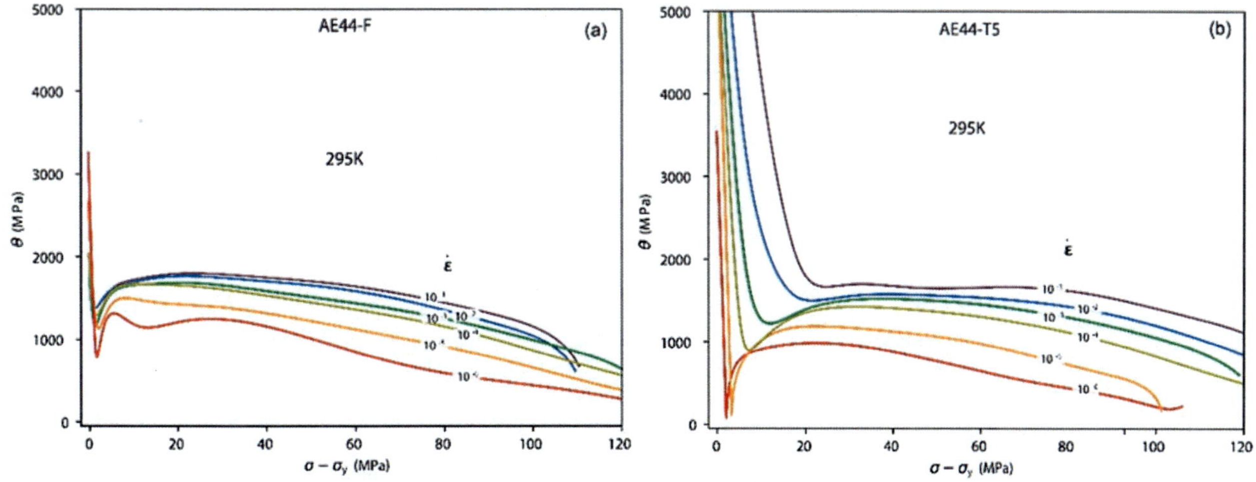

Fig. 7 Work hardening rate ($\Theta = \frac{d\sigma}{d\varepsilon}$) versus stress (relative to proof stress) for (**a**) AE44 and (**b**) AE44-T5 for multiple strain rates at 295 K

reported. The stages of deformation in face-centred cubic (FCC) metals, however, have been studied extensively [19]. In polycrystalline FCC materials, easy glide dislocation slip (referred to as stage I) is prevented by the constraint of adjacent grains. The onset of plastic slip in polycrystalline materials requires cross-slip, a process described as stage II. Stage II is characterised by a nearly constant hardening rate with slip occurring on more than one set of planes [20]. In stage III, the strain hardening rate decreases due to processes known as dynamic recovery [20]. In many cases, stage III begins immediately at the onset of plastic deformation, and so the stage II hardening rate represents only the instantaneous hardening rate at the commencement of plastic deformation. If a metal obeys Voce hardening law, then stage II hardening rate is equivalent to Θ_0. In FCC metals, this hardening rate divided by shear modulus, Θ_0/μ, typically yields a constant value of $\sim$ 0.05 regardless of temperature or strain rate and thus is referred to as the athermal hardening rate, Θ_h [19].

In HCP metals, such as Mg alloys, true stage II deformation does not occur as basal slip only has one plane. That is, the characteristic of stage II, that slip occurs on more than one set of planes, cannot be met by basal slip alone. If the deformation behaviour of Mg is analogous to FCC metals, then there must be some component of prismatic slip in addition to basal slip, so that slip can occur on more than one set of planes. This implies that there is no athermal slip process in Mg alloys with the possible exception of the easy glide stage of basal slip. This is further supported by Fig. 5c which shows that the Θ_0/μ values are clearly not athermal, ranging from 0.18 at 77 K to 0.04 at 473 K.

Previously, the concept of athermal hardening has been applied to Mg alloys to model the deformation behaviour. Caceres and Blake [21] applied Eq. 2 (adapted from [19]) to the room temperature tensile behaviour of Mg castings with a range of grain sizes. Θ_h represents the athermal hardening component and Θ_r accounts for softening due to dynamic recovery. Θ_r is a function of stress, strain rate, and temperature and is the means by which the effects of temperature and strain rate on tensile curves can be modelled. The nearness of Θ_h/μ or Θ_h/E to expected values was cited as evidence supporting this approach.

$$(\sigma - \sigma_y)\frac{d\sigma}{d\varepsilon} = (\sigma - \sigma_y)[\Theta_h - \Theta_r(\sigma, \dot{\varepsilon}, T)] \quad (2)$$

Equation 2 is of the same form as the Voce hardening law (Eq. 1) if we equate Θ_h to Θ_0 and Θ_r to $\Theta_0 \frac{\sigma}{\sigma_v}$. The present study has clearly shown that Θ_0 is not athermal for Mg alloys and that the nearness of Θ_0/μ to expected values only applies to temperatures slightly above room temperature. Instead Θ_0, for the case of Mg alloys, is itself a function of temperature and strain rate and so Eq. 3 is a more appropriate basis for a model.

$$\Theta = \Theta_0(\dot{\varepsilon}, T)\left[1 - \frac{\sigma}{\sigma_v}\right] \quad (3)$$

In both the as-cast and T5-aged conditions, AE44 follows Voce hardening over a wide range of temperatures and strain rates (except AE44-T5 at 295 K), if the knee section is excluded. Modelling the knee section requires a separate approach, and it is still under investigation. A further understanding is also required for AE44-T5 at 295 K which is the temperature where the maximum heat treatment strengthening effect occurs.

Conclusions

The effects of temperature and strain rate on the tensile properties and work hardening behaviour of Mg alloy AE44 have been studied. It is observed that both the AE44-F and AE44-T5 show a continuous decrease in strength and hardening rate with increasing temperature and decreasing strain rate. One significant finding is that the athermal hardening stage, commonly reported for Mg alloys, is not observed for the present AE44 alloy in both the as-cast and T5-aged conditions. The Voce hardening parameter shows a continuous decrease with increasing temperature.

Acknowledgements The authors would like to acknowledge the use of the RMIT Microscopy and Microanalysis Facility (RMMF).

References

1. Ang HQ, Abbott TB, Zhu SM, et al. (2016) Proof stress measurement of die-cast magnesium alloys. Mater. Des. 112: 402–409.
2. Cepeda-Jiménez C, Molina-Aldareguia J, Pérez-Prado M (2015) Origin of the twinning to slip transition with grain size refinement, with decreasing strain rate and with increasing temperature in magnesium. Acta Mater. 88: 232–244.
3. Cepeda-Jiménez C, Molina-Aldareguia J, Pérez-Prado M (2015) Effect of grain size on slip activity in pure magnesium polycrystals. Acta Mater. 84: 443–456.
4. Barnett MR, Bouaziz O, Toth L (2015) A microstructure based analytical model for tensile twinning in a rod textured Mg alloy. Int. J. Plast. 72: 151–167.
5. Ang HQ, Zhu SM, Abbott TB, et al. (2017) Strain-rate sensitivity of die-cast magnesium-aluminium based alloys. Mater. Sci. Eng. A 699: 239–246.
6. Ang HQ, Abbott TB, Zhu SM, et al. (2017) Effect of strain rate on mechanical behaviour of commercial die-cast magnesium alloys. Paper presented at Materials Science and Technology (MS&T) 2017, Pittsburgh, Pennsylvania, 8–12 October 2017: p. 164–170.

7. Su C, Li D, Luo AA, et al. (2019) Quantitative study of microstructure-dependent thermal conductivity in Mg-4Ce-xAl-0.5 Mn alloys. Metall. Mater. Trans. A 50: 1970–1984.
8. Ang HQ, Abbott TB, Zhu SM, et al. (2017) Anelasticity of die-cast magnesium-aluminium based alloys under different strain rates. Mater. Sci. Eng. A 707: 101–109.
9. Ang HQ, Zhu SM, Abbott TB, et al. (2018) Anelastic deformation during cyclic loading-unloading of die-cast magnesium alloys. Paper presented at the 11th International Conference on Magnesium Alloys and Their Applications, Old Windsor, UK, 24–27 July 2018: p. 165–168.
10. Ang HQ (2020) Modelling of the strain hardening behaviour of die-cast magnesium-aluminium-rare earth alloy. Adv. Eng. Forum 35: 1–8.
11. Zhu SM, Abbott TB, Gibson MA, et al. (2016) Age hardening in die-cast Mg–Al–RE alloys due to minor Mn additions. Mater. Sci. Eng. A 656: 34–38.
12. Ang HQ, Zhu SM, Abbott TB, et al. (2022) High-performance Mg–4Al–4RE (RE= cerium and lanthanum) die-casting alloy. Paper presented at the TMS 2022 Annual Meeting & Exhibition, Anaheim, California, 27 February–3 March 2022: p. 61–67.
13. Rzychoń T, Kiełbus A, Dercz G (2007) Structural and quantitative analysis of die cast AE44 magnesium alloy. J. Achiev. Mater. Manuf. Eng 22 (2): 43–46.
14. Zhu SM, Abbott TB, Nie JF, et al. (2021) Re-evaluation of the mechanical properties and creep resistance of commercial magnesium die-casting alloy AE44. J. Magnes. Alloy. 9(5): 1537–1545.
15. Ang HQ (2021) Anelastic behaviour of commercial die-cast magnesium alloys: effect of temperature and alloy composition. Materials 14 (23): 7220.
16. Ang HQ, Abbott TB, Zhu SM, et al (2015) Flaws in standardised proof stress determination methods for magnesium alloys. Paper presented at the Third International Academic Conference of Postgraduates, NUAA, Nanjing, 18–20 November 2015, p. 75–79.
17. Voce E (1955) A practical strain hardening function. Metallurgia 51: 219–226.
18. Ang HQ, Abbott TB, Zhu SM, et al. (2019) An analysis of the tensile deformation behavior of commercial die-cast magnesium-aluminum-based alloys. Metall. Mater. Trans. A 50 (8): 3827–3841.
19. Kocks UF, Mecking H (2003) Physics and phenomenology of strain hardening: the FCC case. Prog. Mater. Sci 48: 171–273.
20. Dieter, GE (1976) Mechanical Metallurgy. McGraw-Hill, New York.
21. Caceres CH, Blake AH (2007) On the strain hardening behaviour of magnesium at room temperature. Mater. Sci. Eng A 462: 193–196.

The Mechanisms to Improve Creep Resistance in a Die-Cast MgREMnAl Alloy

Xixi Dong, Lingyun Feng, Eric A. Nyberg, and Shouxun Ji

Abstract

High-pressure die-cast magnesium (Mg) alloys are required for components working at temperatures above 200 °C. These new alloys are necessary for high-volume applications such as critical parts used for internal combustion (IC) engines in power tools. Here we present new developments of a Mg–RE die-cast alloy that shows excellent ambient and high-temperature tensile strength, creep resistance, stiffness, and thermal conductivity, which are key advantages for alloys used at elevated temperatures. The excellent creep resistance of the die-cast Mg3.5RE (La, Ce, Nd)1.5GdMnAl alloy, in comparison with its counterpart alloy without the Al addition, shows a significant improvement of the steady-state creep rate (SCR) at 300 °C/50 MPa. The synergistic effect of Al, Gd, and Mn has been found to induce a novel and thermally stable AlMnGd ternary short-range order (SRO, 0–2 nm)/cluster (2–10 nm) in the Mg matrix, which is believed to be responsible for the improvement in creep performance.

Keywords

Magnesium alloys • Mechanical properties • Creep resistance • High pressure die casting • Strengthening mechanism

X. Dong · L. Feng · S. Ji (✉)
Brunel Centre for Advanced Solidification Technology (BCAST), Brunel University London, Uxbridge, UB8 3PH, UK
e-mail: Shouxun.Ji@brunel.ac.uk

E. A. Nyberg
Kaiser Aluminum, Spokane Valley, WA 99216, USA

Introduction

Magnesium alloys as lightweight structural materials are attractive in manufacturing powertrain components because of the excellent damping property to achieve significant reduction in vibration [1–3]. This is even more attractive for the application in powered tools as the reduced vibration can increase the comfortability in human operation. However, the inherent features of magnesium alloys are usually not favourable for applications at high temperatures [4, 5].

Generally, having Al in Mg alloys was considered beneficial for castability but detrimental for creep resistance at elevated temperatures [3, 6]. Therefore, the most widely used die-cast Mg–Al-based alloys (AZ91, AM50/60) are not suitable for elevated applications above 120 °C, due to the formation of thermally unstable $Mg_{17}Al_{12}$ phase [7, 8]. The improvement through adding Si, Sr, Ca, Sn, and rare earth (RE) can increase the creep resistance of Mg–Al-based die-cast alloys, but only work well at temperatures up to 150–175 °C [6, 9]. Therefore, further increase of working temperatures has focused on the Mg–RE alloys without Al addition. However, it is unclear whether minor additions of Al can be beneficial for Mg–RE alloys working for higher temperatures.

In this work, it is aimed to study the effect of Al in a die-cast Mg–RE alloy that shows excellent creep resistance at elevated temperatures of 300 °C. The microstructure characterisation is conducted at an atomic level to understand the mechanism of creep resistance. The discussion focuses on the relationship between the microstructure and the creep resistance.

Experimental

Cylindrical creep samples were prepared by high-pressure die casting (HPDC) on a 4500 kN cold chamber machine, and the gauge diameter and length of creep samples were

S. Barela et al. (eds.), *Magnesium Technology 2023*, The Minerals, Metals & Materials Series,
https://doi.org/10.1007/978-3-031-22645-8_29

6.35 mm and 20 mm, respectively. The alloy composition obtained by ICP-OES was Mg1.6La0.9Ce1.0Nd1.5Gd0.3 Zn0.3Mn0.5Al (wt.%). Creep tests were conducted on a lever-type Instron machine following ISO 204:2018. The creep temperatures were set 200–300 °C, and the stress levels of 40–80 MPa were applied for the creep tests. The temperature of the creep samples was controlled by thermocouples with an accuracy of ± 1 °C, and the loaded stresses were lower than the yield stress. One extensometer was attached to the gauge section of samples to measure the strains during creep testing.

Backscattered SEM (BSD-SEM) observation was performed on a Zeiss SUPRA 35VP microscope. Electron backscatter diffraction (EBSD) samples were prepared by electrolytic polishing using picric acid solution (a mixture of 12.6 g picric acid, 30 ml acetic acid, 30 ml water, and 420 ml ethanol) after mechanical polishing. The electrolyte temperature was − 30 °C after liquid nitrogen treatment, and the polishing time was 80–100 s with 20 V voltage. TEM samples were prepared by mechanical polishing followed by final ion thinning. TEM analysis including nano-diffraction (ND), bright-field (BF) imaging, and high-resolution TEM (HRTEM) imaging were conducted using a FEI Tecnai G2 microscope. An in-depth TEM analysis was carried out using a Thermofisher Titan Themis 60–300 microscope at 300 kV, which was equipped with a highly efficient (4 quadrant) energy dispersive X-ray (EDX) system and a probe aberration corrector for atomic and nano-scale characterisation in the scanning TEM (STEM) mode. The collection angle of the high-angle annular dark field (HAADF) detector ranged from 80 to 150 mrad. The probe size was set to 0.1 nm with a convergent semi-angle of 22.5 mrad. Energy dispersive spectrometer (EDS) mapping was also applied under STEM. A 3D visualisation program for structural models (VESTA) was used to construct interfacial atomic structures.

Results

Tensile Creep Properties

Figure 1 shows the creep properties of the die-cast Mg3.5RE1.5GdMnAl alloy, the comparison with the counterpart Al-free Mg3.5RE1.5GdMn alloy and the traditional representative Mg–Al-based commercial die-cast MRI153M, AJ62, MRI230D, and AE44 alloys for elevated applications. As illustrated in Fig. 1a, under the creep temperature of 300 °C and the stress of 50 MPa, the Mg3.5RE1.5GdMnAl alloy went into the tertiary state after creep for 460 h (h) and did not reach the rupture point after 620 h; the counterpart Mg3.5RE1.5GdMn alloy went into the tertiary state after creep for 350 h and ruptured after 450 h, whereas the secondary steady-state creep strain of the Mg3.5RE1.5GdMnAl alloy was much lower than the counterpart Mg3.5RE1.5GdMn alloy and the four commercial alloys. As shown in Fig. 1b, the steady-state creep rate (SCR, s^{-1}) of the MRI153M, AJ62, MRI230D, AE44, Mg3.5RE1.5GdMn, and Mg3.5RE1.5GdMnAl alloys at 300 °C/50 MPa were $(4.58 \pm 1.19) \times 10^{-7}$, $(3.39 \pm 0.88) \times 10^{-7}$, $(3.15 \pm 0.82) \times 10^{-8}$, $(5.69 \pm 1.41) \times 10^{-9}$, $(4.68 \pm 1.23) \times 10^{-10}$, and $(1.35 \pm 0.28) \times 10^{-10}$, respectively. The Mg3.5RE1.5GdMnAl alloy showed a 71% reduction in the SCR compared with the counterpart Mg3.5RE1.5GdMn alloy; it further showed a much-improved creep resistance in contrast to the four typical Mg–Al-based commercial die-cast alloys. In addition, the SCR of the Mg3.5RE1.5GdMnAl alloy was lower than the heat-treated gravity cast Mg alloys (QE22, EQ21, AM-SC1, WE43, WE54, Mg15Gd, Mg8Gd4Y, Mg15Gd2Y, Mg10Gd3Y, and Mg12Y5Gd) for applications at 200–300 °C [10–19]. Therefore, the die-cast Mg3.5RE1.5 GdMnAl alloy has a superior creep resistance that could promote the working temperatures of the die-cast Mg alloys from 120–200 °C [3, 20] to 200–300 °C.

As-Cast Microstructure

Figure 2a–h show the BSD-SEM morphology and SEM–EDS mapping of the counterpart Al-free Mg3.5RE1.5GdMn alloy in the as-cast state. The microstructure comprised the primary α_1-Mg matrix phase with a grain size of ~ 10–30 μm, the secondary α_2-Mg matrix phase with a grain size of ~ 2–10 μm and the intermetallic phase at the grain boundaries (GBs). Previous studies [20, 21] have confirmed that the fine α_2-Mg was formed in the die cavity with a higher cooling rate than the α_1-Mg which was formed in the shot sleeve. The major intermetallic phase at the GBs of the Mg3.5RE1.5GdMn alloy was $Mg_{12}RE$, which was enriched in La, Ce, Nd, and Gd. Figure 2i–q display the BSD-SEM morphology and SEM–EDS mapping of the Mg3.5RE1.5GdMnAl alloy in the as-cast state. The major intermetallic phase at the GBs of the Mg3.5RE1.5GdMnAl alloy was still $Mg_{12}RE$ that was enriched in La, Ce, Nd, and Gd. However, a small amount of fine Al_2RE_3 intermetallic phase formed at the GBs of the Mg3.5RE1.5GdMnAl alloy, which was in blocky morphology. The Al_2RE_3 intermetallic phase was enriched in Al and Gd, as shown in Fig. 2m and n.

Microstructure Evolution at GBs by Creep

Figure 3a and b present the BSD-SEM and STEM morphology of the intermetallic phases at the GBs of the die-cast Mg3.5RE1.5GdMnAl alloy before creep, and the ND

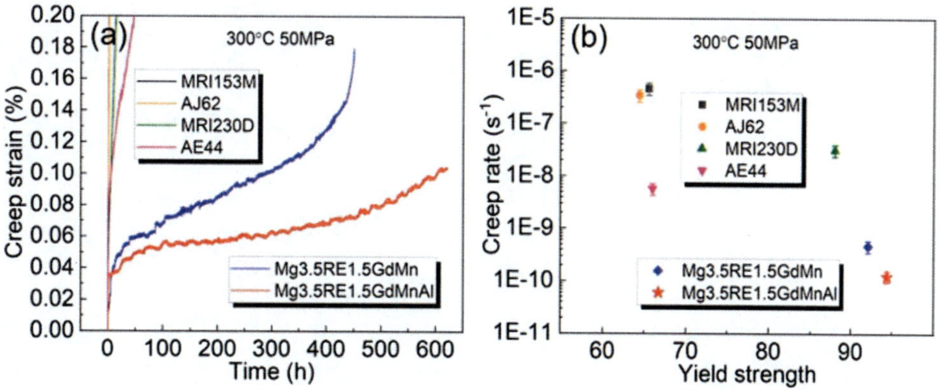

Fig. 1 Tensile creep properties of the die-cast Mg3.5RE1.5GdMnAl alloy under the as-cast condition, compared with the counterpart Al-free Mg3.5RE1.5GdMn alloy and the traditional representative Mg–Al-based commercial die-cast MRI153M, AJ62, MRI230D, and AE44 alloys for elevated applications. **a** Representative creep curves. **b** The steady-state creep rate of the alloys tested at 300 °C with a 50 MPa stress

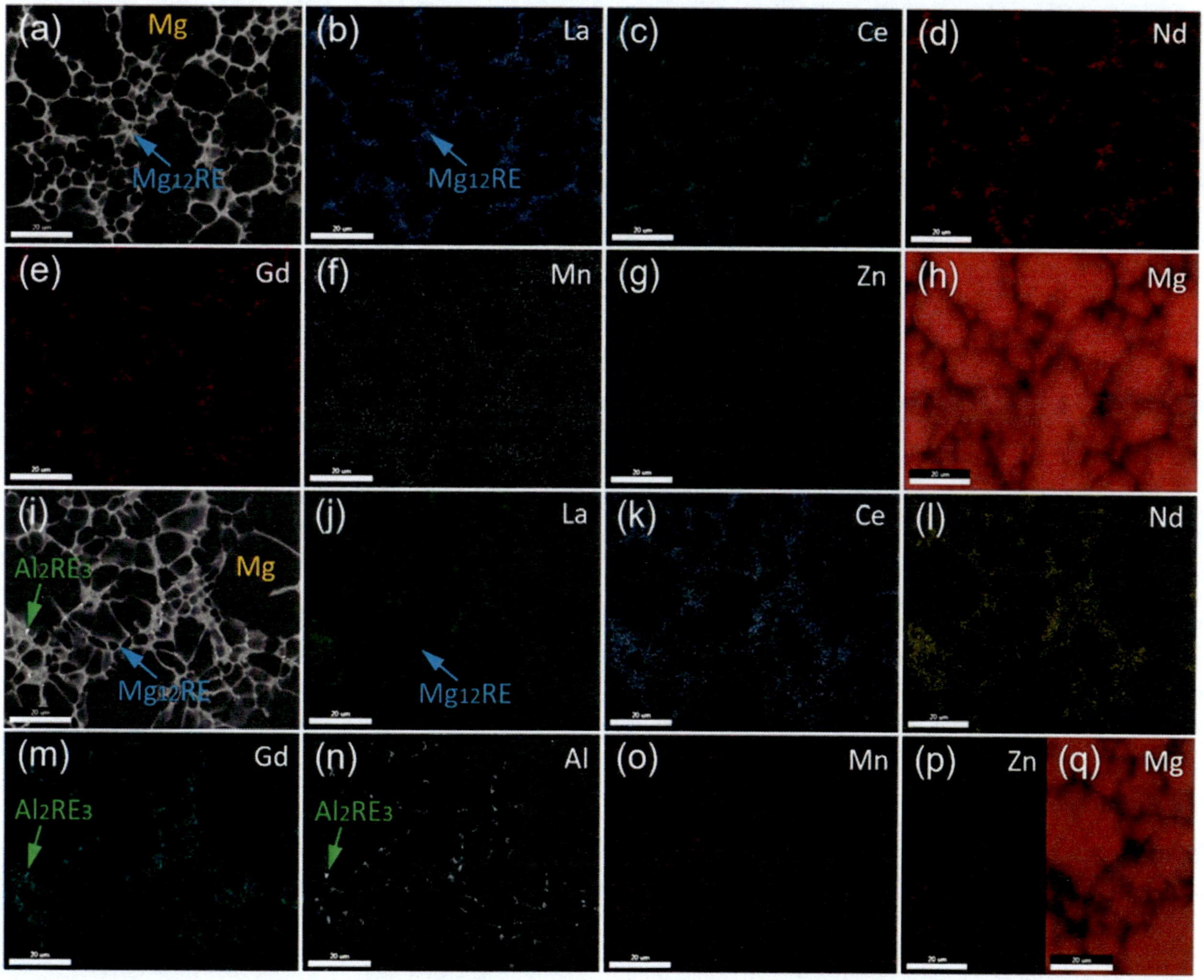

Fig. 2 As-cast microstructure of **a–h** the counterpart Al-free Mg3.5RE1.5GdMn alloy and **i–q** the die-cast Mg3.5RE1.5GdMnAl alloy analyzed by SEM. **a** BSD-SEM morphology of the counterpart Al-free Mg3.5RE1.5GdMn alloy; **b–h** SEM/EDS maps of the elements: **b** La, **c** Ce, **d** Nd, **e** Gd, **f** Mn, **g** Zn, and **h** Mg in **a**; **i** BSD-SEM morphology of the Mg3.5RE1.5GdMnAl alloy; **j–q** SEM/EDS maps of the elements: **j** La, **k** Ce, **l** Nd, **m** Gd, **n** Al, **o** Mn, **p** Zn, and **q** Mg in **i**

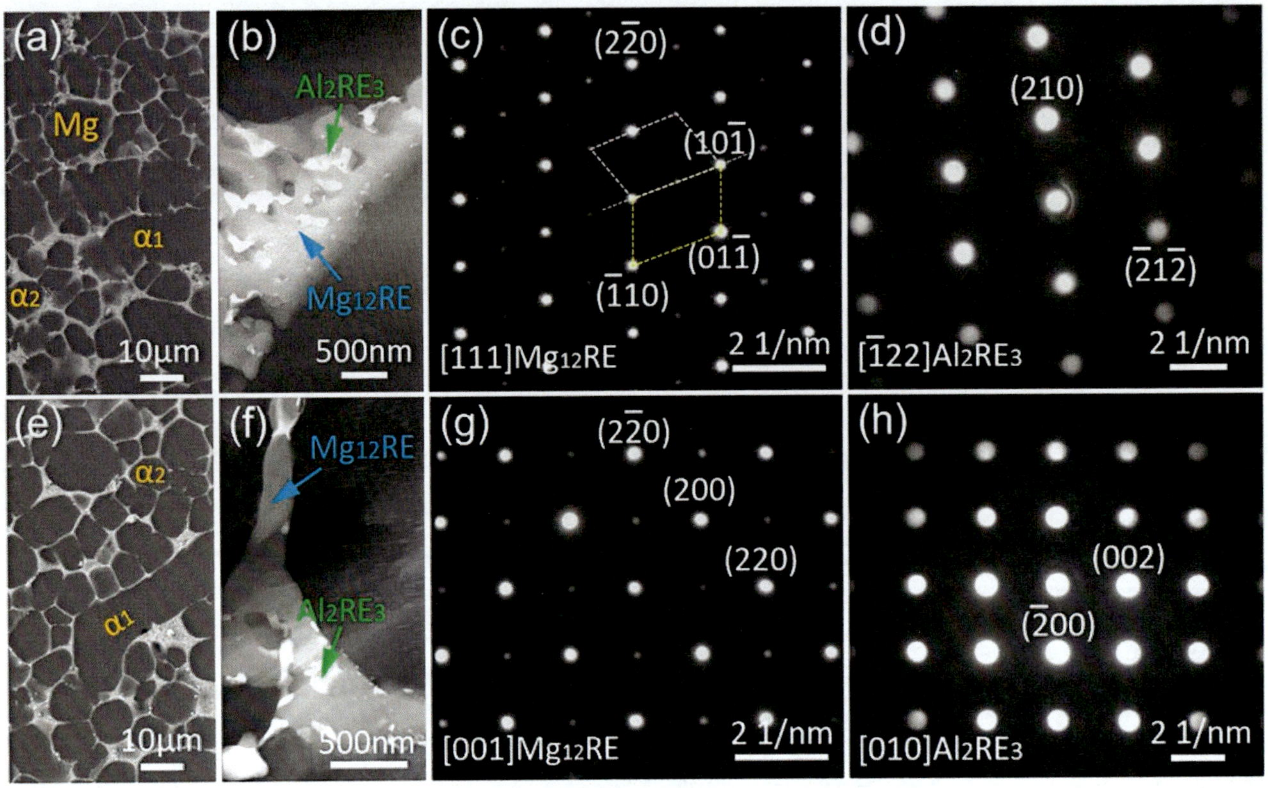

Fig. 3 Microstructure of the as-cast die-cast Mg3.5RE1.5GdMnAl alloy **a–d** before creep and **e–h** after creep at 300 °C/50 MPa for 400 h. **a, e** BSD-SEM morphology; **b, f** STEM images and **c, d, g, h** ND patterns showing the intermetallic phases at the GBs

patterns in Fig. 2c and d verified that the intermetallic phases at the GBs of the Mg3.5RE1.5GdMnAl alloy before creep were $Mg_{12}RE$ and Al_2RE_3. Figure 3e and f show the BSD-SEM and STEM morphology of the Mg3.5RE1.5GdMnAl alloy after creep at 300 °C/50 MPa for 400 h. The $Mg_{12}RE$ and Al_2RE_3 phases at the GBs showed no obvious change of morphology in comparison with that before creep. In addition, the ND patterns in Fig. 3g and h demonstrated that the intermetallic phases at the GBs of the Mg3.5RE1.5GdMnAl alloy after creep were still $Mg_{12}RE$ and Al_2RE_3. Therefore, there were no phase transformations at the GBs after creep, indicating that $Mg_{12}RE$ and Al_2RE_3 were stable and did not degrade at the high creep temperature of 300 °C.

Microstructure Evolution in Mg Matrix by Creep

Figure 4a–e, f and i show the microstructure in the Mg matrix of the Mg3.5RE1.5GdMnAl alloy before and after creep at 300 °C/50 MPa for 400 h, respectively. Isolated solute atoms and a novel ternary AlMnGd SRO/cluster were observed in the Mg matrix of the alloy before creep (Fig. 4a–e). After creep at 300 °C/50 MPa for 400 h, an Al_2Gd precipitate was observed in the Mg matrix (Fig. 4f); isolated solute atoms and the AlMnGd SRO/clusters still existed in the Mg matrix (Fig. 4f–i); the quantity of isolated solute atoms decreased (Fig. 4g), while the size of some clusters increased (Fig. 4f). However, the size of the clusters was under 10 nm after creep at 300 °C/50 MPa for 400 h (Fig. 4f). In this work, the AlMnGd microstructure with the size of 0–2 nm and 2–10 nm was called the SRO and cluster, respectively. In addition, the size of Al_2Gd precipitate was ~ 27–75 nm. Further, the Al_2Gd precipitate exhibited a hexagon-faceted morphology under the projection direction of ⟨001⟩ Mg. Figure 4d and e show the enlarged view of the SRO and cluster in Fig. 4c, h and i show the enlarged view of the SRO and cluster in Fig. 4g. The AlMnGd SRO/clusters occupied the positions of the Mg atoms, both before (Fig. 4d, e) and after creep (Fig. 4h, i). The inserts in Fig. 4d, e, h and i show the corresponding fast Fourier transform (FFT) patterns in the areas, and the FFT patterns of the AlMnGd SRO/clusters were the same as their adjacent areas of the Mg matrix, both before and after creep, which indicated that the AlMnGd SRO/clusters were coherent with the Mg matrix. SRO and clusters were also reported in some Mg alloys [22–24]. Therefore, the microstructure evolution mainly occurred

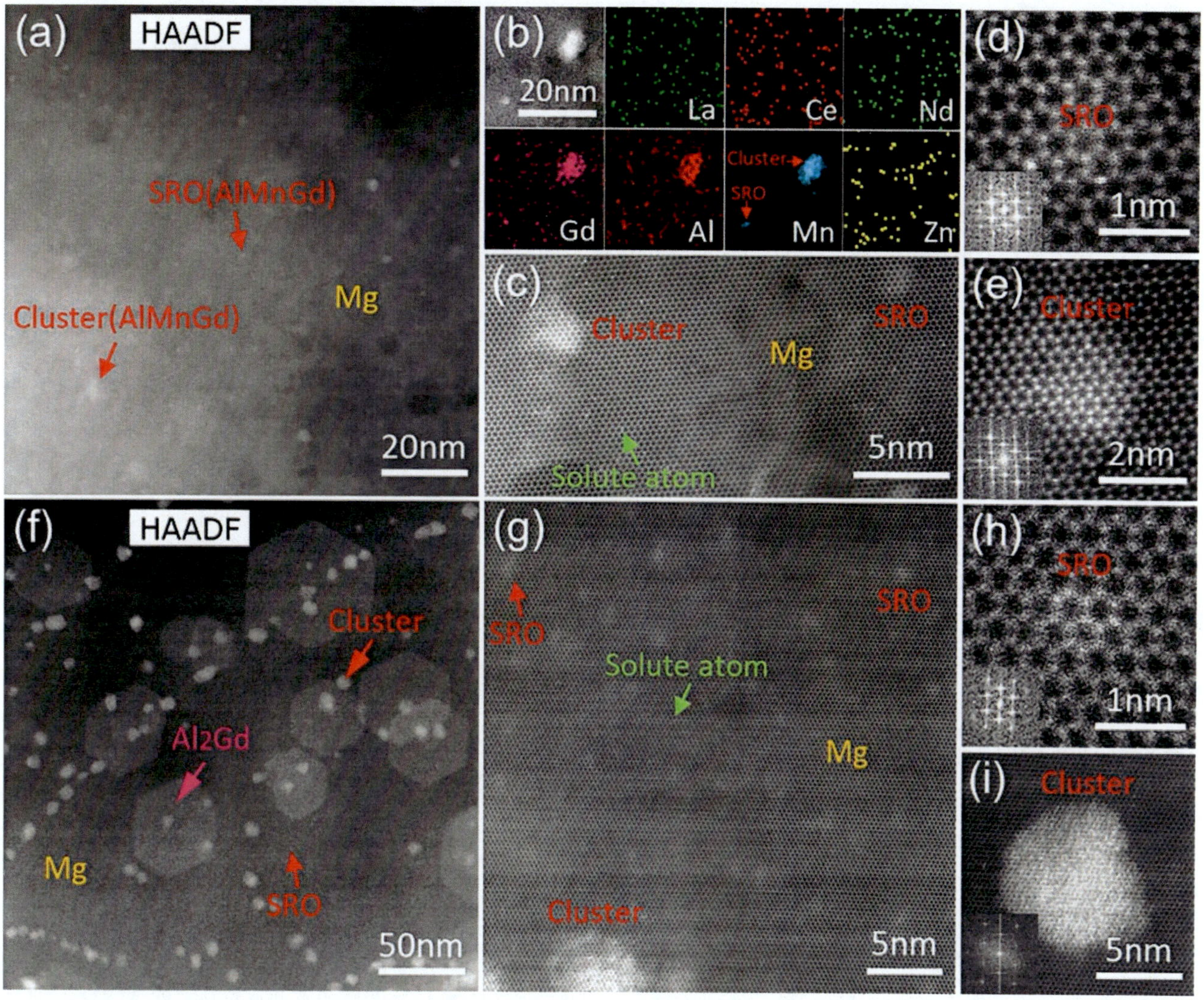

Fig. 4 STEM images showing the microstructure in the Mg matrix of the die-cast Mg3.5RE1.5GdMnAl alloy **a–e** before creep and **f–i** after creep at 300 °C/50 MPa for 400 h, under the projection direction of ⟨0001⟩ Mg. **a**, **c–e** The STEM-HAADF images and **b** STEM/EDS mapping of the solute atoms and AlMnGd SRO/clusters in the Mg matrix before creep; **f–i** the STEM-HAADF images showing the solute atoms, AlMnGd SRO/clusters and Al_2Gd precipitates in the Mg matrix after creep

in the Mg matrix of the Mg3.5RE1.5GdMnAl alloy after creep.

Figure 5 shows the interactions between dislocations and precipitates in the Mg matrix, and between the dislocations and intermetallic phases at the GBs during the steady-state creep of the die-cast Mg3.5RE1.5GdMnAl alloy at 300 °C/50 MPa. Importantly, high-density AlMnGd SRO/clusters were observed for pinning dislocations in the Mg matrix (Fig. 5a). Al_2Gd precipitates can also contribute to dislocation pinning; however, due to the lower number density and larger size, their contribution to the dislocation pinning may not be as significant as the AlMnGd SRO/clusters. The HRTEM images in Fig. 5b–d show the impeding of the dislocations across the GBs by the $Mg_{12}RE$ network.

Discussion

Effects of $Mg_{12}RE/Al_2RE_3$ at GBs on Creep

It was suggested that the formation of the $Mg_{12}RE$ phase at the GBs became possible in Mg–Al–RE alloys when the RE/Al weight ratio was above 1.4 [25]. In differentiation to the divorced or lamellar Al-containing compounds in the die-cast Mg–Al-based alloys, $Mg_{12}RE$ constitutes the

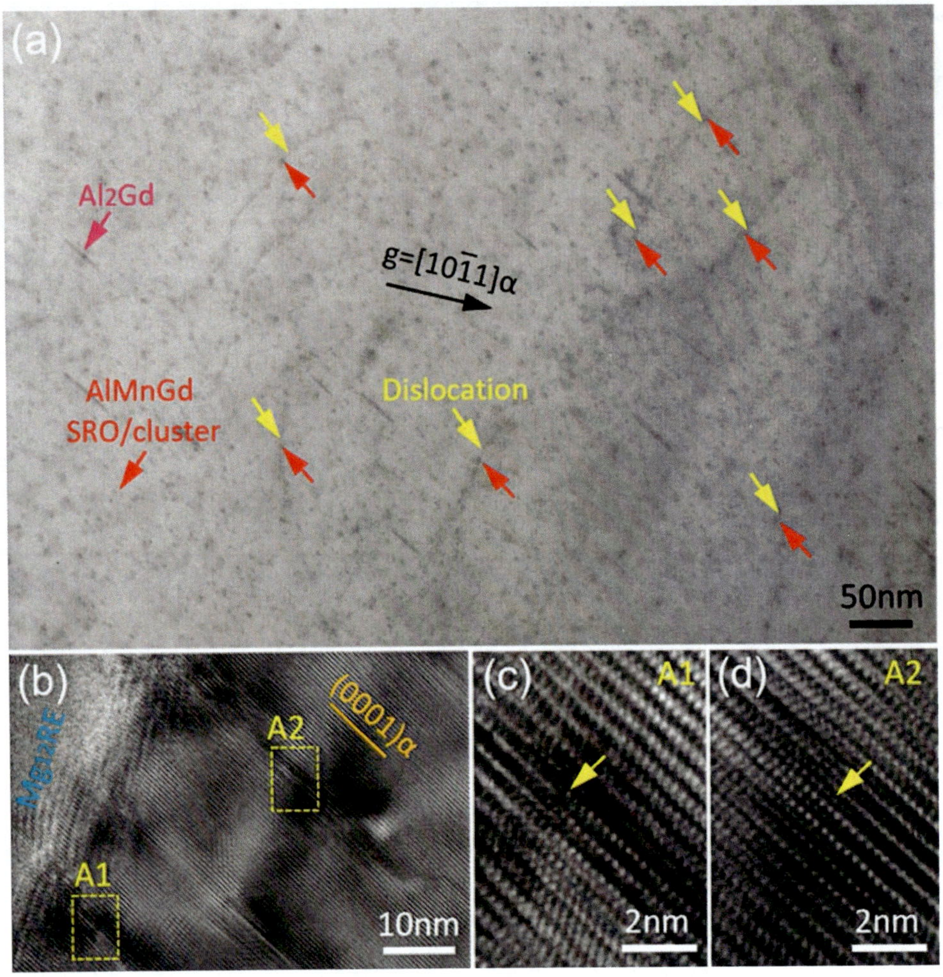

Fig. 5 **a** BF-TEM and **b–d** HRTEM images viewed along the [1 $\bar{2}$ 10] Mg, showing the dislocations in the die-cast Mg3.5RE1.5GdMnAl alloy after creep at 300 °C/50 MPa for 400 h. **a** The dislocations in the Mg matrix far from the GBs and viewed with **g** = [1 0$\bar{1}$ 1] Mg, and dislocations massively pinning by the AlMnGd SRO and clusters; **b** HRTEM image showing the dislocations very near the GBs; **c** and **d** enlarged views of the dislocation areas marked by A1 and A2 in **b**, respectively

major intermetallic phase at the GBs of the Al-free die-cast Mg-RE alloys [26, 27]. More importantly, $Mg_{12}RE$ can form a network. With the increase of the RE/Al weight ratio, there is a transition from the Al-rich $Al_{11}RE_3$ to the Al-free $Mg_{12}RE$ at the GBs. In theory, the transition sequence is $Al_{11}RE_3 \rightarrow Al_2RE \rightarrow AlRE \rightarrow Al_2RE_3 \rightarrow AlRE_2 \rightarrow Mg_{12}$ RE [28]. As the die-cast Mg3.5RE1.5 GdMnAl alloy has a RE/Al weight ratio of ~ 10:1, the $Mg_{12}RE$ network can form at the GBs (see Figs. 2 and 3). However, Al is tied up by RE, especially Gd, to form the divorced Al_2RE_3 compound at the GBs (see Fig. 2). $Mg_{12}RE$ was reported to be more stable than $Al_{11}RE_3$, which can degrade to Al_2RE via a releasing of Al [29, 30]. The diffraction analysis after creep (see Fig. 3) also supported the stability of $Mg_{12}RE$. It was further reported [31] that the increase of the interconnectivity of the intermetallic phase at the GBs was important for the enhancement of creep resistance as it can shield more load from the matrix. Regarding Al_2RE_3, it is highly rich in RE, rather than Al, and it has the only theoretical possibility of decomposition into $AlRE_2$. However, this decomposition would not occur (see Fig. 3) even at 300 °C/50 MPa as the later lanthanides including Gd had a strong combination with Al. Moreover, Al_2RE_3 has a melting point of more than 1000 °C [28]. Thus, in terms of the intermetallic phases at the GBs, the addition of Al here did not deteriorate the creep properties, as Al was tied up by Gd to form the thermally stable Al_2RE_3 compound. Considering that the major phase at the GBs was the thermally stable network of $Mg_{12}RE$, $Mg_{12}RE$ should also play a significant part in the excellent creep resistance in the die-cast Mg3.5RE1.5GdMnAl alloy via impeding the movement of dislocations across the GBs (see Fig. 5b–d). This is in comparison to the traditional Mg–Al-based commercial die-cast alloys (see Fig. 1) for elevated temperature applications. However, for the counterpart Al-free Mg3.5RE1.5GdMn alloy, the addition of 0.5wt.% Al should have a very limited effect on the improvement of creep resistance at the GBs since it hardly changed the major $Mg_{12}RE$ network at the GBs (see Figs. 2 and 3).

Effects of Precipitates and SRO/Clusters on Creep

Dynamic precipitates were traditionally thought to contribute to the creep resistance of Mg alloys via dislocation pinning [32–34]. However, in some alloys such as Mg-0.45La-1.2Y (in at.%), SRO/clusters can play an important role for creep resistance [22]. There have been some intense debates about the effects of the dynamic precipitates and SRO/clusters on creep resistance [3]. Some researchers thought that SRO/clusters were thermally unstable and disappeared after ageing and/or creep [35, 36]. Clearly, massive amounts of the AlMnGd SRO/clusters still existed in the Mg matrix of the die-cast Mg3.5RE1.5GdMnAl alloy, after creep at 300 °C/50 MPa for a duration of 400 h, which can be attributed to the high thermal stability of the AlMnGd SRO/clusters [37]. SRO/clusters could also hinder dislocation movement and contribute to creep resistance [23, 24]. Here, both the Al_2Gd precipitates and the AlMnGd SRO/clusters can contribute to creep resistance. However, the number density and thermal stability [37] of the AlMnGd SRO/clusters were much higher than the Al_2Gd precipitates, so the AlMnGd SRO/clusters should play an important role for the creep resistance of the Mg3.5RE1.5GdMnAl alloy. In addition, SRO/clusters were hardly observed in the Mg matrix of the crept counterpart Mg3.5RE (La, Ce, Nd)1.5GdMn alloy, and REZn, ZnGd, and MnGd precipitates with larger size dominated the creep of the alloy rather than SRO/clusters. Moreover, the singular addition of Al into traditional Gd-free Mg-RE (La, Ce, Nd) alloys deteriorated creep resistance [26, 30]. Therefore, it is not the singular effect of Al or Gd that provides the exceptional improvement of creep resistance; it is the synergistic effect of Al, Gd, and Mn to form the novel thermally stable and creep-resistant ternary AlMnGd SRO/clusters in the Mg matrix that provides the exceptional improvement in the creep resistance in the Mg3.5RE (La, Ce, Nd) 1.5GdMnAl alloy. This is in contrast to the counterpart Al-free Mg3.5RE (La, Ce, Nd) 1.5GdMn alloy.

Conclusions

(1) The die-cast Mg3.5RE1.5GdMnAl alloy offers a superior low steady-state creep rate of 1.35×10^{-10} s^{-1} under the critical condition of 300 °C/50 MPa, which is 71% lower than the counterpart Al-free Mg3.5RE1.5GdMn alloy, and much lower than the traditional Mg–Al-based commercial die-cast alloys for elevated temperature applications; this enables die-cast Mg alloys to work at the higher temperatures of 200–300 °C.

(2) The addition of Al induces the formation of a novel thermally stable ternary AlMnGd short-range order/cluster in the Mg matrix of the Gd-containing Mg3.5RE1.5GdMnAl alloy. The AlMnGd short-range order/clusters still exist after creep at 300 °C/50 MPa for 400 h. Moreover, high-density AlMnGd short-range order/clusters can be observed for dislocation pinning during creep. It is the synergistic effect of Al, Gd, and Mn to form the thermally stable and creep-resistant AlMnGd short-range order/clusters that provides the exceptional improvement of creep resistance in the Mg3.5RE1.5GdMnAl alloy in comparison to the counterpart Al-free Mg3.5RE1.5GdMn alloy; this alters the traditional understanding of the disadvantage of Al on the creep resistance of Mg alloys.

(3) A thermally stable network of the $Mg_{12}RE$ (La, Ce, Nd) phase forms at the grain boundaries of the Mg3.5RE1.5GdMnAl alloy; it also impedes the further motion of dislocations across the grain boundaries, which plays a significant role in achieving the super creep resistance in the Mg3.5RE1.5GdMnAl alloy, in contrast to the traditional Mg–Al-based commercial die-cast alloys for elevated temperature applications.

Acknowledgements This work has been supported by Innovate UK (Project reference: 10004694).

References

1. Luo AA (2013) Magnesium casting technology for structural applications. J. Magnes. Alloy. 1:2–22.
2. Feng LY, Dong XX, Cai Q, Wang B, Ji S (2022) Effect of Gd on the microstructure and mechanical properties of high-pressure die-cast Mg-La-Ce alloys at ambient and elevated temperatures. J. Alloys Compd. 923:166364.
3. Mo N, Tan QY, Bermingham M, Huang YD, Dieringa H, Hort N, Zhang MX (2018) Current development of creep-resistant magnesium cast alloys: A review. Mater. Des. 155:422–442.
4. Dong XX, Nyberg EA, Ji S (2020) A die-cast magnesium alloy for applications at elevated temperatures. Magnes. Technol. 31–36.
5. Dong XX, Feng LY, Wang SH, Wang F, Ghasemi R, Ji G, Nyberg EA, Ji S (2022) A quantitative strategy for achieving the high thermal conductivity of die-cast Mg-Al-based alloys. Materialia 22:101426.
6. Pekguleryuz M, Celikin M (2010) Creep resistance in magnesium alloys. Int. Mater. Rev. 55:197–217.
7. Weiler JP, Wood JT (2012) Modeling the tensile failure of cast magnesium alloys. J. Alloys Compd. 537:133–140.
8. Luo AA (2004) Recent magnesium alloy development for elevated temperature applications. Int. Mater. Rev. 49:13–30.

9. Powell BR, Rezhets V, Luo AA, Bommarito JJ, Tiwari BL (2001) Creep resistant magnesium alloy die casting. US Patent, 6264763.
10. Yang Q, Guan K, Li BS, Lv SH, Meng FZ, Sun W, Zhang YQ, Liu XJ, Meng J (2017) Microstructural characterizations on Mn-containing intermetallic phases in a high-pressure die-casting Mg–4Al–4RE–0.3Mn alloy. Mater. Charact. 132:381–387.
11. Almgren M, Assarsson H, Hjalmarsson S, Dong XX, Ji S, Nyberg EA, Orestig P (2020) WIPO Patent, WO2020171758A1.
12. Momma K, Izumi F (2011) *VESTA 3* for three-dimensional visualization of crystal, volumetric and morphology data. J. Appl. Crystallogr. 44:1272–1276.
13. Sklenicka V, Kucharova K, Svoboda M, Kvapilova M, Dvorak J, Kra P (2020) Creep behaviour of unreinforced and reinforced QE22 magnesium alloy: A comparison. Kovove Mater. 58:71–82.
14. Kielbus A, Rzychon T (2011) Mechanical and creep properties of Mg-4Y-3RE and Mg3Nd-1Gd magnesium alloy. Procedia Eng. 10:1835–1840.
15. Bettles CJ, Gibson MA, Zhu SM (2009) Microstructure and mechanical behaviour of an elevated temperature Mg-rare earth based alloy. Mater. Sci. Eng. A 505:6–12.
16. Mordike BL (2002) Creep-resistant magnesium alloys. Mater. Sci. Eng. A 324:103–112.
17. Smola B, Stulıková I, Pelcova J, Mordike BL (2004) Significance of stable and metastable phases in high temperature creep resistant magnesium–rare earth base alloys. J. Alloys Compd. 378:196–201.
18. Janik V, Yin DD, Wang QD, He SM, Chen CJ, Chen Z, Boehlert CJ (2011) The elevated temperature mechanical behavior of peak-aged Mg–10Gd–3Y–0.4 Zr alloy. Mater. Sci. Eng. A 528:3105–3112.
19. Anyanwu IA, Kamado S, Kojima Y (2001) Creep properties of Mg-Gd-Y-Zr alloys. Mater. Trans. 42:1212–1218.
20. Zhu SM, Easton MA, Abbott TB, Nie JF, Dargusch MS, Hort N, Gibson MA (2015) Evaluation of magnesium die-casting alloys for elevated temperature applications: microstructure, tensile properties, and creep resistance. Metall. Mater. Trans. A 46A:3543–3554.
21. Dong XX, Feng LY, Wang SH, Nyberg EA, Ji S (2021) A new die-cast magnesium alloy for applications at higher elevated temperatures of 200–300° C. J. Magnes. Alloy. 9:90–101.
22. Gavras S, Zhu SM, Nie JF, Gibson MA, Easton MA (2016) On the microstructural factors affecting creep resistance of die-cast Mg–La-rare earth (Nd, Y or Gd) alloys. Mater. Sci. Eng. A 675:65–75.
23. Abaspour S, Cáceres CH (2015) Thermodynamics-based selection and design of creepresistant cast Mg alloys. Metall. Mater. Trans. A 46:5972–5988.
24. Li J, Wu JL, Jin L, Celikin M, Wang FH, Dong S, Dong J (2021) The role of dislocation-solute interactions on the creep behaviour of binary Mg–RE alloys. Sci. Rep. 11:2860.
25. Pettersen G, Westengen H, Hoier R, Lohne O (1996) Microstructure of a pressure die cast magnesium-4wt.% aluminium alloy modified with rare earth additions. Mater. Sci. Eng. A 207:115–120.
26. Easton M, Gibson MA, Zhu SM, Abbott T, Nie JF, Bettles CJ, Savage G (2018) Development of magnesium-rare earth die-casting alloys. Magnes. Technol. 329–336.
27. Zhu SM, Nie JF, Gibson MA, Easton MA, Bakke P (2012) Microstructure and creep behavior of high-pressure die-cast magnesium alloy AE44. Metall. Mater. Trans. A 43A:4137–4144.
28. Razazi M, Amini RN, Eshov BB, Badalov AB (2012) Preparation of physical and chemical and thermodynamic properties of aluminum alloys-cerium. Mat. Sci. Res. India 9:1–7.
29. Moreno IP, Nandy TK, Jones JW, Allison JE, Pollock TM (2003) Microstructural stability and creep of rare-earth containing magnesium alloys. Scr. Mater. 48:1029–1034.
30. Moreno IP, Nandy TK, Jones JW, Allison JE, Pollock TM (2001) Microstructural characterization of a die-cast magnesium-rare earth alloy. Scr. Mater. 45:1423–1429.
31. Amberger D, Eisenlohr P, Göken M (2012) On the importance of a connected hard-phase skeleton for the creep resistance of Mg alloys. Acta Mater. 60:2277–2289.
32. Choudhuri D, Jaeger D, Gibson MA, Banerjee R (2014) Role of Zn in enhancing the creep resistance of Mg–RE alloys. Scr. Mater. 86:32–35.
33. Nie JF, Oh-ishi K, Gao X, Hono K (2008) Solute segregation and precipitation in a creep-resistant Mg–Gd–Zn alloy. Acta Mater. 56:6061–6076.
34. Zhu SM, Abbott TB, Gibson MA, Nie JF, Easton MA (2017) The influence of minor Mn additions on creep resistance of die-cast Mg–Al–RE alloys. Mater. Sci. Eng. A 682:535–541.
35. Nie JF, Wilson NC, Zhu YM, Xu Z (2016) Solute clusters and GP zones in binary Mg–RE alloys. Acta Mater. 106:260–271.
36. Matsuoka Y, Matsuda K, Watanabe K, Nakamura J, Lefebvre W, Nakagawa D, Saikawa S, Ikeno S (2014) Precipitation sequence in the Mg–Gd–Y system investigated by HRTEM and HAADF-STEM. Mater. Trans. 55:1051–1057.
37. Dong XX, Feng LY, Wang SH, Ji G, Addad A, Yang HL, Nyberg EA, Ji S (2022) On the exceptional creep resistance in a die-cast Gd-containing Mg alloy with Al addition. Acta Mater. 232:117957.

Part V

Primary Production and Recycling/Alloy Development

Condensation Behavior of Magnesium in Horizontal Furnace in Argon and Vacuum by Inert Gas Condensation Method

Jibiao Han, Quan Yang, Xianglin Bai, Daxue Fu, Junhua Guo, and Ting'an Zhang

Abstract

The condensation behavior of magnesium vapor in vacuum and argon was studied by inert gas condensation method. The effects of heat source temperature 1273–1473 K on the macro morphology and condensation temperature of magnesium vapor condensation were obtained in vacuum and 0.2m^3/h argon flow. The results show that magnesium vapor condenses in a small area in vacuum, while the existence of argon would prevent the condensation process of magnesium vapor, resulting in the increase of the condensation area. The initial condensation temperature of magnesium vapor carried by argon is 974.0–1159.8 K, which is higher than 781.8–1034.8 K in vacuum; The condensed products with dense combination can be obtained in vacuum, while the condensed products with different particle sizes can be obtained in argon flows. A new idea is provided for the manufacture of magnesium powder and magnesium alloy through the condensation experiment of magnesium by inert gas condensation method.

Keywords

Magnesium vapor • Condensation behavior • Vacuum • Argon flow • Inert gas condensation method

J. Han · Q. Yang (✉) · X. Bai
Sino-Platinum Metals Resources (Yimen) Co. Ltd, Yuxi, 653100, China
e-mail: yq@ipm.com.cn

J. Han · X. Bai
Precious Metal Laboratory Co., Ltd. of Yunnan, Kunming, 650000, China

J. Han · D. Fu · J. Guo · T. Zhang (✉)
School of Metallurgy, Northeastern University, Shenyang, 110819, China
e-mail: zta2000@163.com

Introduction

Magnesium is the lightest structural metal; it is widely used in aerospace, automobile, rail transit, 3C, metallurgical industry and civil fields because of its excellent performance in engineering [1–3]. While the magnesium powder has more active chemical properties, which can be used in hot metal desulfurization, metal reductant, aerospace booster fuel additives, lighting flares, and other aspects, and has high application value. There are many methods to produce metal powder, such as ball milling [4, 5], grinding [6], electrodeposition [7], granulation, and atomization [8, 9]. For metal magnesium powder, it is generally prepared by ball milling, grinding, or atomization. It is common to use grinding method. However, due to the active chemical properties of magnesium, the products obtained by physical and chemical methods of mechanical crushing or smelting are easy to be oxidized, and the purity is not high. Therefore, it is necessary to adopt new methods to prepare metal magnesium powder or magnesium alloy materials, so as to reduce process pollution and improve production effect.

Inert gas condensation (IGC) is one of the main physical methods for preparing clean powders or materials because of its advantages of less preparation procedures, easy collection of powders, and high particle purity. It is applied to the preparation of various types of powders or materials such as metals, alloys, and metal oxides [10–13]. The IGC method promotes the volatilization of raw materials by heating the heat source in the evaporation chamber and fills the condensation chamber with inert gas, so that the metal, alloy, or compound to be prepared can be gasified and then condensed to form powder or material, which can provide a method for the clean preparation and production of magnesium powder.

Therefore, this study was designed to compare the condensation behavior of Mg vapor in 0.2 m^3/h argon (Ar) flow at 1273–1473 K with its behavior under a vacuum, and the condensation law of magnesium vapor in the process of

S. Barela et al. (eds.), *Magnesium Technology 2023*, The Minerals, Metals & Materials Series,
https://doi.org/10.1007/978-3-031-22645-8_30

condensation was obtained. Mg condensation, the condensation versus temperature trends, and microcondensation morphology of pure Mg were obtained and examined. Through Mg condensation experiments under vacuum and with Ar, the condensation conditions for Mg and its condensation behavior were obtained, with the goal of providing a new strategy for the production of Mg powders through these condensation experimental results.

Experiment and Method

Mg ingot (99.9%) was used as the raw material, and volatilization and condensation experiments were carried out under vacuum and Ar flow to observe the condensation behavior of Mg vapor under different conditions. Vacuum conditions were created by vacuum pump and Ar (99.99%) controlled by flowmeter. Experiments were carried out in a horizontal furnace (SK-G03143, Zhonghuan, Tianjing, China) (Fig. 1), a stainless steel tube, and furnace temperature set at 1273, 1323, 1373, 1423, and 1473 K. After reaching temperature, 30 g (± 1 g) of Mg in a graphite crucible was placed in the furnace center and the chamber evacuated less than 100 Pa or a flow of 0.2 m^3/h of Ar introduced to the chamber for 15 min to ensure the complete Mg volatilization. The Mg sample volatilized at the corresponding temperature and Mg vapor moved to the condensation area. Graphite foil (0.2 mm thick) acted as a condenser for Mg vapor. When the temperature was < 373 K, the condenser was removed and Mg condensation products collected. These products were characterized by X-ray diffraction analysis (XRD; D8 Advance, Bruker Corp., Billerica, MA, USA) and scanning electron microscopy (SEM; Zeiss EVO 18, Carl Zeiss AG, Oberkochen, Germany).

Results and Discussion

Description of Experimental Macroscopic Phenomena of Magnesium Vapor Condensation

Results from Mg vapor condensation at 1273–1473 K showed that the volatilization and condensation of metals are closely related to temperature and vapor pressure (Fig. 2). Due to increased particle internal energy, molecular thermal movement of Mg particles allowed them to move from the liquid metal surface into space at the furnace temperature and then moved into the furnace tube. When the tube temperature had reduced to a certain temperature, the Mg vapor pressure was higher than the saturated Mg vapor pressure at this temperature, reaching a supersaturated state. Mg atoms continuously collided with other atoms and the condensation surface. After further energy loss the Mg vapor condensed, the atoms adsorbed onto the condenser surface to form surface aggregations, subsequently nucleating and growing.

Observations of Mg vapor condensation under vacuum conditions showed that condensed products formed in a small area, which decreased with increased furnace temperature (Fig. 2a–e). Condensation occurred in larger areas under 0.2 m^3/h Ar flow, which was different from vacuum results (Fig. 2f–j). These differences were due to the existence of Ar in the Mg/Ar mixed gas, which collided with Mg vapor reduces the collision between Mg and Mg atoms such that the this noncondensable and nonreactive gas weakened the Mg vapor condensation process, an effect that has been reported in other references [14, 15]. Moreover, Ar has a directional flow, which contributed kinetic energy to Mg atomic vapor, such that they moved with the flow before condensing in the condenser, thus producing longer condensation zones. With decreased condensation temperature, there were three forms of condensation products in Ar flow,

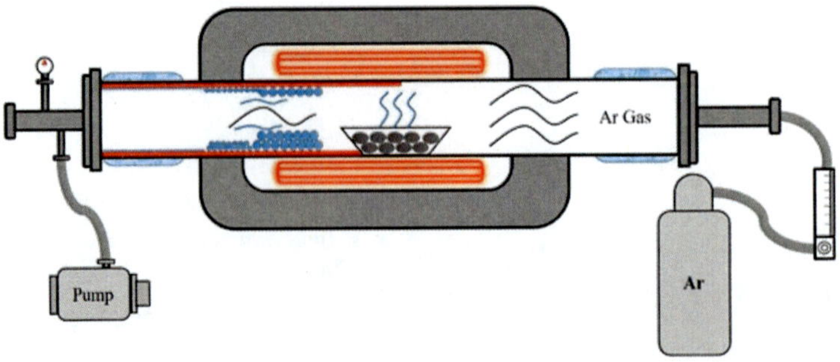

Fig. 1 Schematic diagram of magnesium vapor condensation device

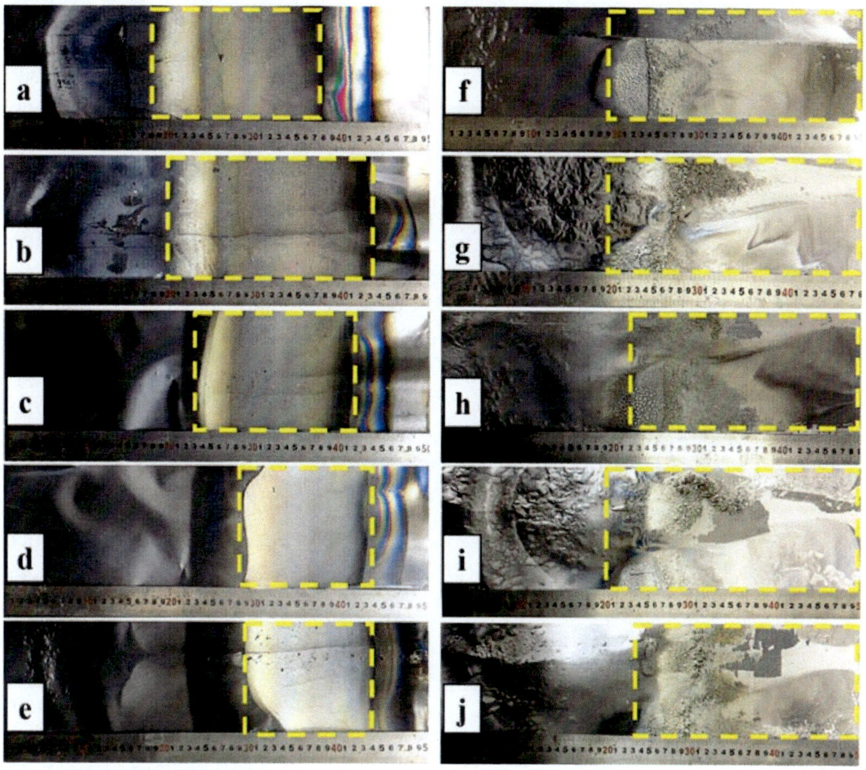

Fig. 2 Macromorphology of Mg vapor condensation under vacuum and Ar. Condensation products in yellow dashed boxes: 1273, 1323, 1373, 1423, and 1473 K with vacuum (**a–e**) and 0.2 m^3/h Ar flow (**f–j**, respectively)

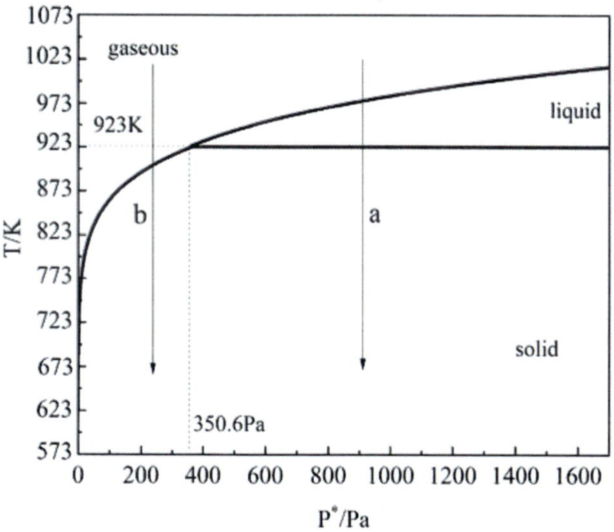

Fig. 3 Gas–liquid–solid phase diagram of Mg

including large spherical, transitional, and powder particles. According to the gas–liquid–solid phase diagram of Mg (Fig. 3), with decreased temperature and vapor pressure, Mg vapor followed different condensation paths and resulted in different macroscopic morphologies [16, 17].

Condensation Temperature of Magnesium Vapor in Vacuum and Argon Flow

The condensation temperature is the main factor affecting Mg vapor condensation. The distribution of condensation temperature at 1273–1473 K in vacuum and 0.2 m^3/h Ar flow showed that, when the condensation end temperature of Mg vapor in vacuum was 520.4–571.3 K, changes in furnace temperature did not affect the end temperature of Mg vapor condensation, which was consistent with the results of Xiong Neng's study on Mg vapor in vacuum (Fig. 4) [18]. However, this was different from Ferguson's results, because in his experiment, vapor flowed vertically upward and this buoyancy plume had a great influence on the experimental temperature gradient, thus affecting condensation temperature [19]. The higher the furnace temperature was, the greater the energy carried by Mg vapor. When volatilizing to the condensation area, a certain residence time was required to fully exchange heat with outside materials before crystallization nucleation and condensation could occur. When the vapor pressure in the tube increased, the vacuum pump affected vapor movement and made it move to the pipe outlet. Therefore, with increased furnace temperature, the initial condensation temperature decreased. The initial condensation temperature of Mg vapor in vacuum was 781.8–1034.8 K.

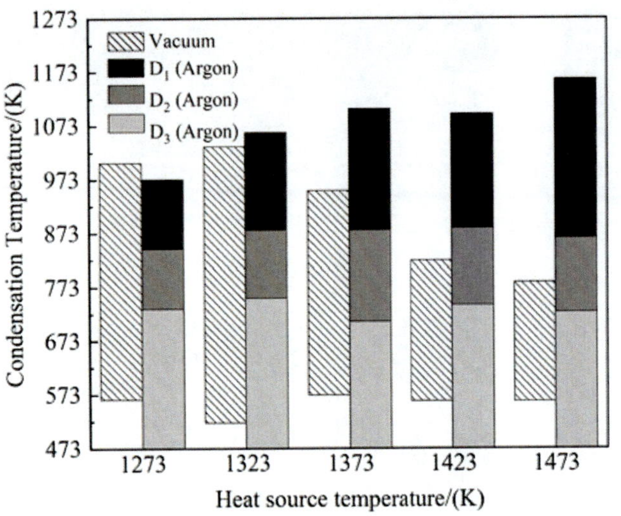

Fig. 4 Condensation temperature distribution in vacuum and Ar. Condensation range of large spherical, transitional, and powder particles, D_1–D_3, respectively

When the furnace temperature increased, Mg atoms volatilized more rapidly and were carried by Ar gas to the cooler area to condense. This resulted in higher Mg vapor pressure and thus higher partial pressure in the Ar/Mg mixed gas, increasing the number of collisions between atoms and thus reducing the critical nucleation radius of Mg vapor. This situation was conducive to the aggregation and growth of Mg particles, such that the initial condensation temperature increased. The initial condensation temperature of Mg vapor in 0.2 m^3/h Ar at 1273–1473 K was 974.0–1159.8 K, which was different from that under vacuum conditions. According to the three-phase diagram of Mg, the reason for these results was that the Mg vapor partial pressure under Ar flow was greater than under vacuum and, therefore, when Mg vapor was supersaturated, it condensed at higher temperature. The end temperature could not be determined due to the presence of Mg in the furnace outlet under Ar flow. The observed results were due to gas–liquid–solid and gas–solid phase transitions. With the melting point of Mg at 923 K, when the temperature of the condensation area was lower than the melting point, the Mg vapor pressure was still in a supersaturated state. At this time, the gas–solid phase transition of Mg vapor occurred and liquid phase particles did not aggregate and grow, directly becoming powder particles.

SEM and XRD of Condensation Products of Magnesium Vapor

SEM patterns of condensation products in vacuum and 0.2 m^3/h Ar flow at a heat source temperature of 1273–1473 K (Fig. 4). Products of Mg vapor condensation under vacuum showed the phenomenon of sheet growth, with metal blocks bonded with each other and condensed into metal Mg sheets on graphite paper in the temperature range of 1273–1473 K (Fig. 4a). This was consistent with Yang's SEM results of Mg vapor condensation under vacuum [17]. However, results were different between condensation under Ar and vacuum. The condensation products in D_1 area were found to be spherical, which was due to the vapor–liquid–solid phase change during the condensation process and the existence of liquid surface tension, such that spherical droplets with small surface area were obtained (Fig. 5b). SEM analysis of D_1, D_2, and D_3 areas at 1273, 1373, and 1473 K showed that the particle sizes of D_2 and D_3 were significantly reduced. In D_2 area, there were transitional particles with different sizes. Powder particles were found to have hexagonal structure, which indicated that Mg vapor easily formed hexagonal structures when the temperature was lower than the metal's melting point in an Ar flow. The results showed the Mg particle in D_1 area had particle size of 0.15–1.04 mm, Mg particle in D_2 area had particle size of 2–35 μm, and Mg particle in D_3 area had particle size of 0.5–8 μm.

XRD patterns of condensation products showed that heat source temperature had an increased preference for the (102) and (103) crystal orientation but inhibited the preference for the (110) orientation under vacuum (Fig. 6). However, Mg vapor condensation under Ar was beneficial for the (102) orientation. The reason for this difference between vacuum and Ar was that the Mg vapor condensation process under Ar had a single flow effect. For condensation products in D_1, D_2, and D_3 condensation zones at 1373 K, XRD results were basically consistent and the results showed that product purity was high and not easily oxidized.

Conclusion

Mg vapor condensation products carried by Ar were different from that under vacuum, with the former being large spherical, transitional, and powder particles. The initial condensation temperature of Mg vapor at 0.2 m^3/h was 974.0–1159.8 K, which was higher than 781.8–1034.8 K under vacuum, was caused by different vapor pressures.

SEM results showed that Mg vapor condensation products in vacuum were blocky, but also different under Ar. In the higher temperature region, the product was spherical due to the influence of surface tension. When the temperature was lower than the melting point, particle size was decreased and with hexagonal structure.

Due to the influence of single direction Ar flow, Mg vapor was encouraged to the growth of (102) crystal faces.

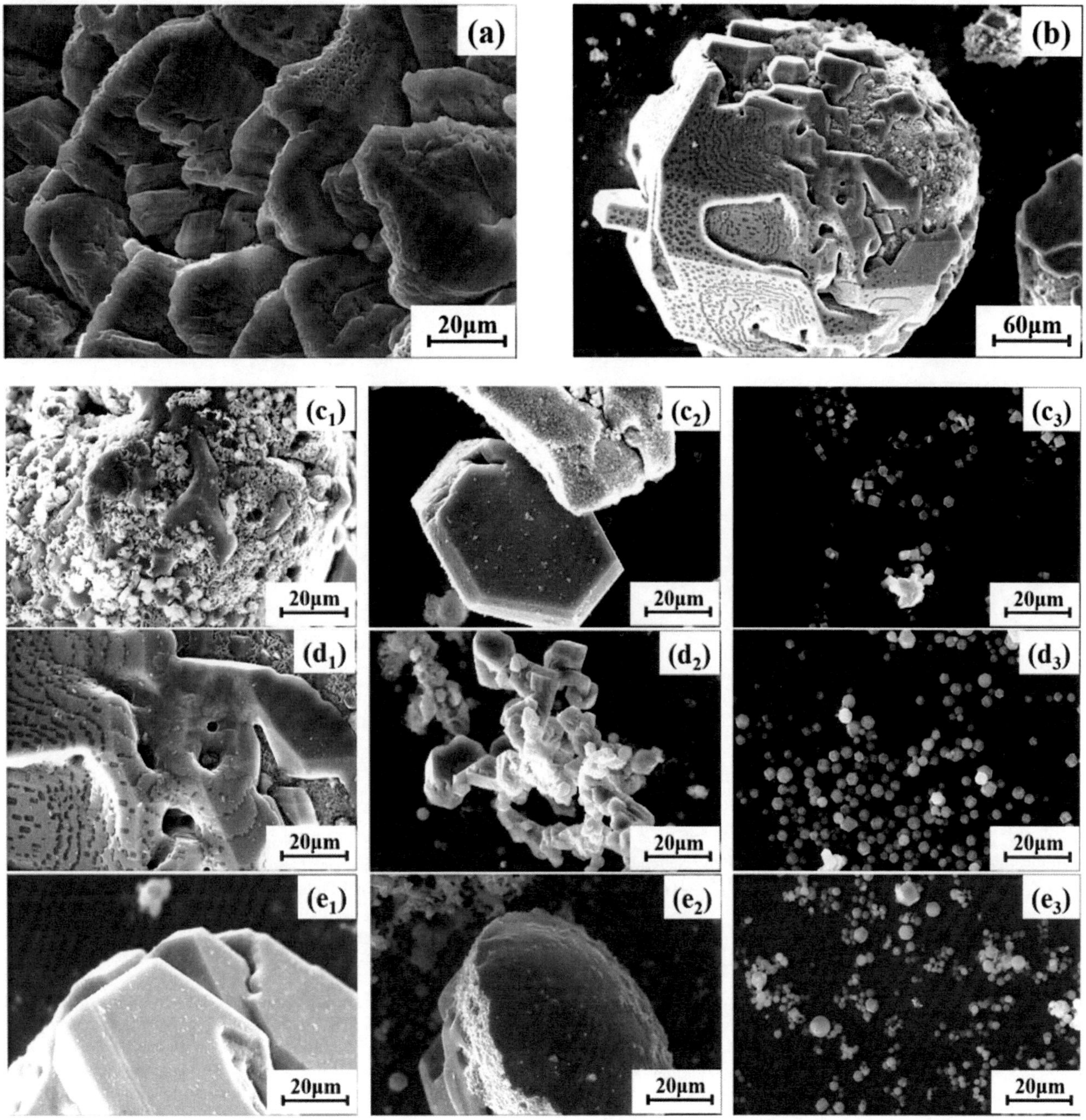

Fig. 5 SEM analysis of condensation products from Mg vapor. Here, 1473 K and vacuum (**a**), 0.2 m^3/h Ar flow at 1373 K (**b**), D_1, D_2, and D_3 in 0.2m^3/h Ar flow at 1273 (**c_1–c_3**), 1373 (**d_1–d_3**), and 1473 K (**e_1–e_3**, respectively)

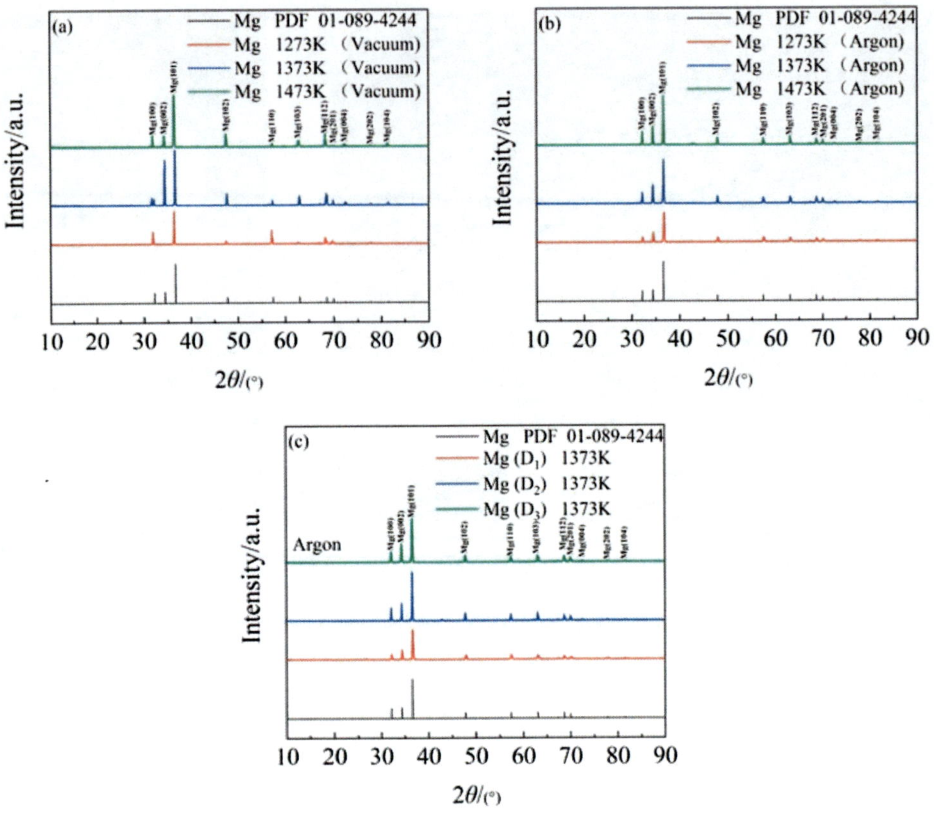

Fig. 6 XRD patterns of condensation products under vacuum and Ar. Vacuum (**a**), Ar flow of 0.2 m^3/h (**b**), and condensation products in different condensation areas under 0.2 m^3/h Ar flow at 1373 K (**c**)

High purity crystalline Mg was obtained in both vacuum and Ar flow conditions.

Acknowledgements This work was supported by National Natural Science Foundation of China (51504058; U1508217; 51404054; 51374064).

References

1. Hirsch J, Al-Samman T (2013) Superior light metals by texture engineering: Optimized aluminum and magnesium alloys for automotive applications. Acta Materialia 61(3): 818–843.
2. Froes F, Eliezer D, Aghion E (1998) The science, technology, and applications of magnesium. JOM 50(9): 30–34.
3. Fu D, Ji Z, Guo J, et al. (2020) Diffusion and phase transformations during the reaction between ferrosilicon and CaO center dot MgO under vacuum. Journal of Materials Research and Technology-Jmr&T 9(3): 4379–4385.
4. Alizadeh M, Sharifian-Jazi F, Haghshenas-Jazi E, et al. (2015) Production of Nanosized Boron Oxide Powder by High-Energy Ball Milling. Synthesis and Reactivity in Inorganic 45(1): 11–14.
5. Enayati M, Aryanpour G, Ebnonnasir A (2009) Production of nanostructured WC–Co powder by ball milling. International Journal of Refractory Metals and Hard Materials 27(1): 159–163.
6. Gan L, Xiao Z, Wang A, et al. (2021) Efficient preparation of ultrafine powder from waste cellulose by physicochemical method. Powder Technology 379: 478–484.
7. Xu J, Yu G, Hu B, et al. (2014) Preparation of copper coated tungsten powders by intermittent electrodeposition. Powder Technology 264: 561–569.
8. Jayasinghe S, Dorey R, Edirisinghe M, et al. (2005) Preparation of lead zirconate titanate nano-powder by electrohydrodynamic atomization. Applied Physics A 80(4): 723–725.
9. Liu T, Sakurai O, Mizutani N, et al. (1986) Preparation of spherical fine ZnO particles by the spray pyrolysis method using ultrasonic atomization techniques. Journal of Materials Science 21 (10): 3698–3702.
10. Ahn J, Kim J, Park J, Huh M (2004) Microstructure and gas-sensing properties of thick film sensor using nanophase SnO2 powder. Sensors and Actuators B-Chemical 99(1): 18–24.
11. Haas V, Birringer R, Gleiter H (1998) Preparation and characterisation of compacts from nanostructured powder produced in an aerosol flow condenser. Materials Science and Engineering a-Structural Materials Properties Microstructure and Processing 246(1–2): 86–92.
12. Huh M, Kim J, Ahn J (2002) IGC synthesis of tin nanophase particles Part I: Study on convection gas currents in IGC chamber. Powder Metallurgy 45(2): 154–159.
13. Gleiter H (1989) Nanocrystalline Materials, Springer Press, Berlin.
14. Hu H, Tang G, Niu D (2015) Experimental investigation of condensation heat transfer on hybrid wettability finned tube with large amount of noncondensable gas. International Journal of Heat and Mass Transfer 85: 513–523.
15. Yi Q, Tian M, Yan W, et al. (2016) Visualization study of the influence of non-condensable gas on steam condensation heat transfer. Applied Thermal Engineering 106: 13–21.
16. Hischier I, Chubukov B, Wallace M, et al. (2016) A novel experimental method to study metal vapor condensation/oxidation:

Mg in CO and CO2 at reduced pressures. Solar Energy 139: 389–397.
17. Yang C, Tian Y, Qu T, et al. (2014) Magnesium vapor nucleation in phase transitions and condensation under vacuum conditions. Transactions of Nonferrous Metals Society of China 24(2): 561–569.
18. Xiong N, Tian Y, Yang B, et al. (2018) Volatilization and condensation behaviours of Mg under vacuum. 156: 463–468.
19. Ferguson F, Nuth J, Lilleleht L (1996) Experimental studies of the vapor phase nucleation of refractory compounds .4. The condensation of magnesium. Journal of Chemical Physics 104(9): 3205–3210.

Design of the Continuous Gravity-Driven Multiple-Effect Thermal System (G-METS) for Efficient Low-Cost Magnesium Recycling

Daniel McArthur Sehar, Gabriel Espinosa, Armaghan Ehsani Telgerafchi, Chinenye Chinwego, Keira Lynch, Benjamin Perrin, and Adam Powell

Abstract

Recent global supply shortages, combined with need for light-weight structural components for efficient transportation, have led to very high prices for magnesium due to its low density and high part stiffness/weight. This study continues prior work on magnesium recycling using multiple-effect distillation with gravity as the parameter to create the pressure difference between effects, called G-METS. The long-term vision is for continuous distillation, but experiments to date have only used batch distillers with maximum rate of 1 kg/h. and minimum energy consumption of 2 kWh/kg. This study presents the design of laboratory-scale continuous distiller including a custom-made furnace to sustain a high-power density and temperature gradient for rapid boiling and efficient recovery of magnesium.

Keywords

Magnesium • Distillation • Recycling • Secondary recovery • Extraction • Processing

Introduction

With the significant demand for magnesium metal, there are certain obstacles to the prevalent usage of magnesium. Based on an increase in the global population, industrialization and digitalization headed the pressure on the raw material resources drastically in current trends. According to a report in the EU list of critical raw materials (2020), OECD forecasts that the demand for these critical materials will double from 79 billion tons in 2020 to 167 billion tons by the end of 2060. It has been said that the critical raw materials' extent of dependence will be substituted with today's needful reliance on oil [1]. The latest 2022 U.S Geographical Survey (USGS) and EU reports on the list of critical materials represent the economic importance of magnesium metal along with the high supply risk [2, 3] (Fig. 1).

It is a well-known fact that China is the monopoly epicenter for the magnesium market in the supply chain, around 87% of the world's magnesium market. However, due to the uncertain circumstances of COVID-19, the production of magnesium in China has reduced significantly and thereby resulted in very high of magnesium metal 5–7 times higher than when compared to the earlier period [4].

We have presented the concept of Gravity-Driven Multiple-Effect Thermal System (G-METS) in Magnesium Technology symposium at the TMS annual meeting in 2021 and 2022 [5, 6]. This paper provides the information of partial results of the one-effect distillation experiments. We presented five batch distillation experiments with the similar design described in the previous published paper on the TMS magnesium symposium on 2022 [6]. Currently, we are working on designing the custom-made bottom and side heating elements for enhancing the uniform temperature gradient between the evaporator and condenser of the distiller. In this paper, we have discussed a detailed proposed plan to design manufacturing of semi-continuous one-effect

D. McArthur Sehar · C. Chinwego · A. Powell (✉)
Materials Science and Engineering Program, Department of Mechanical Engineering, Worcester Polytechnic Institute, Worcester, MA, USA
e-mail: acpowell@wpi.edu

G. Espinosa · A. E. Telgerafchi · K. Lynch · B. Perrin
Department of Mechanical Engineering, Worcester Polytechnic Institute, Worcester, MA, USA

A. Powell
100 Institute Rd, Worcester, MA 01609, USA

S. Barela et al. (eds.), *Magnesium Technology 2023*, The Minerals, Metals & Materials Series,
https://doi.org/10.1007/978-3-031-22645-8_31

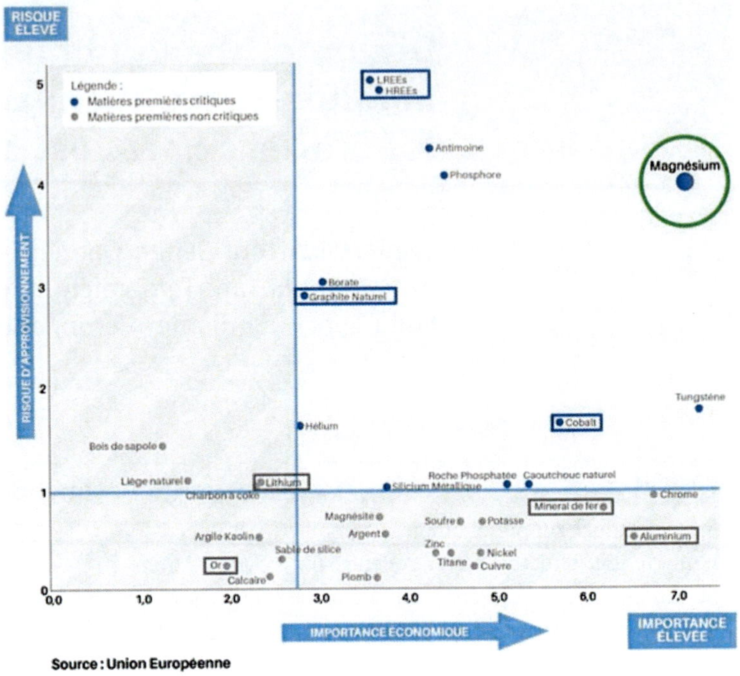

Fig. 1 European Union critical materials assessment including magnesium [3]

distiller and custom furnace. Also, we have included some of the same figures presented in the prior paper as a reference set of figures for this paper [5–8].

G-METS (Gravity Multiple-Effect Thermal System) Magnesium Distiller

Magnesium has played a prominent role in the commercial field from the manufacturing of automobile vehicles to defense artillery and ground-terrain vehicles. For economic and environmental sustainability, there is an essential need the recycling magnesium scraps for the better welfare of the environment. There are several refining methods have been done based on either flux or flux-less refining of the accumulated magnesium scraps. However, vacuum distillation methodology under flux-less refining is a very much efficient process in extracting magnesium from scrap materials at remarkably high purity.

Vapor Compression Distillation (VCD) can significantly reduce energy consumption, the capital cost of magnesium distillation, and other operating costs. VCD provides the enthalpy of vaporization to the evaporator. The compressor in the system raises the pressure of the condenser above that of the evaporator. The boiling point of the condenser is significantly higher than the evaporator, thereby providing the hotter heat flow from the condenser to the cooler evaporator. This method helps in refining the distilled magnesium to about 90% when compared to batch distillation. G-METS uses a similar principle with gravity as the factor to create the pressure difference between the evaporator and condenser. G-METS has no internal moving element and making it more energy efficient when compared to VCD with multiple moving systems of varying pressure differences. G-METS distiller will likely use well below 1 kW/kg of magnesium element and produces an extraordinary record during the continuous distillation process, thereby substantially reducing the capital cost. This distiller can potentially play a prominent role in recycling magnesium alloys from rare-earth magnets [9] to the primary production of magnesium elements [10]. The prominent role of the Mg distillation can lower the concentrations of impurity than the conditions of the low-corrosion alloys like AZ91E [11]. The process of distillation facilitates the direct reduction of the magnesium oxide (MgO) in the Hall-Heroult cell with a dense reactive liquid cathode (i.e., tin metal) to produce an alloy. From then on, the pure Mg can be separated from the formed alloy using G-METS and tin restored to the host metal in the electrolysis cell [10, 12–16] (Fig. 2).

The description of the G-METS multiple-effect magnesium distiller design is elaborated briefly in the previous published article at the TMS magnesium symposium [5, 6]. This design method of multiple-effect distillation is similar to the concept of the multiple-effect distillation apparatus patented by Charles W. Lotz in 1970, initiated the design to covert the saline water to fresh water, by evaporating the fresh water and leaving the saline or distilland salts in the evaporating surface [17]. Similarly, water distillation multiple-effect evaporation system was patented by John P. Petrek in the year 1992 [18]. It uses the similar idea

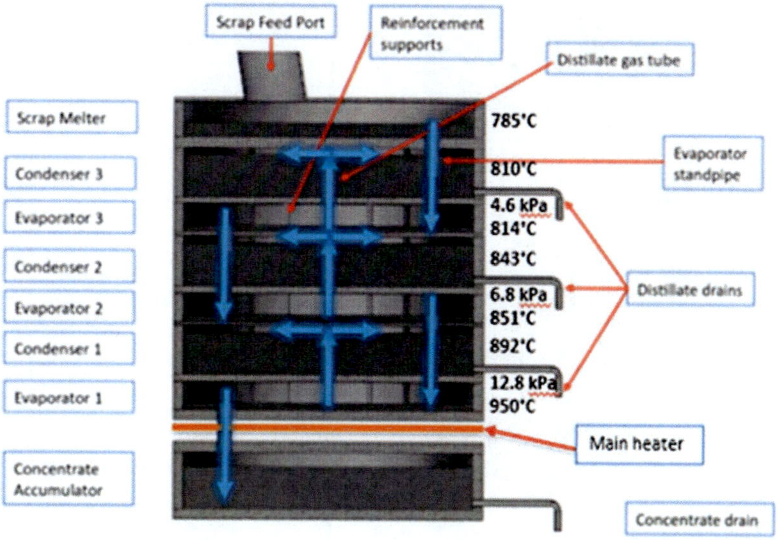

Fig. 2 G-METS three-effect magnesium distiller preliminary design, showing temperatures and pressures predicted by a reduced-order model [5, 7]

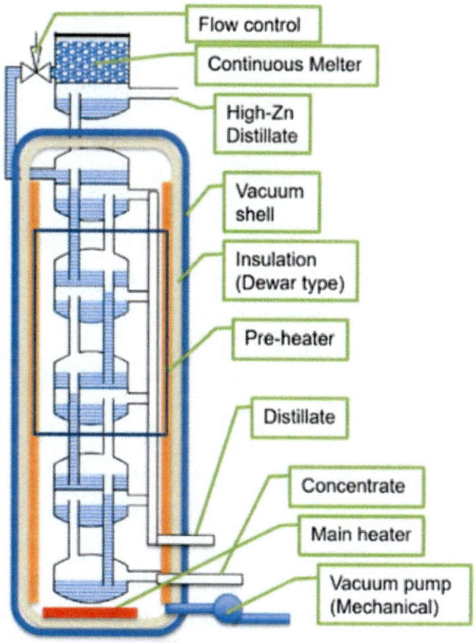

Fig. 3 Schematic diagram of four-effect G-METS distiller

of vertical drainpipes between the two-effects to procure the desired pressure difference between the evaporator-condenser pair (Fig. 3).

Experiments

One-Effect Distiller

As mentioned prior, each effect consists of a condenser and evaporator combined with the standpipe between them. The term "one-effect" distiller means the interaction between the condenser-evaporator system. We conducted five rounds of experiments with the dimensions and specifications as elaborated previously [5]. Initially, for the first three experiments, we faced several challenges as in the due to improper vacuum conservation in the distiller led to the formation of magnesium nitride—leading to the formation of ammonia and thereby inhibiting the evaporation process. Also, even though the temperature we maintained furnace temperature to about 1100 °C, there was an insufficient temperature gradient to distill magnesium. Therefore, there were some changes made in the design parameter as we continued to conduct the batch distiller experiments. In the 4 and 5th experiments, we increased the diameter of the distiller plates from 7 to 10 inches and decreased the pipe height by 1.75 inches in the condenser and 3 inches in the evaporator. In this newer design, we believe that the heat transfer is enhanced better than in the previous model to maintain a proper temperature gradient for the magnesium to evaporate and condense at the condenser. However, we are still facing some trials in accomplishing the magnesium to distill in the condenser (Fig. 4).

Two-Effect Distiller

The next planned experiment is to conduct a two-effect distiller as shown in Fig. 5. Wherein it consists of each two effects (condenser-evaporator) vertically arranged in the successive pattern of the distiller. The working is very much similar to the one-effect distiller. In this working process, the condensation heat provides a significant amount of heat for the evaporation process. This principle can be achieved by

Fig. 4 One-effect distiller—AE41 Mg-Al-Nd alloy distillation

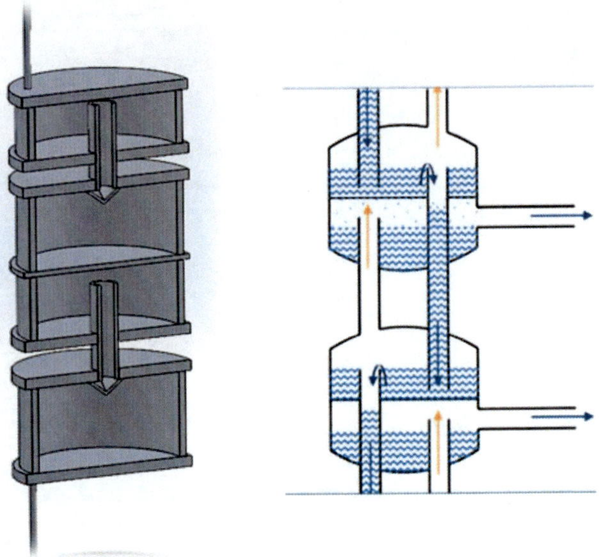

Fig. 5 Two-effect distiller 3-D CAD model (left); detailed working of G-METS magnesium distillation (right) [5, 6]

building higher pressure in the condenser than in the evaporator. Wherein the boiling point of the condenser is considerably higher than the evaporator such that heat flows effortlessly from the higher condenser to the lower evaporator temperature. Another interesting aspect is the standpipe wherein the liquid magnesium weight creates the pressure difference, as the weight of the liquid metal in the standpipe column causes to enhance the pressure in the evaporating chamber which is successively arranged below one another. Thereby the condenser connected to the lower part of the evaporator also raises pressure significantly (Fig. 5).

Semi-Continuous One-Effect Distiller

The design of this model is quite different with the same setup as of a one-effect distiller but has a new part called "melter" with the walls of the same 10" nominal schedule 40 pipes of 0.5" thick A572 mild carbon steel. This part is attached right above the condenser. This newer part consists of a scheduled pipe diameter of 6" inside the 10" pipe and consists of a long melter standpipe connecting the evaporator and the melter chamber as seen in Fig. 6. The main purpose of the standpipe is to maintain the pressure flow of the liquid magnesium into the evaporator chamber. To avoid the moving parts inside the system, we use argon (Ar) gas as a piston or pushing agent to compress the adequate amount of magnesium liquid into the evaporator via a standpipe. The pushing action of the magnesium liquid in the melter by the argon gas is done by actuating the valves at intermittent time intervals. There are two holes proposed in the top plate of the melter. The central hole into the 6" pipe embedded inside the 10" pipe is the mainstream flow for argon gas and vacuum line. The other hole vacuumed is to depressurize in the area inside the area between the 6" and 10" scheduled pipe to avoid the leakage of argon gas from the 6" pipe chamber into the standpipe, which can disrupt maintaining the pressure flow of the magnesium into the evaporator. The two streamlines of the argon gas and the vacuum line are controlled via a valve as shown in Fig. 6.

Furnace

The furnace is a custom build with a bottom and side heater to alter the heat supply accordingly to sustain a proper temperature gradient for distillation to take place. The standard side heaters and the insulations are obtained from Thermcraft company. However, the bottom heater is custom built using insulating firebrick, ceramic pegs, and Cr-Al-Fe coils from Kanthal wires wound around the pegs. The main purpose of building the furnace is to create a sufficiently high-power density for the bottom heater. Also, the idea of this furnace is to create a proper temperature gradient for magnesium to distill into the condenser from the evaporator. Initially, the side and bottom heaters are started together for the solid magnesium to melt into liquid magnesium completely. Once the magnesium is melted to liquid, only the bottom heater is operated continuously maintain continuous distillation of magnesium. The whole furnace is covered with insulation to minimize heat loss for better efficiency of the distillation process (Fig. 7).

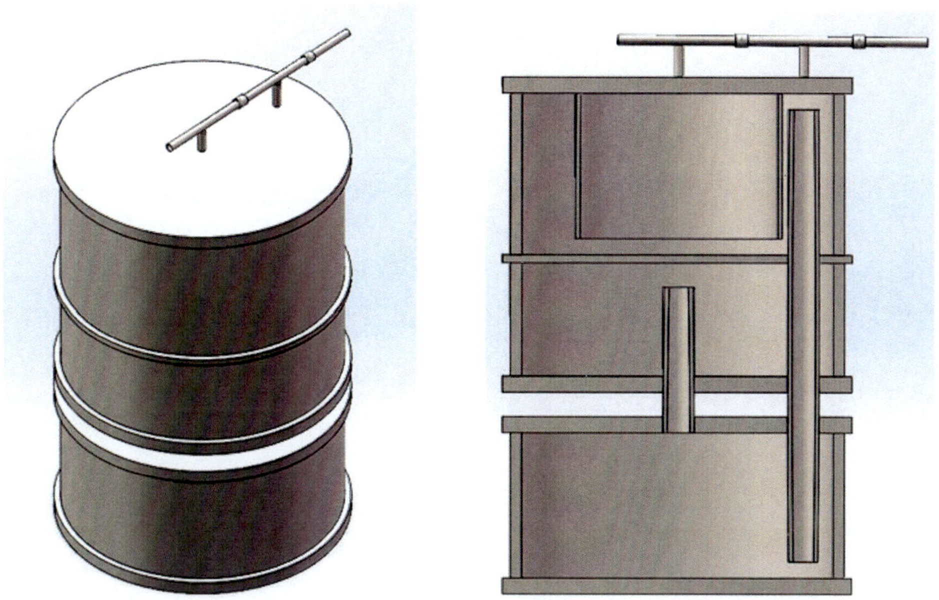

Fig. 6 Solid CAD image of one-effect continuous distiller, two valves for vacuum and argon gas inlet each—full view (left); sectional view (right)

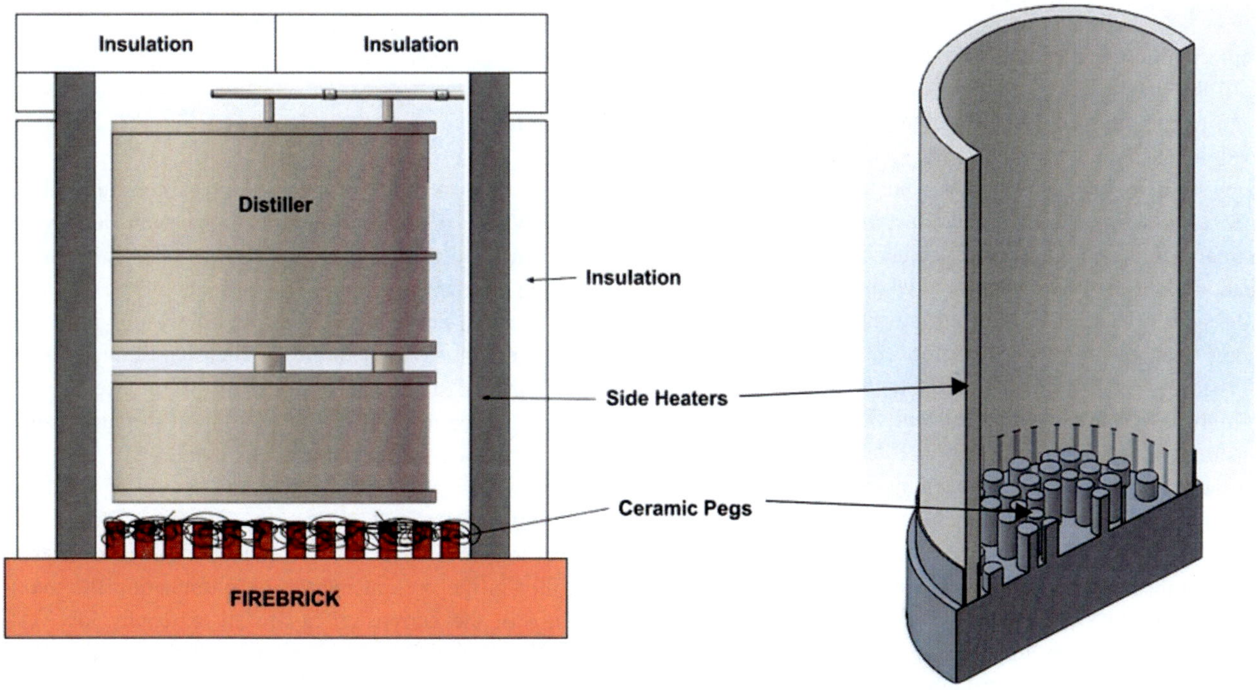

Fig. 7 Schematic illustration of side and bottom heater furnace with insulation (left); sectional solid CAD image of the furnace (right)

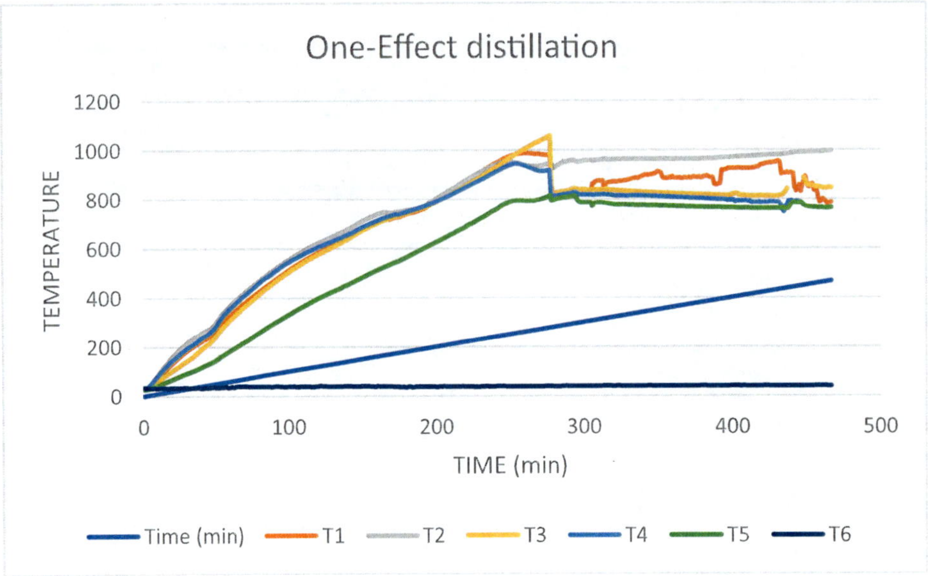

Fig. 8 Temperature–time graph for one-effect distillation experiment

Results

From conducting the one-effect distillation experiments to distill the magnesium from the evaporator to the condenser at the temperature of 1100 °C, there were some challenges faced in each experiment for the complete distillation of magnesium to happen. Some of the major challenges we faced are holding the inert vacuum environment inside the distiller while crimping process, wherein the presence of a small amount of air can inhibit the whole distillation process and lead to the formation of strong pungent smelling ammonia (magnesium nitride). Moreover, the main reason for the issues in the experiments is due to the insufficient supply of high-power density. Inconsistent fluctuations of power densities from the existing furnace led to failures in the distillation experiments. However, in the 5th experiment of the one-effect distillation, which was conducted for about 8 h at 1150 °C. There were some accumulations of the magnesium metal deposition clustered on the top of the standpipe in the condenser section. The reason we believe this situation is due to clogging of magnesium on the standpipe due to insufficient temperature gradient. We have recorded the temperature–time graph of each top and bottom plate of the condenser and evaporator with the help of thermocouples from the data logger. The recorded temperature–time graph of the fifth experiment is shown in Fig. 8.

Conclusion

Therefore, the concluding analysis of the successful working of the G-METS distillation requires high-power density for the efficient distill of magnesium, as was used in Infinium distillation experiments [5, 6]. The next step of the plan is to build a custom-made furnace as described previously to maintain a perfect temperature gradient in future experiments. With the successful building of the furnace, we can expect a positive result in the distillation of the magnesium in all one, two-effect, and continuous distillers.

If all succeeds, detailed thermal models will be validated for all the effects and continuous distiller. This would enhance the growth of the G-METS distiller for a great possibility in the present market by plummeting the maximum use of the current energy to an extremely low 0.5–1 kWh/Kg. Especially, the continuous distiller will open a new door for removing the barrier and will enable a low-energy efficient recycling process, mainly for the low-grade challenging post-consumer scraps such as Cu and Ni contaminated materials. This G-METS methodology will then be able to distill high-purity magnesium from low-grade scraps.

References

1. Study on the EU's list of Critical Raw Materials (2020) Final Report; September 2020; https://doi.org/10.2873/11619.
2. USGS- Federal Register, https://www.govinfo.gov/content/pkg/FR-2022-02-24/pdf/2022-04027.pdf.
3. Alliance Magnesium- Critical materials, https://alliancemagnesium.com/en/magnesium/critical-materials/.
4. Magnesium supply shortage: cross-industry coalition and workers warn against catastrophic impact, https://www.acea.auto/news/magnesium-supply-shortage-cross-industry-coalition-and-workers-warn-against-catastrophic-impact/.
5. E. Telgerafchi, G. Espinosa, M. Rutherford, A. Powell, and D. Dussault, "Efficient Low-Cost Gravity-Driven Multiple Effect Thermal System (G-METS) Distillation of Magnesium," in Magnesium Technology 2021, Mar. 2021, pp. 145–152. https://doi.org/10.1007/978-3-030-65528-0_22.
6. Design of Efficient Low-Cost Recycling of Magnesium Using Gravity-Driven Multiple Effect Thermal System (G-METS), https://doi.org/10.1007/978-3-030-92533-8_22, The Minerals, Metals & Materials Society 2022.
7. A.C. Powell, "A Magnesium Clean Energy Ecosystem Vision," in Magnesium Technology 2022.
8. C. Powell, D. M. Dussault, M. R. Earlam, A. Tajima, and C. Raymes, "Method and Apparatus for Efficient Metal Distillation and Related Primary Production Process," U.S. Patent Application 16/944,147 [Online]. https://patents.google.com/patent/US20210040633A1/en.
9. Chinenye Chinwego, Hunter Wagner, Emily Giancola, Jonathan Jironvil, and Adam Powell "technoeconomic analysis of rare-earth metal recycling using efficient metal distillation"; JOM, vol. 74, no. 4, 2022 https://doi.org/10.1007/s11837-021-05045-7
10. M. Rutherford, A. E. Telgerafchi, G. Espinosa, A. C. Powell, and D. Dussault, "Low-Cost Magnesium Primary Production Using Gravity-Driven Multiple Effect Thermal System (G-METS) Distillation," in Magnesium 2021, Orlando, Florida, Mar. 2021, pp. 139–144. https://doi.org/10.1007/978-3-030-65528-0_21.
11. Ditze and C. Scharf, Recycling of Magnesium. 2008.
12. L. A. Yerkes, "Electrolytic Method for Producing Magnesium," US 2,342,723, Dec. 02, 1947 Accessed: Feb. 20, 2020. [Online]. https://patents.google.com/patent/US2431723A/en.
13. J. Kang, J.-Y. Lee, and T.-H. Lee, "Metal refining method by using liquid metal cathode," KR 10–2004902, Jul. 29, 2019 Accessed: Aug. 26, 2020. [Online]. https://patents.google.com/patent/KR102004920B1/en.
14. T.-H. Lee, T. H. Okabe, J.-Y. Lee, Y. M. Kim, and J. Kang, "Molten Salt Electrolysis of Magnesium Oxide Using a Liquid–Metal Cathode for the Production of Magnesium Metal," Metall. Mater. Trans. B, vol. 51, no. 6, pp. 2993–3006, Dec. 2020, https://doi.org/10.1007/s11663-020-01976-9
15. T.-H. Lee, T. H. Okabe, J.-Y. Lee, Y. M. Kim, and J. Kang, "Development of a novel electrolytic process for producing high-purity magnesium metal from magnesium oxide using a liquid tin cathode," J. Magnes. Alloys, p. S2213956721000220, Feb. 2021, https://doi.org/10.1016/j.jma.2021.01.004.
16. D.-H. Lee et al., "Scale-Up Study of Molten Salt Electrolysis using Cu or Ag Cathode and Vacuum Distillation for the Production of High-Purity Mg Metal from MgO," J. Sustain. Metall., vol. 7, no. 3, pp. 883–897, Sep. 2021, https://doi.org/10.1007/s40831-021-00367-x
17. Charles W. Lotz, "Multiple Effect Distillation Apparatus", US 3,356,591; patented on 1970 [Online]. https://patents.google.com/patent/US3536591A/en?inventor=Charles+W+Lotz
18. J. P. Petrek, "Multiple-Effect Evaporation System and Process," US 5,094,721, Mar. 10, 1992 Accessed: Jul. 28, 2020. [Online]. https://patents.google.com/patent/US5094721A/en.

Development of Compound Vertical Retort Technology for Magnesium Production and Its Application

Fengqin Liu, Shaojun Zhang, Rongbin Li, Michael Ren, Peixu Yang, Jinhui Liu, and Zegang Wu

Abstract

This paper introduces a new silicothermic process, compound vertical retort technology for magnesium production. The workflow and key equipment and technologies were introduced. With the new technology, a demonstration unit was established in China. For the demonstration unit, the new equipment of lined steel retort was developed to solve the problem of "adhesion and glaze," realizing the mechanical releasing slag; the new compound magnesium crystallizer with radiation heating surface and with function of sectional crystallization was developed to obtain magnesium metal of higher purity 99.8%; the new dual-regenerative furnace with combustion mode of double-deck, double-direction, hedging, suction, and reducing atmospheric firing was developed to realize temperature uniformity, and to further improve the combustion efficiency and heat efficiency. According to the actual production, these new technology's economic and technical indices meet or even exceed the design expectations, significantly better than the Pidgeon process.

Keywords

Silicothermic process • Vertical retort • Magnesium

F. Liu · R. Li · Z. Wu
School of Metallurgical and Ecological Engineering, University of Science and Technology Beijing, Beijing, 100083, China

S. Zhang (✉) · P. Yang · J. Liu
Henan Province Industrial Technology Research Institute of Resources and Materials, Zhengzhou University, Zhengzhou, 450001, China
e-mail: zhangs@zzu.edu.cn

M. Ren
Sunlightmetal Consulting Inc., Beijing, China
e-mail: michael.ren@sunlightmetal.ca

Introduction

The global prominence of climate problems has prompted the widespread application of lightweight materials in the fields of automobiles, aerospace, and electronic products [1, 2]. As the lightest structural materials, magnesium and magnesium alloys are increasingly favored. The amount of magnesium and magnesium alloys is increasing year by year, but the growth of demand is limited due to the production of primary magnesium. In fact, due to the lack of major breakthroughs in magnesium production technology, the development of the magnesium industry has slowed down in recent years.

At present, there are two kinds of magnesium production technology used in industrial production in the world: silicothermic reduction process and electrolysis process. The United States, the United Kingdom, Australia, Israel, CIS countries and other countries use the electrolysis process, France and Brazil use slag conductive semi-continuous silicothermic process, and Canada used the electrolysis method before and has now changed to the horizontal retort silicothermic method (Pidgeon method). South Korea introduced the Chinese vertical retort silicothermic process, China mainly uses horizontal retort silicothermic (Pidgeon process) [3]. Because the energy consumption and cost of electrolytic magnesium production greatly exceed that of silicothermic process, the global annual output of electrolytic method is less than 100,000 tons (see Fig. 1), and most of them have been stopped or semi-suspended. In recent years, the United States [4], Australia [5], and other countries have been vigorously researching silicothermic process.

Due to the abundant mineral resources such as magnesite and dolomite and cheap labor, China has vigorously developed the Pidgeon process since 2000 and gradually promoted its application. Up to now, the annual output of magnesium metal is about 800,000 tons, accounting for 80% of the world's magnesium production above. However, for a long time, the Pidgeon process has problems such as low

S. Barela et al. (eds.), *Magnesium Technology 2023*, The Minerals, Metals & Materials Series,
https://doi.org/10.1007/978-3-031-22645-8_32

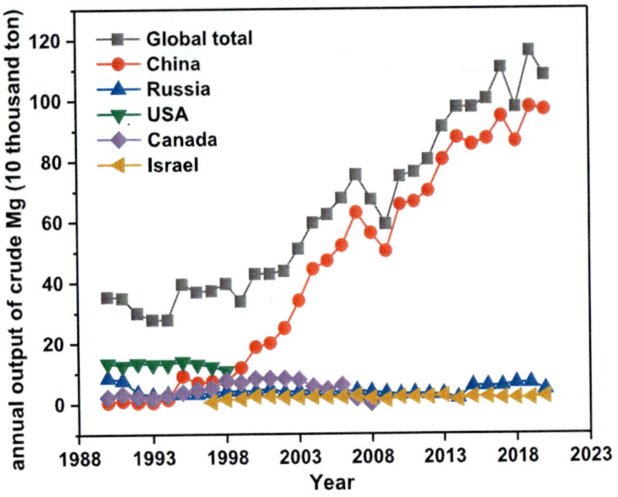

Fig. 1 Primary magnesium output yearly

production efficiency, high energy consumption, high carbon emissions, and high labor intensity; especially, the production method of horizontal retort is difficult to achieve mechanization. Vertical Retort magnesium production technology has been developed in recent years [6].

This paper introduces a new type of silicothermic magnesium production technology, namely the compound vertical retort magnesium production technology. The main technical process, key technology research and development, construction and operation of demonstration plant of the compound vertical retort magnesium production technology are mainly introduced. And through the comparison and analysis of the technical and economic indicators of the compound-vertical process with the Pidgeon process and Magnetherm process, the superiority and advancement of the compound vertical retort magnesium production technology are discussed.

The Compound Vertical Retort Technology for Magnesium Production

Working Flow

By integrating the self-developed compound vertical retort, compound crystallizer, compound reduction furnace, reduction slag waste heat recovery device, and supporting auxiliary systems, the compound vertical retort magnesium production technology process has been formed. The main process flow of the double vertical retort magnesium production technology is shown in Fig. 2, which is mainly composed of raw material production/transportation system, magnesium reduction production system, magnesium/slag product transportation system, and supporting auxiliary systems. The magnesium reduction production system is the core part of the magnesium production process. It consists of the upper compound magnesium crystallizer and mechanized feeding device, the middle regenerative compound reduction furnace body (including the compound vertical retort), the lower mechanized slag removal device and the high-temperature slag waste heat boiler and other equipment.

Fig. 2 Working flow of compound vertical retort technology

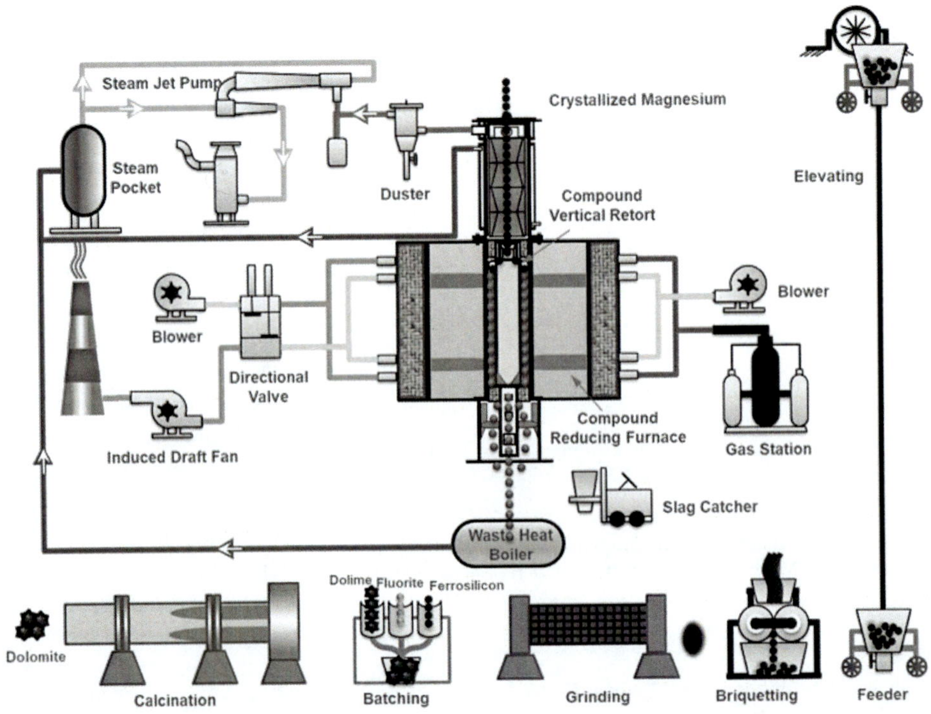

During the production operation, the raw material dolomite is calcined in a rotary kiln to produce calcined white, and the calcined white is mixed with ferrosilicon and fluorite according to the proportion and then finely ground in a ball mill, and the powder is sieved and pressed into pellets; the storage equipment containing the pellets is lifted to the top of the compound vertical retort by the lifting device, and the pellets are loaded into the compound vertical retort by a mechanized feeding device; vacuum system and regenerative combustion system are operated to make the temperature and pressure conditions in the retort meet the conditions required for production and start the magnesium reduction process; the magnesium vapor flows upward through the central cylinder of the compound vertical retort and enters the crystallizer for condensation, the water jacket absorbs the crystallization heat, and the generated steam is directly supplied to the vacuum steam jet pump for use; the reduction residue is discharged from the bottom of the retort and enters the waste heat boiler through a closed slag-catching device. After the slag is cooled, it is conveyed to the slag bin for subsequent processing through an automatically closed conveyor. For the mechanical compound vertical technology, the manhours for completing the feeding and discharging of one retort were largely shorten to around 15 min compared to the Pidgeon process.

Research and Development of Key Technique and Equipment

Firstly, from the thermophysical and chemical changes and gas, liquid and solid multiphase changes that occur under the conditions of silicothermic magnesium production, combined with experimental verification and analysis, the internal reaction mechanism of melting and bonding of materials is clarified, the root cause of "sticking pot" and "glaze formation" is found, and effective measures to solve the problems of "sticking pot" and "glaze formation" are proposed [7]. Based on this, a vertical pot with inner liner structure is developed. In addition, through the study of thermodynamics and dynamics mechanism of the solid-phase reaction of magnesium production by silicothermic process, a mathematical model that can accurately describe the magnesium reduction process is constructed, and the structure of the compound vertical retort is optimized with the help of numerical simulation technology [8–11], and reasonable structural parameters of the compound vertical retort with a central cylinder are given; combined with the research results of fusion bonding and the experimental research on the creep properties of high-temperature heat-resistant steel materials, the composite mode between the body of the compound vertical retort and the inner liner and the center cylinder are given [12].

Secondly, the magnesium yield of a single retort of a compound vertical retort is higher than that of the traditional horizontal retort by the Pidgeon process. With the increase in the output of crystalline magnesium, the slow heat transfer rate of the traditional crystallizer and the congestion and overflow of magnesium vapor in the crystallizer have also emerged; the traditional structure and size of crystallizer can no longer meet the demand of compound vertical retort magnesium production. The compound crystallizer is significantly larger than the traditional crystallizer in height and diameter, the heat transfer area is increased, and the heat transfer efficiency is improved by welding the radiation heat transfer plate inside the outer condensate jacket and the outside of the inner crystallizer one by one [13]. This "fin type" structure design can increase the heat transfer area between the crystallization cylinder and the condensate jacket by more than two times, and the crystallization volume of magnesium can reach about 10 kg. m^{-2}. h^{-1}, which is more than three times of the traditional crystallizer.

Furthermore, the regenerative combustion reduction furnace is generally used in the industrial production of silicon thermal magnesium production. The retort filled with filler balls is placed in the furnace for heating. The external heat flow field of the retort includes the convection exchange of high-temperature flue gas flowing through the retort wall, radiation of high-temperature flue gas to the retort wall. By establishing a coupling model of the calculation model of the flow around the retort and the magnesium reduction model in the retort, our team simulated the temperature field and heat flow field distribution in the silicothermal magnesium retort. By optimizing the nozzle structure, fuel, and air operating parameters, the temperature difference inside the furnace chamber of the single-module reduction furnace is $\leq$ 50 °C. The numerical simulation results were used to predict the change of magnesium reduction rate in a single reduction retort in the reduction furnace and the change characteristics of the total reduction rate [14]. Based on simulation, the design of the compound reduction furnace adopts double-layer/bidirectional/interleaved/hedged/jet X flame, etc., to fully realize the uniform temperature in the vertical and horizontal directions in the reduction furnace and to improve the combustion efficiency and thermal efficiency of the reduction furnace [13]. In addition, the negative pressure reducing atmosphere combustion technology adopted can significantly reduce the oxidation corrosion rate of the surface of the reduction retort, and the life of the reduction furnace body and the reduction retort have been improved.

The Demonstration Unit and Its Working Status

On the basis of the comprehensive completion of the design, verification, and development of the key equipment and supporting equipment for magnesium production in

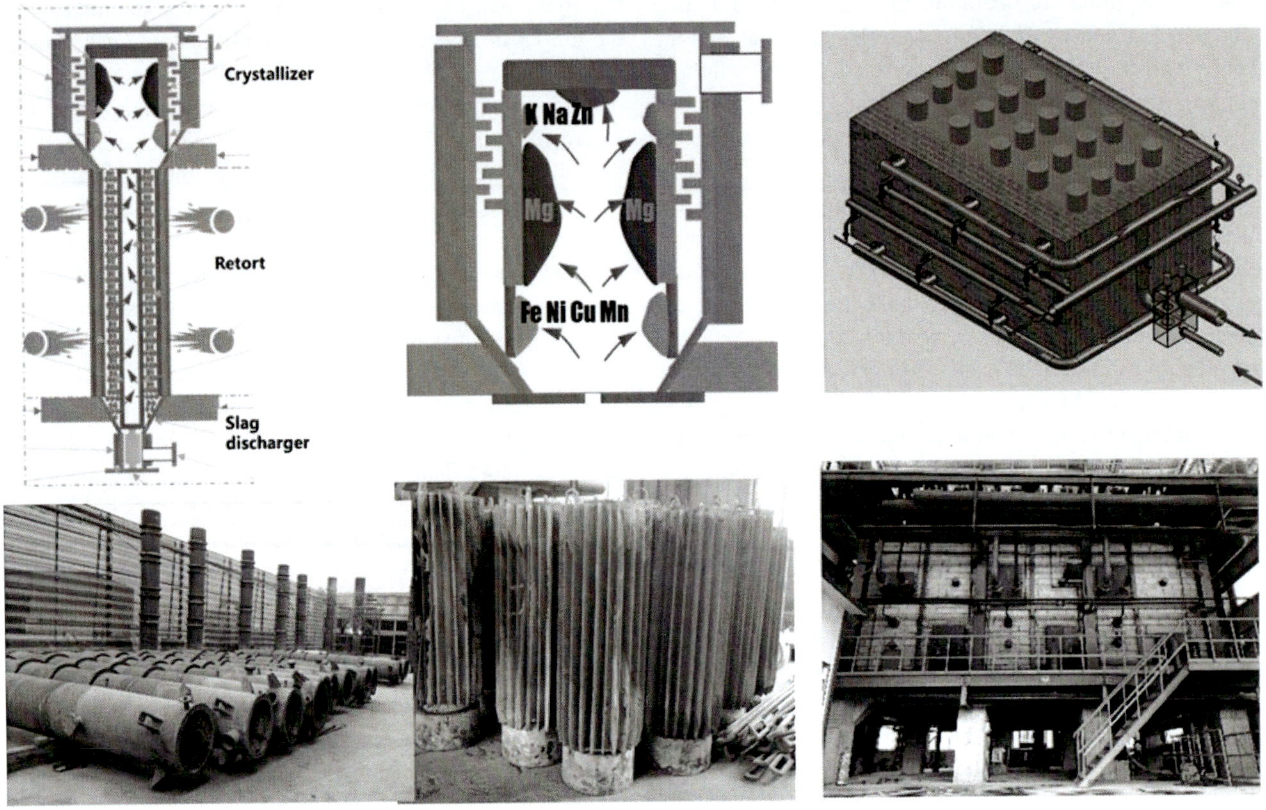

Fig. 3 Research and development of key equipment of compound vertical retort technology

compound vertical retort, we have built a pilot plant for magnesium production with such a vertical retort in China see Fig. 3. The core equipment, key supporting equipment, and supporting process system of the double vertical retort magnesium production technology have been fully verified in actual production operation. After the pilot plant was put into operation, the problems such as excessive water and tar in the external gas supply system, insufficient power of external spiral slag dispenser, and material backflow and jamming were found and solved successively. The influence of external factors and supporting devices on the overall system operation was eliminated. The actual operation shows that the pilot production unit can run continuously and stably for a long time, and all the economic and technical indexes have reached/exceeded the design criteria.

With the material-magnesium ratio of 6.2 as the measurement benchmark, the main technical and economic indicators of compound vertical retort magnesium production technology are listed in Table 1 in this paper according to the operational indicators of the pilot production unit. The comparison shows that all the technical and economic indicators of the compound vertical retort are significantly better than that of the horizontal retort of the Pidgeon process and Magnetherm process.

Table 1 Main technical and economic indicators of compound vertical retorts

No.	Index	Compound vertical retort	Pidgeon process	Magnetherm process
1	Reduction cycle (h)	**12**	12	16–24
2	Magnesium yield per retort (kg)	**~100**	20–30	~3000 (one batch)
3	Reduction furnace energy consumption (tce/t-Mg[a])	**>1.0**	~1.6	~2.4
4	Service life of retort (day)	**~180**	60–90	–
5	Purity of rough magnesium	**> 99.8%, dense solid**	< 99.5%, not dense solid	< 99.5%, liquid

[a]tce/t-Mg index represents the energy consumption for producing 1 tons magnesium in reduction stage, the energy consumption for raw materials of dolime and ferrosilicon not included

Summary

This paper comprehensively introduces a new magnesium production technology, the compound vertical retort magnesium production technology, and introduces in detail the basic theoretical problems and solutions of engineering thermophysics and thermochemistry in the research and development of the compound vertical retort magnesium production technology. The main process flow and technical characteristics of the double vertical retort magnesium production technology are introduced. Through the comparative analysis with the technical and economic indicators of the horizontal retort Pidgeon method, this paper discusses the superiority and advancement of the composite vertical retort magnesium production technology.

A new integrated calcination and silicothermic reduction short process for magnesium production was investigated in detail in this paper. Deep research on this technology was introduced, including fundamental theoretical problems and solutions of engineering thermophysics and thermochemistry in the research and development of the compound vertical retort magnesium production technology, the construction and operation of the scale-up experiment and the pilot plant and the main process flow and technical characteristics of the compound vertical retort magnesium production technology. Through the comparative analysis with the technical and economic indicators of the horizontal retort Pidgeon process, this paper discusses the superiority and advancement of the composite vertical retort magnesium production technology.

Acknowledgements The authors gratefully acknowledge the financial support from the Fundamental Research Funds for the Central Universities (2302018FRF-TP-18-095A1) and National Key Research and Development Project (2016YFB0301101).

References

1. Yang LX, Huang YD, et al (2022) Applications of magnesium alloys in aerospace and aviation. Aerospace Shanghai (Chinese and English) (01): 31–45+91.
2. Huang YD, Jiang B (2022) Editorial for special issue on developments of magnesium alloys for structural and functional applications. IJMMM 29(07): 1307–1309.
3. Che YS, Du SM, et al (2022) Research status and progress of novel technology for magnesium production. Transactions of Nonferrous Metals Society of China 32(06): 1719–1733.
4. Chubukov BA, Rowe SC, et al (2019) Investigation of continuous carbothermal reduction of magnesia by magnesium vapor condensation onto a moving bed of solid particles Powder Technol. 365: 2–11.
5. CSIRO (2017) Australia's "magnesium sonic" brings energy-saving and environmentally friendly magnesium production new technology. China Nonferrous Metallurgy. 46(06): 63.
6. Yu A, Hu H, Li NY (2002) Mathematical modelling of magnesium reduction in a novel vertical Pidgeon process. Model. Simul. Mater. Sc. 10 (4): 413–423.
7. Li RB, Chen M, et al (2022) Catalytic mechanism and control of fusion and agglomeration in indirectly heated silicothermic reduction. Nonferrous Metals (extractive metallurgy) (06): 30–36.
8. Li RB, Zhang SJ, et al. (2013) Numerical study of magnesium (Mg) production by the Pidgeon process: impact of heat transfer on Mg reduction process. Int. J. Heat Mass Trans. 59: 328–337.
9. Li RB, Zhang C, et al. (2014) Experimental and numerical modeling studies on production of Mg by vacuum silicothermic reduction of CaO·MgO. Metall. and Mater. Trans. B 45(1): 236–250.
10. Wang C, Zhang C, et al. (2015) The effect of CaF_2 on the magnesium production with silicothermal process. Int. J. Miner. Process. 142: 147–153.
11. Zhang C, Wang C, et al. (2015) The effects of hydration activity of calcined dolomite (HCD) on the silicothermic reduction process. Int. J. Miner. Process. 142: 154–160.
12. Li RB, Chen X, et al. (2020) Structure design and parameters optimization of retorts for magnesium production via silicothermic process. Light metal 12: 34–39.
13. Zhang SJ (2014) Theoretical and Experimental Research on New Type of Silicothermic Reduction of Magnesium and Development of Key Technologies and Equipment. Ph.D. thesis, Xi'an Jiaotong University.
14. Li RB (2013) A Research on the Experimental and Theoretical Modeling of the New-type Silicothermic Reduction to Magnesium and its Engineering Demonstration. Ph.D. thesis, Xi'an Jiaotong University.

Development of Magnesium-Strontium/ Calcium (Mg-Sr/Ca)-Based Alloys with Improved Sinterability for Next-Generation Biomedical Implants

Mert Celikin, Ava Azadi, Hyeonseok Kim, Ted Vaughan, and Eoin O'Cearbhaill

Abstract

The use of biodegradable magnesium (Mg) alloys for bone fixation devices has potential to improve patients' quality of life by avoiding the necessary secondary operations conducted regularly for the removal of implants fabricated from conventional non-resorbable alloys. Mg-alloys have excellent biocompatibility and biodegradability along with a low modulus of elasticity which will decrease bone-shielding effects. However, low corrosion resistance and relatively poor mechanical performance limit the use of Mg-based alloys for biomedical applications. This study focuses on the processing of Mg-Ca- and Mg-Sr-based alloys via powder metallurgical route. Thermodynamic calculations are used to predict the liquid phase fractions in order to optimise sinterability and porosity levels. Materials characterisation was conducted to validate the thermodynamic modeling results using optical and scanning electron microscopy (SEM/EDS) as well as X-ray Diffraction (XRD).

Keywords

Magnesium alloys • Biodegradable implants • Thermodynamic calculations • Microstructural characterisation • Sinterability

M. Celikin (✉) · A. Azadi · H. Kim · E. O'Cearbhaill
School of Mechanical and Materials Engineering, University College Dublin, Dublin, Ireland
e-mail: mert.celikin@ucd.ie

T. Vaughan
Biomedical Engineering, School of Engineering, College of Science and Engineering, Biomechanics Research Centre (BioMEC), University of Galway, Galway, Ireland

M. Celikin · A. Azadi · H. Kim
Materials Design and Processing Laboratory, University College Dublin, Dublin, Ireland

Introduction

Additive Manufacturing (AM) technology can produce patient-specific implants with very complex geometries [1, 2]. Magnesium (Mg) alloys produced by high-temperature AM techniques such as Laser Powder Bed Fusion (L-PBF) have been extensively investigated for biomedical applications [3–5]. However, the use of high power sources such as laser and electron beam raises safety concerns due to the intrinsic properties of Mg: high affinity to oxygen, low boiling temperature, and high vapour pressure [6–9].

Low-temperature AM techniques such as fused deposition modeling (FDM) have high potential to be used for processing Mg-based alloys targeting biomedical applications which can avoid issues related to PBF-based AM techniques. FDM is a cost-efficient 3D-printing technique commonly used to produce polymer-based components from filaments. Employing FDM that operates at low temperatures (<200 °C) can offer key technological advancement in the customization of patient-specific biodegradable Mg implants with maximum design flexibility. However, the application of FDM technology is currently not widely adopted for Mg-alloys due to the low sinterability of Mg. Binder jetting and material extrusion have been mainly investigated for producing Mg alloys. Salehi et al. achieved high and isotropic mechanical properties of ZK51 produced by binder jetting showing around 177 MPa of ultimate compressive strength (UCS) [10]. Furthermore, specific sintering processes such as microwave sintering and two-step sintering process enhance the mechanical properties of Mg alloys produced by Binder Jetting [10, 11]. However, optimizing porosity levels without sacrificing structural integrity during the sintering process is the main challenge.

Mg has high oxygen affinity, and, therefore, upon exposure to air, a thin stable oxide layer (MgO) is formed on the particle surface acting as a barrier to diffusion and thereby strongly inhibits sintering [12]. There exist some practical

S. Barela et al. (eds.), *Magnesium Technology 2023*, The Minerals, Metals & Materials Series,
https://doi.org/10.1007/978-3-031-22645-8_33

Table 1 Results of particle size analysis (laser diffraction)

Alloy	d10 (μm)	d50 (μm)	d90 (μm)
Mg-20Sr	87.44	207.40	277.50
Mg-20Ca	63.40	144.40	271.70

solutions with which this challenge can be lowered and accordingly improve the sintering behaviour of Mg, including the addition of the alloying elements to the pure Mg along with the control over the sintering atmosphere. For example, sintering in the presence of a high purity inert gas to protect parts from oxygen and/or to use oxygen getters inside the crucible during heat treatment are reported to reduce oxidation [12–14]. In this regard, Wolff et al. [14] added two binary alloys, Mg-7Ca and Ca-18 Mg (in wt.%), along with the calcium hydride (CaH_2), to pure Mg powders during powder blending step and sample preparation for the powder metallurgy process. In their work, the addition of the master alloys or calcium hydride increased the sinterability of Mg, and it was concluded that the highest mechanical properties were corresponded to Mg-0.6wt.% Ca blend. Furthermore, Shaper et al. [15] investigated the mechanical properties and microstructural analysis of AZ81 Mg alloy parts manufactured via metal injection moulding (MIM) followed by T4 heat treatment. It was reported that AZ81 alloy demonstrated higher mechanical properties and significantly shorter sintering time (4 h at 605 °C) as compared to Mg-0.9Ca. Also, the T4 post-heat treatment at 420 °C for 10 h did not improve the mechanical properties of the MIM part.

Liquid-phase sintering (LPS), where a liquid phase forms just below the melting temperature, will allow enhanced sinterability due to solid wetting, providing a capillary force that pulls the grains together [16]. Furthermore, supersolidus LPS, where sintering is conducted between the solidus and liquidus, allows liquid to form between grain boundaries, is effective for the sintering of pre-alloyed powders [17]. Sintering behaviour affects the final porosity and degree of shrinkage, determining the final component's mechanical properties and structural integrity.

The objective is study is to investigate the sinterability of two binary Mg-20Ca and Mg-20Sr alloys where particles are obtained via slow speed dry milling. 525 and 575 °C were selected as sintering temperatures based on thermodynamic calculations to understand the contributions from LPS and solid-state sintering.

Experimental Procedure

As-received master alloys (Shanghai Xinglu Chemical Tech.) with nominal compositions of Mg-20Sr and Mg-20Ca (% in weight) were used as the base materials. Magnesium alloy powder was obtained from the ingots via slow speed milling without using any cutting fluid. Dry milling of the ingots was conducted using Emill milling equipment in automatic feed mode and NDrill with parameters: depth of cut 1 mm, length of travel 200 mm/min where 1 revolution took 24 s/5 mm. During drilling, a maximum temperature increase of 3 °C was determined using an infrared thermometer.

Mg alloy powder was then sieved to separate particles coarser than 300 microns, and all the particles were determined to be larger than 25 microns (i.e. delimited particle size (x) range: 25 μm < x < 300 μm). Particle size distributions (*d10, d50, and d90* are defined as 10%, 50%, and 90% of the particles smaller than x, respectively) were determined via laser diffraction for both Mg-20Ca and Mg-20Sr master alloys (Table 1). Microtrac S3500 dry dispersion laser diffractometer was used to analyse the particle size.

Particles were compacted in a cylindrical die using a cold pressing setup (PA 260, Josef Lucas LTD Birmingham) with an applied load of 30–50 kN and dwell time of 15–20 s, and thereby Mg-20Ca and Mg-20Sr discs with a diameter of 25 mm were manufactured. Sintering was conducted in a tube furnace (Lenton Tube Furnace 1600) under argon (Ar) as a protective atmosphere using two different sintering temperatures, 525 and 575 °C, with the following parameters: 5 °C/min heating to aforementioned temperatures, holding for 2 h at each temperature, and subsequent cooling to ambient temperature with 5 °C/min cooling rate. It should be mentioned that even though the cooling rate is set as 5 °C/min, the actual cooling rate may vary. Thereafter, specimens were extracted from the sintered discs and cross sections were metallographically prepared following the standard procedure, finishing with a 1 μm diamond paste polishing step.

For the assessment of liquid phase fractions and the equilibrium phases, ThermoCalc thermodynamic software (TCS Magnesium-based alloys database) was used. The calculations were based on Mg-Ca and Mg-Sr binary phase diagrams.

Morphology and size of raw particles were analysed before and after sieving using optical microscopy (Buehler Coventry) and scanning electron microscopy (SEM, TM4000Plus Tabletop Hitachi, Japan), and the porosity percentage was calculated from the optical micrographs using the Image J software (an open platform for scientific image analysis, https://imagej.net/Welcome). Chemical

Fig. 1 SEM images of raw particles after milling

Fig. 2 SEM images of raw particles after sieving

analysis of the phases was performed using energy dispersive spectroscopy (EDS). X-ray diffraction (XRD) analysis was conducted using Cu-K_α source (Siemens D500 X-ray diffractometer) with a step size of 0.02° and dwell time of 3 s.

Results and Discussion

Raw particle morphology of the as-milled samples was found to be irregular with sharp edges (Fig. 1). Additionally, large differences in particle sizes were observed for both alloys. While some particles with sizes over 6 mm were seen, there were also numerous particles as small as 6 µm in size. This type of irregular morphology with sharp edges is expected for particles obtained via milling.

Sieving was conducted using mesh sizes of 300 and 25 µm in order to remove large particles and decrease the difference in particle size. After sieving, particle sizes were analysed using SEM along with laser diffractometer (size distribution–Table 1). Smaller variation in particle sizes than as-milled condition was evident for both alloys (Mg-20Sr and Mg-20Ca) shown in Fig. 2. It was determined that Mg-20Ca alloy has smaller particles compared to Mg-20Sr; 50% of the particles (d50 value) were below 144.4 µm and 207.4 µm for Mg-20Ca and Mg-20Sr, respectively (Table 1). Based on the EDS analyses, obtained particles possess close to nominal composition; Ca content was found to be around 20.8%, and Sr content was approximately 19% for the binary alloys. As both elements have high affinity to oxygen, around 3–5% oxygen (in wt.%) was present as surface oxide.

XRD analyses was carried out on the milled and sieved particles for both alloys. According to the binary Mg-Ca and Mg-Sr phase diagrams [18], the most stable secondary phases were Mg_2Ca and $Mg_{17}Sr_2$. Consistently, it was determined via XRD analysis that the same equilibrium phases were present in as-milled and sieved particles obtained from the as-cast master alloys (Fig. 3). This finding is consistent with the author's previous studies on Mg-Sr- and Mg-Ca-Sr-based alloys where Mg_2Ca and $Mg_{17}Sr_2$ intermetallics are the main interdendritic phases formed upon casting [19–21].

Thermodynamic calculations were done using Thermo-Calc software (TCS Magnesium-based alloys database). Based on the calculations, liquidus temperatures were determined: 518.3 °C (Mg-20Ca) and 588.2 °C (Mg-20Sr). Two (2) separate sintering treatments were selected as 525 °C and 575 °C. For Mg-20Ca alloy, liquid phase sintering (LPS) should dominate both at low and at high temperatures; however, only solid-state sintering is expected to be active during sintering process for Sr-containing alloy as liquidus temperature (588.2 °C) is higher than both 525 °C and 575 °C.

Figure 4 shows SEM images obtained from both Mg-20Sr and Mg-20Ca alloys upon sintering conducted at 525 °C for 2 h. Image analyses conducted on several images indicate lower porosity levels for Mg-20Ca (22.5%) in comparison to Mg-20Sr alloy (47.5%). This is probably due to the lower liquidus temperature of Mg-20Ca alloy where partial liquid formation occurred during sintering as observed via SEM analysis.

In the Ca-containing alloy, dendritic formation of Mg-rich region is evident (Fig. 4). This has been observed partially in the structure as a result of liquid to solid phase transformation. Moreover, porosity levels are not homogeneous throughout the structure; there are parts of the sample with higher porosity levels. As thermodynamic calculations

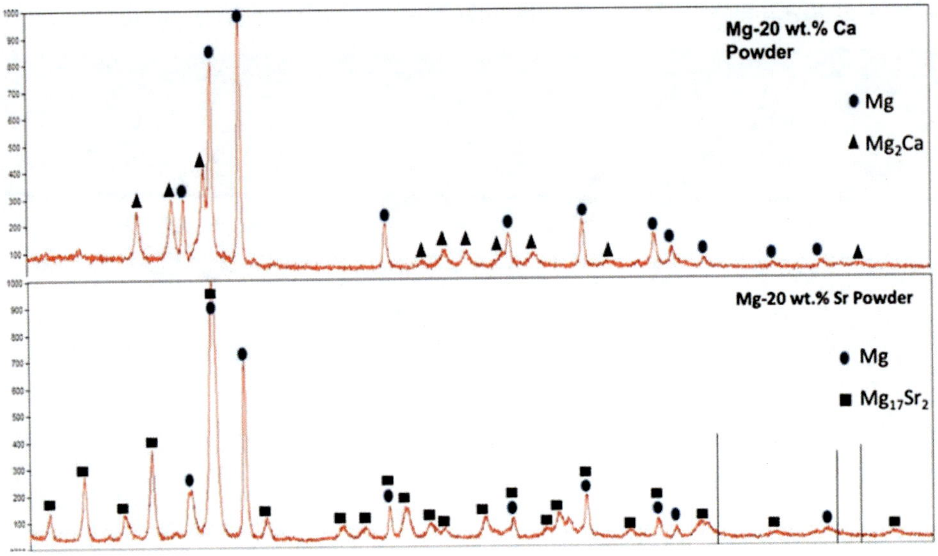

Fig. 3 XRD spectra of raw particles (upon sieving) from master alloys (Mg-20Ca and Mg-20Sr)

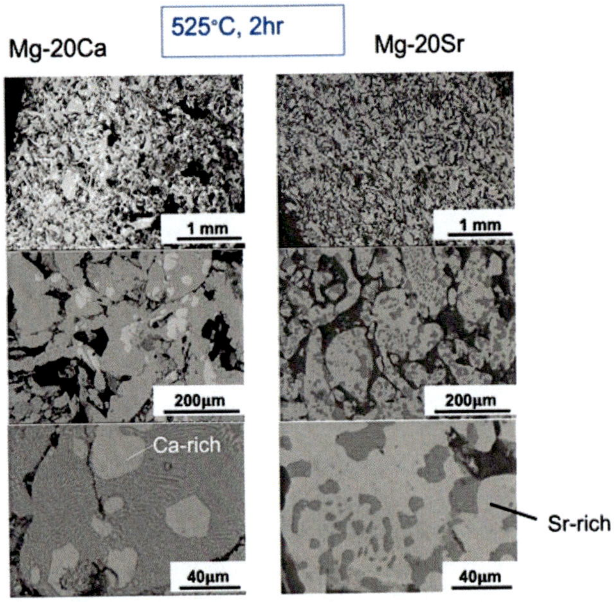

Fig. 4 SEM images of alloys after sintering at 525 °C for 2 h

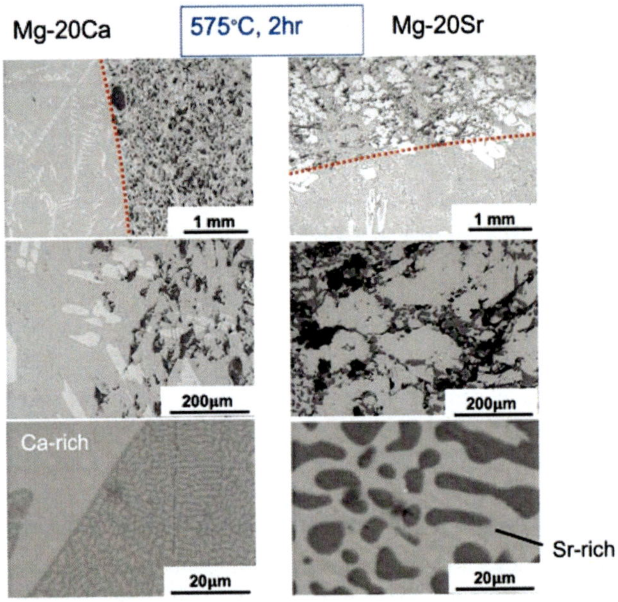

Fig. 5 SEM images of alloys after sintering at 575 °C for 2 h

do not involve the kinetics of phase transformation, further in-depth studies are needed to link the amount of liquid phase fraction and sintering conditions. On the other hand, in Sr-containing alloy no dendritic phase formation was observed upon sintering at 525 °C, consistent with dominant solid-state sintering mechanism predicted by thermodynamic calculations. Limited sinterability was determined without the liquid phase formation (47.5% porosity). Additionally, non-spherical particle morphology may have also decreased the area of contact between separate particles reducing sinterability; however, more detailed analysis should be conducted to understand the effect of particle morphology on sinterability.

Figure 5 shows final microstructures of both compositions upon high-temperature sintering process (575 °C for 2 h). In contrast to low-temperature sintering, both alloys have shown the partial activation of LPS which created a clear interface between LPS only region and the region that is formed by LPS and solid-state sintering (shown as dashed line). It is determined that LPS only region has 0% porosity and possess typical cast microstructure. However, the regions where both LPS and solid-state sintering mechanisms are active have higher porosity levels; 14.9% (Mg-20Sr) and 20.5% (Mg-20Ca). It should also be stated that porosity levels decreased significantly when sintering temperature increased from 525 to 575 °C.

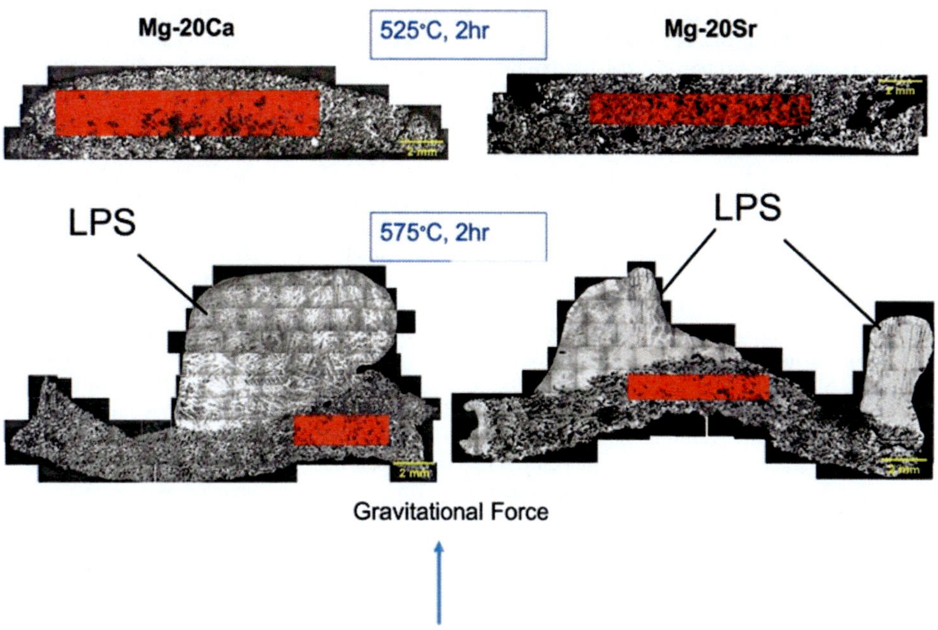

Fig. 6 Macrostructures of Mg-20Sr and Mg-20Ca alloys sintered at 525 and 575 °C (2 h) exhibiting LPS only and LPS/solid-state sintering regions

Macrostructures of both alloys (Fig. 6) exhibited loss of structural integrity upon sintering at 575 °C after 2 h due to the formation of LPS controlled regions. The liquid only regions which have solidified upon cooling are mainly at the bottom of the samples (consistent with gravitational force). Boxes in the figures are representing the areas involved solid-state sintering where porosity analyses were done. In order for structural integrity to be kept without creating high level of porosity, sintering temperature, and time should be optimised. Future work will involve the use of cryo-ball milling technique to further reduce the size and to obtain more spherical particles which will also improve sinterability between magnesium particles.

Conclusion

Dry milling and sieving processes have been successfully used to obtain particles of Mg-20Sr and Mg-20Ca alloys with sizes $25 < x < 300$ µm. Particles that had irregular morphology and sharp edges were compacted to form cylindrical samples. Thermodynamic calculations were used to assess liquidus temperatures, equilibrium phases as well as fraction of liquid formation. However, it should be noted that kinetic considerations were not involved for the calculations. This data was used to select two separate sintering temperatures where for Ca-containing system LPS is expected to dominate sintering, whereas solid-state sintering mechanism should control sintering of Sr-containing samples. Preliminary data indicated that at high sintering temperatures LPS mechanism is activated leading to not only lower levels of porosity but also the loss of structural integrity. The results also indicated that additional optimisation is necessary to link thermodynamic calculations with sintering conditions (temperature and duration). This is part of an on-going research where kinetics of phase transformation is studied to optimise the amount of liquid formed without losing the structural integrity. The future work involve the development of new Mg alloy compositions with superior sinterability which will shed light into the use of low temperature AM for processing biomedical Mg alloys.

Acknowledgements Authors acknowledge the EU financial support received under Marie-Sklodowska Curie Fellowship (AMBIT 101029651). This publication has also emanated from research conducted with the financial support of Science Foundation Ireland under grant number 20/FFP-P/8868. For the purpose of Open Access, the author has applied a CC BY public copyright licence to any Author Accepted Manuscript version arising from this submission. Authors also thank South Eastern Applied Materials (SEAM) Research Centre for their support in conducting laser diffraction experiments.

References

1. S. Piazza, B. Merrigan, D.P. Dowling, M. Celikin, 'The effects of geometry and laser power on the porosity and melt pool formation in additively manufactured 316L stainless steel', International Journal of Advanced Manufacturing Technology, v.111, n. 5-6, p. 1457 – 14701, 2020
2. F.R. Kaschel, D.P. Dowling, M. Celikin, 'In-situ XRD study on the effects of stress relaxation and phase transformation heat treatments on mechanical and microstructural behaviour of additively manufactured Ti-6Al-4V', Materials Science and Engineering A, v. 819,141514, 2021
3. H. Hyer, L. Zhou, G. Benson, B. McWilliams, K. Cho, Y. Sohn, 'Additive manufacturing of dense WE43 Mg alloy by laser powder bed fusion', Additive Manufacturing, v.33, 101123, 2020.

4. H. Hyer, L. Zhou, Q. Liu, D. Wu, S. Song, Y. Bai, et al. 'High strength WE43 microlattice structures additively manufactured by laser powder bed fusion' Materialia, v.16, 101067, 2021.
5. J. Liu, B. Yin, Z. Sun, P. Wen, Y. Zheng, Y. Tian, 'Hot cracking in ZK60 magnesium alloy produced by laser powder bed fusion process', Materials Letters, v.301, 130283, 2021.
6. J. Dong, Y. Li, P. Lin, M.A. Leeflang, et al., 'Solvent-cast 3D printing of magnesium scaffolds', Acta Biomaterialia, v.114, p. 497-514, 2020.
7. C. Liu, S. Lu, Y. Fu, H. Zhang, 'Flammability and the oxidation kinetics of the magnesium alloys AZ31, WE43, and ZE10', Corrosion Science, v.100, p.177-185, 2015.
8. N.V. Ravi Kumar, J.J. Blandin, M.Suéry, E. Grosjean, 'Effect of alloying elements on the ignition resistance of magnesium alloys', Scripta Materialia, v.49, n3, p.225–230, 2003.
9. N. Wegner, D. Kotzem, Y. Wessarges, N. Emminghaus, C. Hoff, J. Tenkamp, et al. 'Corrosion and Corrosion Fatigue Properties of Additively Manufactured Magnesium Alloy WE43 in Comparison to Titanium Alloy Ti-6Al-4V in Physiological Environment', Materials (Basel), v.12, n.18, 2019.
10. M. Salehi, H.L. Seet, M. Gupta, H. Farnoush, S. Maleksaeedi, M. L.S. Nai, 'Rapid densification of additive manufactured magnesium alloys via microwave sintering', Additive Manufacturing, v.37, 2021.
11. C. Su, J. Wang, H. Li, Z.You, J. Li, 'Binder-jetting additive manufacturing of Mg alloy densified by two-step sintering process', Journal of Manufacturing Processes, v.72, p.71-79, 2021.
12. R. Karunakaran, S. Ortgies, A. Tamayol, F. Bobaru, and M. P. Sealy, 'Additive manufacturing of magnesium alloys', Bioactive materials, v.5, n.1, p. 44–54, 2020, doi: https://doi.org/10.1016/j.bioactmat.2019.12.004.
13. P. Burke, G. J. Kipouros, D. Fancelli, and V. Laverdiere, 'Sintering Fundamentals of Magnesium Powders', Canadian Metallurgical Quarterly, v.48, n.2, p.123–132, 2009, doi: https://doi.org/10.1179/cmq.2009.48.2.123.
14. M. Wolff, T. Ebel, and M. Dahms, "Sintering of Magnesium," Adv. Eng. Mater., v.12, n.9, p. 829–836, 2010
15. J. G. Schaper, M. Wolff, B. Wiese, T. Ebel, and R. Willumeit-Römer, 'Powder metal injection moulding and heat treatment of AZ81 Mg alloy', Journal of Materials Processing Technology, v. 267, p. 241–246, 2019
16. R.M. German, P. Suri, S.J. Park, Review: liquid phase sintering. Journal of Materials Science,*v.*44, p.1–39, 2009
17. J Liu, A Lal, R.M German, 'Densification and shape retention in supersolidus liquid phase sintering', Acta Materialia, v.47, n.18, p. 4615–4626, 1999
18. ASM Handbook Volume 3 'Alloy Phase Diagrams', ASM International, 1992
19. M. Celikin, R. Gauvin, M. Pekguleryuz 'Dynamic co-precipitation of α-Mn / Mg12Ce in creep resistant Mg-Sr-Mn-Ce alloys', Materials Science and Engineering A, v.719, p.199 – 205, 2018
20. M. Bornapour, M. Celikin, M. Pekguleryuz, 'Thermal exposure effects on the in vitro degradation and mechanical properties of Mg–Sr and Mg–Ca–Sr biodegradable implant alloys and the role of the microstructure', Materials Science and Engineering C, v.46, p.16-24, 2015.
21. M. Celikin, M. Pekguleryuz, 'The role of α-Mn precipitation on the creep mechanisms of Mg–Sr–Mn', Materials Science and Engineering A, v.556, p.911– 920, 2012

Development of Mg-Based Superelastic Alloy Through Aging Heat Treatment

Keisuke Yamagishi, Yukiko Ogawa, Daisuke Ando, and Yuji Sutou

Abstract

The effect of aging heat treatment on microstructure, hardness, and superelasticity at room temperature was investigated herein for Mg-18.8 at.% Sc alloy. The hardness of the alloy is increased via aging heat treatment at temperatures between 423 and 523 K, from ~ 90 Hv (as-quenched condition) to a maximum of 180 Hv. Aging heat treatment at a higher temperature reduces the incubation time before the onset of age hardening. Scanning electron microscopy observations and X-ray diffraction analysis showed that the precipitation of hexagonal close-packed (α) phases within the body-centered cubic (β) matrix phase causes age hardening and the hardness value almost depends linearly on the volume fraction of α precipitates. Furthermore, the α precipitates formed via aging heat treatment can be deformed along with β matrix phase upon stress-induced martensitic transformation in a sample with ~ 10% volume fraction of α precipitates, resulting in a slight reduction in stress hysteresis and a minor increase in superelastic recovery compared with the as-quenched condition.

Keywords

Bcc-magnesium alloy • Shape memory alloy • Aging • Superelasticity

Introduction

Shape memory alloys (SMAs) have attracted attention in several fields owing to their unique properties, such as large recoverable strain and flexibility. The large recoverable strain in SMAs can be obtained upon heating them after unloading (shape memory effect) and/or simply upon unloading (superelasticity) due to reversible martensitic transformation; the strain value generally reaches several percent [1, 2]. Since the discovery of Au–Cd alloy [3], various SMAs have been discovered and developed, e.g., Cu—[4–6], NiTi—[1, 2, 7], Fe—[8–10], and Ti-based alloys [7, 11, 12]; however, Mg-based SMA was not discovered until we first reported the superelasticity of a Mg–Sc alloy [13].

According to reported phase diagrams, the Mg–Sc alloy is the only Mg-based binary alloy whose microstructure can be controlled, depending on heat treatment conditions, between a low-temperature hexagonal close–packed (α) phase and a high-temperature body centered cubic (β) phase at the same Sc content [14, 15]. The unique characteristic of this alloy appears to enable it to show thermoelastic martensitic transformation between β and α″ phases (orthorhombic structure), and we have demonstrated that a Mg-20.5 at.% Sc alloy with β single-phase exhibits superelasticity at 123 K owing to stress-induced martensitic transformation [13]. The working temperature range in which superelasticity occurs depends on the Sc content, and we recently discovered that the Sc content of ~ 19 at.% is suitable for room temperature superelasticity [16, 17]. Because the density of the superelastic Mg–Sc alloy is ~ 2 g/cm^3, which is approximately one-third of

K. Yamagishi · D. Ando (✉) · Y. Sutou
Department of Materials Science, Graduate School of Engineering, Tohoku University, 6-6-11 Aoba, Aramaki, Aoba-Ku, Sendai, 980-8579, Miyagi, Japan
e-mail: daisuke.ando.c4@tohoku.ac.jp

K. Yamagishi
e-mail: keisuke.yamagishi.p2@dc.tohoku.ac.jp

Y. Sutou
e-mail: ysutou@material.tohoku.ac.jp

Y. Ogawa
Research Center for Structural Materials, National Institute for Materials Science, 1-2-1 Sengen, Tsukuba, 305-0047, Ibaraki, Japan
e-mail: OGAWA.Yukiko@nims.go.jp

Y. Sutou
Advanced Institute for Materials Research, Tohoku University, 2-1-1 Katahira, Aoba-Ku Sendai, Miyagi, 980-8577, Japan

S. Barela et al. (eds.), *Magnesium Technology 2023*, The Minerals, Metals & Materials Series,
https://doi.org/10.1007/978-3-031-22645-8_34

conventional NiTi-based superelastic alloys, the alloy may attract attention in automobile and aerospace field from the perspective of fuel efficiency. With its low Young's modulus and high biocompatibility [18, 19], the Mg–Sc superelastic alloy can be applied in medical field as bone plates, stents, etc. Recently, Li et al. implanted a Mg-30 wt.% Sc alloy (equivalent to 18.8 at.% Sc) into rats and showed that the alloy exhibits acceptable biocompatibility and biodegradability [20].

Although the Mg–Sc superelastic alloy exhibits the abovementioned unique properties, it still faces some challenges that need to be addressed. The most critical of which is its low strength and eventual low superelastic recovery at room temperature [16, 17]. Although the recovery can be considerably increased by coarsening the grain size, similar to the case of Cu- and Fe-based shape memory alloy [5, 6, 10], maximum superelastic recovery obtained in Mg−Sc alloy at room temperature was only 3% [16] because slip deformation can occur easily at the same time with stress-induced martensitic transformation. From this perspective, an increase in strength of the β matrix phase is strongly desired to further improve superelasticity.

Herein, we focused on aging heat treatment because age hardening is a well-known strengthening mechanism in various alloys, including Mg-based alloys. The effect of age hardening on superelasticity is also reported in SMAs, such as NiTi—[2, 7], Fe—[8, 9, 21, 22], Ti—[7, 11, 12], and Cu-based alloys [23, 24], where the effect can be either positive or negative depending on the amount and/or size of the aging products. For example, it is reported that nanosized precipitation due to aging heat treatment effectively strengthens the matrix phase of the Fe−Mn−Al−Ni SMA and realizes superior superelasticity, whereas an excess increase in size of the precipitates deteriorates the superelasticity of the alloy [9, 21, 22].

In the Mg−Sc alloy, aging heat treatment is also effective in increasing the hardness and tensile strength owing to the precipitation of the α phase within the β phase [25–27]. However, the Sc content of the measured samples was either 16.8 at.% or > 20 at.% and the effect of aging on superelasticity was not investigated. In this study, we investigated the effect of age hardening on the superelasticity of the Mg −Sc alloy at room temperature.

Procedures

A Mg–Sc ingot with the nominal composition of Mg-18.8 at. % Sc was prepared by OSTECH Co. Ltd. and Hunan Oriental Scandium Co. Ltd. A block was cut from the ingot and hot rolled at 923 K. The resulting sheet was homogenized at 873 K for 24 h before being cold rolled with intermediate annealing at 873 K for 15 min until its thickness reached ~ 1.3 mm. For each measurement, a specific size of sample was cut from the cold rolled sheet; subsequently, the sample was heat treated and then aged as mentioned below.

The changes in hardness and microstructure were measured using samples with the size of ~ 5 × 5 mm^2. The samples were solution heat treated at 963 K for 30 min, followed by quenching into iced water to obtain β single-phase, and finally aged at 423, 448, 473, or 523 K for different durations. Hardness was measured using a Vickers hardness tester with an applied load of ~ 10 N. The hardness of a sample was calculated from the average of 10 points. The microstructure was observed using a scanning electron microscope (SEM) with an accelerated voltage and probe current of 15 kV and 10 μA, respectively. The observed samples were polished using SiC papers and diamond pastes of different grit sizes (3, 1, 0.25 μm) and chemically etched in acid solution for 5 s to obtain a mirror-like surface. X-ray diffraction (XRD) analysis was performed to investigate the constituent phases in measured samples.

Samples with a length and width of approximately 40 and 5 mm were prepared to evaluate superelasticity at room temperature. Cyclic heat treatment (CHT) was applied to the samples to obtain the grain size of ~ 1 mm because a small grain size of about several hundred micrometers considerably deteriorates superelasticity [16, 17] and therefore makes it difficult to determine the effect of aging on superelasticity. Specifically, the samples were solution heat treated at 963 K for 30 min, followed by air cooling to room temperature. In the following cycle, they were again solution heat treated at 963 K for 30 min, followed by quenching into iced water to obtain β single-phase. This type of abnormal grain growth (AGG) was first observed by Omori et al. in the Cu-based SMA [28], where subgrains seemed to be formed by the precipitation and dissolution of the second phase via CHT, and eventually, the subgrain boundary energy acted as an additional driving force for grain growth, resulting in AGG [28, 29]. The second phase is the α phase in the case of the Mg–Sc alloy [16], and the obtained average grain size was ~ 980 μm in the current study. The detailed mechanism underlying the AGG is under investigation and will be reported in the future study. The sample that underwent CHT was finally aged at 423 K for 4 ks, and a cyclic tensile test was conducted at room temperature to evaluate the superelasticity of the alloy. The tested sample was loaded to *i*% strain and then unloaded to zero stress at the *i*th cycle (i = 1, 2, 3, etc.). The test was repeated until fracture. The strain rate was set to 10^{-3} s^{-1}, and sample thickness and gauge length were approximately 0.5 and 15 mm, respectively. A sample that underwent CHT without aging heat treatment (as-quenched sample, hereafter) was also tested at room temperature.

Results and Discussion

Figure 1 shows the plots of Vickers hardness versus aging time at different temperatures, where the hardness of the as-quenched sample was ~ 90Hv (shown with a gray shaded line). At each aging temperature, the hardness increases considerably with aging heat treatment and exhibits a sigmoidal change against aging time. Incubation time before the onset of hardening becomes longer with decreasing aging temperature. By contrast, the maximum hardness values of ~ 180 Hv are almost identical within the range of measured aging time.

A microstructural change observed using an SEM is shown in Fig. 2. The as-quenched sample has a β single-phase with a small amount of α phase at the grain boundaries of the β phase (Fig. 2a). These α phases at grain boundaries should be formed during quenching and were also reported in previous studies [13, 15]. Inclusions shown in white in the figure were measured using SEM–EDX (energy dispersive X-ray spectroscopy) and were found to be Sc_2O_3 that could be formed during the melting process. These inclusions are also seen in all the aged samples; therefore, we believe they do not affect the age hardening behavior. Figure 2b presents the microstructure after aging at 423 K for 4 ks, showing the presence of needle-like precipitates within the β matrix phase. These precipitates grow along their longitudinal direction with aging time, and the volume fraction increases, as shown in Fig. 2c and 2d. The same microstructural changes were also seen in the samples aged at 448, 473, and 523 K.

XRD analysis was performed on the samples aged at 423 K to confirm the phase of the precipitates, and the results are shown in Fig. 3. In the XRD pattern, reflection peaks appear at ~ 36.5° after aging for 4 ks, and the peak intensity increases while that of β phase decreases with

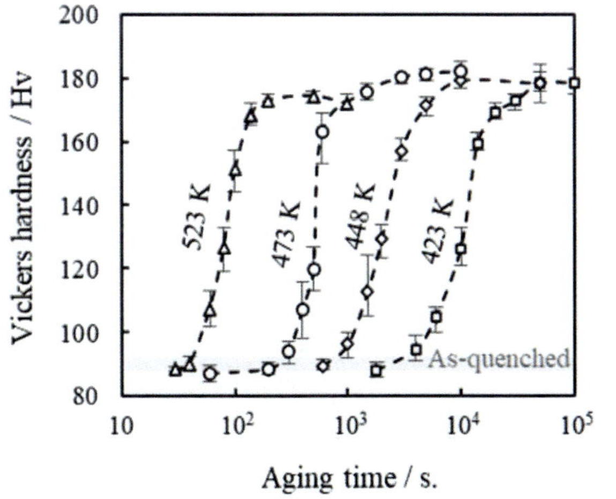

Fig. 1 Vickers hardness as a function of aging time at different temperatures

aging time, indicating that aging products shown in Fig. 2b–d are α phases. Hardness is also plotted in Fig. 4 as a function of the volume fraction of the α phase. The volume fraction was measured using the ImageJ software using the SEM images of the sample aged under different conditions. It is clear from Fig. 4 that the change in hardness due to aging heat treatment is related to the change in the volume fraction of α phase, and an increase in volume fraction of the α phase leads to age hardening in the studied alloy. The same results have been reported in our previous works, where Vickers hardness increases with the volume fraction of needle-like α precipitates due to aging [25–27].

Arrhenius plots are presented in Fig. 5 to investigate the kinetics of age hardening behavior, where t_c denotes the incubation time, as defined in the inset of the figure, and T represents the aging temperature. According to the Arrhenius equation,

$$\ln\left(\frac{1}{t_c}\right) = -\frac{Q}{RT} + C,$$

activation energy Q is equivalent to the slope of the Arrhenius plots, where R and C denote the gas constant and a constant value, respectively. The Q value is calculated as ~ 83.9 kJ/mol and is almost the same as the reported one (~ 83.5 kJ/mol) in a previous report which suggests the precipitation of the α phase is dominated by interface diffusion in the present Mg–Sc alloy [27]. Meanwhile, interface diffusion is also dominant in the precipitation of bainite plates in the case of the Cu-based alloy that shows age hardening owing to bainitic precipitation [30]. Furthermore, notably, if the volume fraction is below 30%, the bainite plates can be deformed together with the matrix phase upon stress-induced martensitic transformation and hence do not deteriorate the superelastic recovery at all despite a slight increase in the hardness of the alloy owing to bainite plates [23]. From this perspective, it is expected that α precipitates could also be deformed in the present Mg–Sc alloy upon stress-induced martensitic transformation and, in turn, could improve the superelasticity of the alloy at room temperature if the volume fraction of α precipitates is in a proper range.

To investigate the effect of α precipitates on the superelasticity of the alloy, we conducted cyclic tensile tests at room temperature. The grain size of the tested sample was increased to ~ 980 μm using AGG induced via CHT, as mentioned in Section Procedures. The aging condition was set to 423 K × 4 ks, where the volume fraction of α precipitates was ~ 10%. Notably, the aged samples with higher volume fraction, i.e., the ones aged at 423 K for 10 and 50 ks, were too fragile to undergo cyclic tensile tests. Figure 6a shows the cyclic stress–strain curves obtained using the as-quenched and aged samples. Both samples

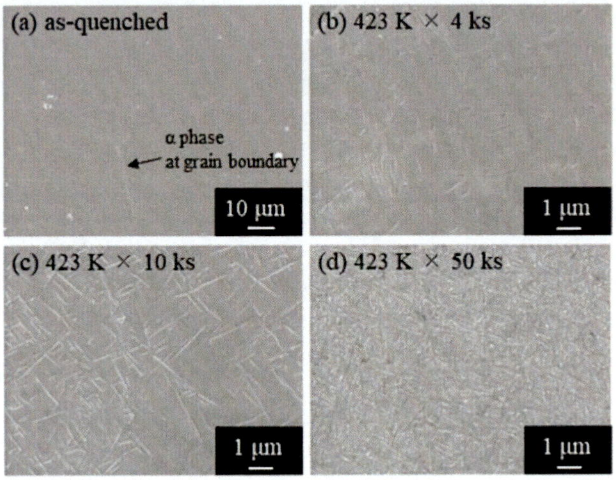

Fig. 2 Secondary electron (SE) images of the samples; **a** as-quenched, **b** after 423 K × 4 ks, **c** after 423 K × 10 ks, and **d** after 423 K × 50 ks

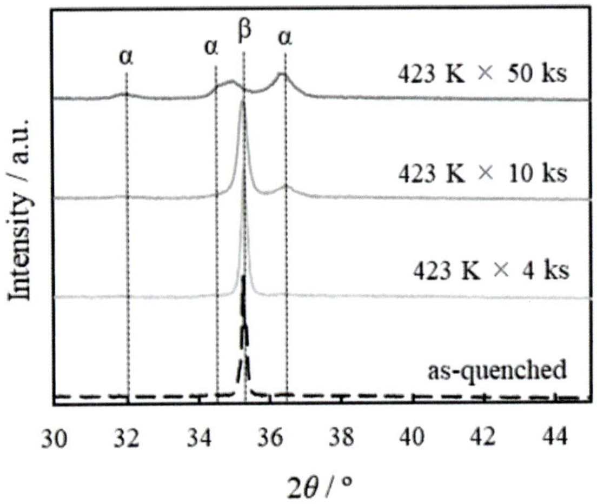

Fig. 3 X-ray diffraction patters obtained from the as-quenched and aged samples

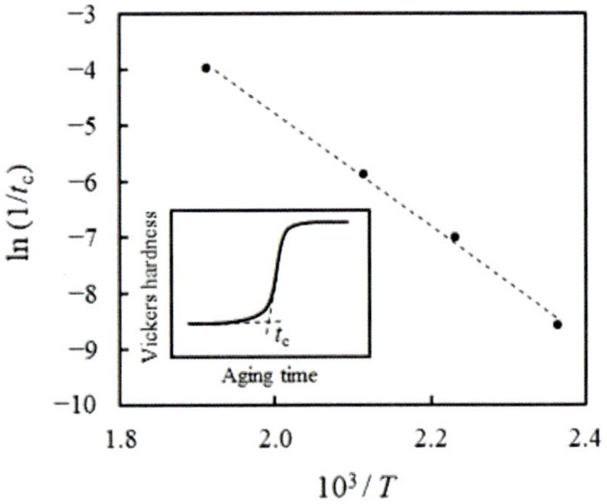

Fig. 5 Arrhenius plots of In $(1/t_c)$ versus $1/T.t_c$ denotes the incubation time as defined in the inset of the figure and T represents the aging temperature

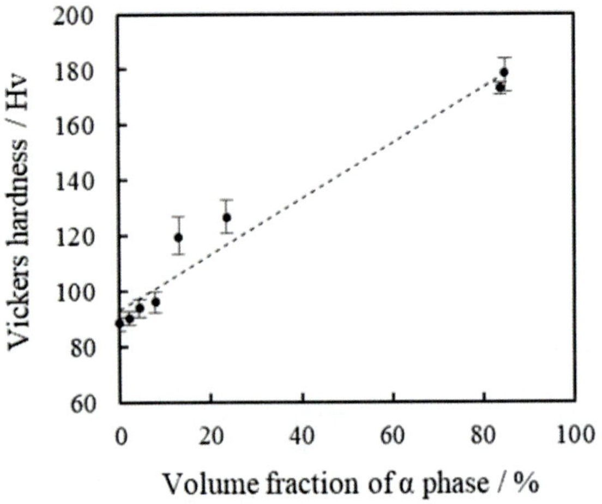

Fig. 4 Vickers hardness as a function of volume fraction of α phase regardless of the aging condition

show clear stress hysteresis during loading and unloading and exhibit superelasticity at room temperature. To compare these two samples quantitatively, superelastic strain (ε^i_{SE}) and applied strain (ε^i_a) at the ith cycle were defined as shown in Fig. 6a. Stress hysteresis (σ^i_{hys}) was also defined as the difference between stresses during loading and unloading at i/2% strain (Fig. 6a). The plots of ε^i_{SE} and σ^i_{hys} as the functions of ε^i_a are presented in Fig. 6b and c, respectively, showing that the maximum superelastic recovery in the aged sample (~ 1.1%) is higher than that in the as-quenched sample (~ 0.8%), whereas σ^i_{hys} is slightly decreased in the aged sample. To investigate the possible causes of the differences, SEM observation was performed on the surface of the aged sample after a fracture. Figure 7a and b presents the obtained microstructures, showing that martensite plates have been formed across α precipitates all over the observed

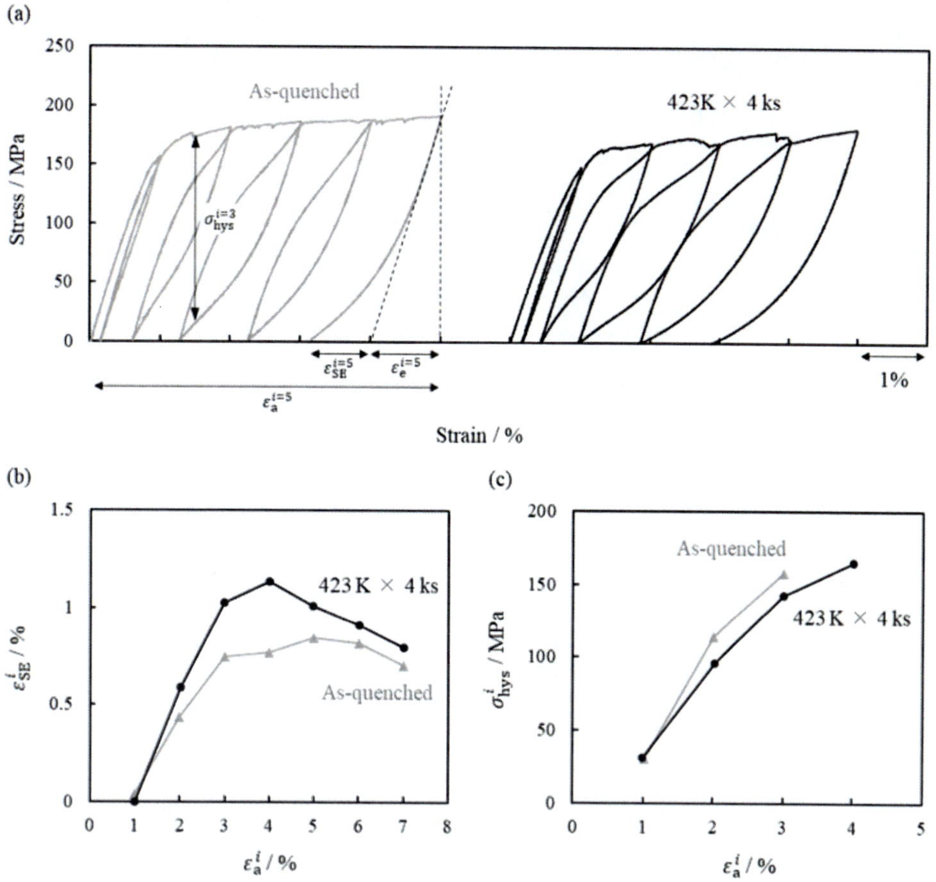

Fig. 6 **a** Comparison of cyclic stress–strain curves at room temperature between as-quenched and aged samples at 423 K for 4 ks. Superelastic strain (ε^i_{SE}), elastic strain (ε^i_e) applied strain (ε^i_a), and stress hysteresis (σ^i_{hys}) at *i*th cycle are also defined in the figure. **b** ε^i_{SE}, and **c** σ^i_{hys} of the samples as the function of ε^i_a

area and the precipitates seem to be deformed along with the β matrix phase upon stress-induced martensitic transformation, as in the case of the foregoing Cu-based superelastic alloy [23, 30]. It is therefore suggested that α precipitates that had been deformed along with β matrix phase cause back stress to assist the reverse transformation of the stress-induced martensite phases. The back stress and a slight increase in hardness might result in decreased stress hysteresis and slightly larger superelastic recovery.

Conclusions

The effect of aging heat treatment on the hardness, microstructure, and superelasticity of the Mg-18.8 at.% Sc alloy at room temperature was investigated herein, and the following conclusions were drawn:

(1) Aging heat treatment at temperatures between 423 and 523 K effectively increased the Vickers hardness of the alloy from ~ 90 Hv (as-quenched condition) to the maximum value of 180 Hv.
(2) SEM observations and XRD analysis revealed that the precipitation of the α phases within the β matrix phase caused age hardening and the hardness value after aging almost linearly depended on the volume fraction of the α phase regardless of the aging condition.
(3) The cyclic tensile test of the sample that had been aged at 423 K for 4 ks revealed that it was possible for the α precipitates with a volume fraction of ~ 10% to be deformed along with the β matrix phase of the alloy upon stress-induced martensitic transformation, which in turn caused back stress during reverse transformation. The back stress and a slight increase in hardness of the aged sample might result in slightly smaller stress hysteresis and minor improvement in superelastic recovery at room temperature compared with those of the as-quenched sample.

Acknowledgements This work was supported by JSPS KAKENHI [grant numbers 21J12146, 18H01691] and the Adaptable and Seamless Technology Transfer Program through target-driven R&D (A-STEP) from Japan Science and Technology Agency (JST) [grant number JPMJTR20TJ].

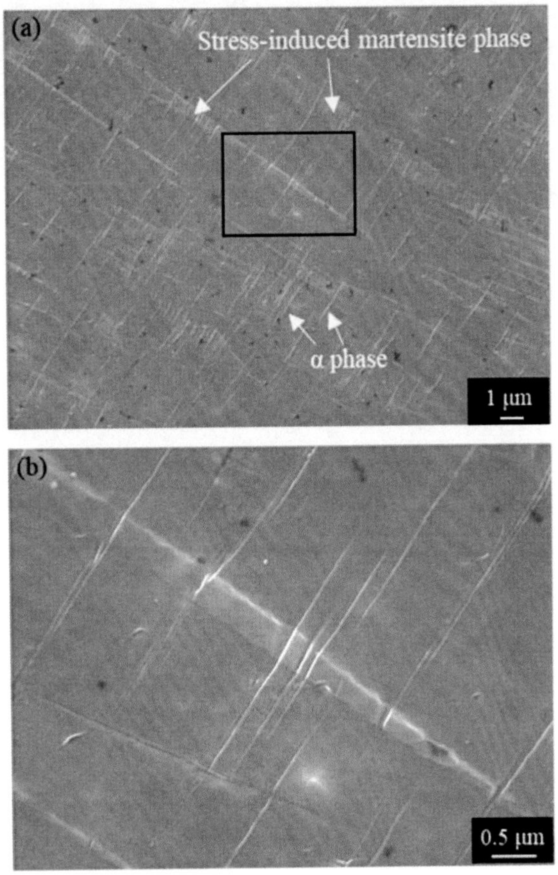

Fig. 7 Secondary electron (SE) images of the aged samples at 423 K for 4 ks after fracture: **a** low and **b** high magnification images

References

1. Otsuka K, Shimizu K (1986) Pseudoelasticity and shape memory effects in alloys. Int. Met. Rev. 31:93–114.
2. Miyazaki S, Otsuka K (1989) Development of shape memory alloys. ISIJ Int. 29:353–377.
3. Ölander A (1932) An electrochemical investigation of solid cadmium-gold alloys. J. Am. Chem. Soc. 54:3819–3833.
4. Otsuka K, Wayman CW, Nakai K, Sakamoto H, Shimizu K (1976) Superelasticity effects and stress-induced martensitic transformations in Cu-Al-Ni alloys. Acta Metall. 24:207–226.
5. Sutou Y, Omori T, Yamauchi K, Ono N, Kainuma R, Ishida K (2005) Effect of grain size and texture on pseudoelasticity in Cu-Al-Mn-based shape memory wire. Acta Mater. 53:4121–4133.
6. Sutou Y, Omori T, Kainuma R, Ishida K (2013) Grain size dependence of pseudoelasticity in polycrystalline Cu-Al-Mn-based shape memory sheets. Acta. Mater. 61:3842–3850.
7. Miyazaki S (2017) My experience with Ti-Ni-based and Ti-based shape memory alloys. Shap. Mem. Superelasticity 3:279–314.
8. Tanaka Y, Himuro Y, Kainuma R, Sutou Y, Omori T, Ishida K (2010) Ferrous polycrystalline shape-memory alloy showing huge superelasticity. Science 327:1488–1490.
9. Omori T, Ando K, Okano M, Xu X, Tanaka Y, Ohnuma I, Kainuma R, Ishida K (2011) Superelastic effect in polycrystalline ferrous alloys. Science 333:68–71.
10. Omori T, Okano M, Kainuma R (2013) Effect of grain size on superelasticity in Fe-Mn-Al-Ni shape memory alloy wire. APL Mater. 1:032103.
11. Miyazaki S, Kim HY, Hosoda H (2006) Development and characterization of Ni-free Ti-base shape memory and superelastic alloys. Mater. Sci. Eng. A 438–440:18–24.
12. Kim HY, Miyazaki S (2015) Martensitic transformation and superelastic properties of Ti-Nb base alloy. Mater. Trans. 56:625–634.
13. Ogawa Y, Ando D, Sutou Y, Koike J (2016) A lightweight shape-memory magnesium alloy. Science 353:368–370.
14. Beaudry BJ, Daane AH (1969) A study of the scandium-magnesium system from 0 to 60 at.% scandium. J. Less-Common Met. 18:305–308.
15. Ogawa Y, Ando D, Sutou Y, Yoshimi K, Koike J (2016) Deformation of α/β phase boundaries and mechanical characterization of Mg-Sc binary alloys. Mater. Sci. Eng. A 670:335–341.
16. Yamagishi K, Ogawa Y, Ando D, Sutou Y, Koike J (2019) Room temperature superelasticity in a lightweight shape memory Mg alloy. Scr. Mater. 168:114–118.
17. Yamagishi K, Ando D, Sutou Y, Ogawa Y (2020) Texture formation through thermomechanical treatment and its effect on superelasticity in Mg-Sc shape memory alloy. Mater. Trans. 61:2270–2275.
18. Zheng YF, Gu XN, Witte F (2014) Biodegradable metals. Mater. Sci. Eng. R 77:1–34.
19. Haghshenas M (2017) Mechanical characteristics of biodegradable magnesium matrix composites: A review. J. Magnes. Alloys 5:189–201.
20. Liu J, Lin Y, Bian D, Wang M, Lin Z, Chu X, Li W, Liu Y, Shen Z, Liu Y, Tong Y, Xu Z, Zhang Y, Zheng Y (2019) *In vitro* and *in vivo* studies of Mg-30Sc alloys with different phase structure for potential usage within bone. Acta. Biomater. 98:50–66.
21. Ozcan H, Ma J, Wang SJ, Karaman I, Chumlyakov Y, Brown J, Noebe RD (2017) Effects of cyclic heat treatment and aging on superelasticity in oligocrystalline Fe-Mn-Al-Ni shape memory alloy wires. Scr. Mater. 134:66–70.
22. Tseng LW, Ma J, Hornbuckle BC, Karaman I, Thompson GB, Luo ZP, Chumlyakov YI (2015) The effect of precipitates on the superelastic response of [100] oriented FeMnAlNi single crystals under compression. Acta Mater. 97:234–244.
23. Sutou Y, Koeda N, Omori T, Kainuma R, Ishida K (2009) Effects of aging on stress-induced martensitic transformation in ductile Cu-Al-Mn-based shape memory alloys. Acta Mater. 57:5759–5770.
24. Kainuma R, Yoshinaka Y, Omori T (2018) Cyclic properties of superelasticity in Cu-Al-Mn single-crystalline sheets with bainite precipitates. Shap. Mem. Superelasticity 4:428–434.
25. Ando D, Ogawa Y, Suzuki T, Sutou Y, Koike J (2015) Age-hardening effect by phase transformation of high Sc containing Mg alloy. Mater. Lett. 161:5–8.
26. Ogawa Y, Ando D, Sutou Y, Koike J (2016) Aging effect of Mg-Sc alloy with α+β two-phase microstructure. Mater. Trans. 57:1119–1123.
27. Ogawa Y, Sutou Y, Ando D, Koike J (2018) Aging precipitation kinetics of Mg-Sc alloy with bcc+hcp two-phase. J. Alloys Compd. 747:854–860.

28. Omori T, Kusama T, Kawata S, Ohnuma I, Sutou Y, Araki Y, Ishida K, Kainuma R (2013) Abnormal grain growth induced by cyclic heat treatment. Science 341:1500–1502.
29. Kusama T, Omori T, Saito T, Kise T, Tanaka T, Araki Y, Kainuma R (2017) Ultra-large single crystals by abnormal grain growth. Nat. Commun. 8:354.
30. Sutou Y, Koeda N, Omori T, Kainuma R, Ishida K (2009) Effects of ageing on bainitic and thermally induced martensitic transformations in ductile Cu-Al-Mn-based shape memory alloys. Acta Mater. 57:5748–5758.

Processing Map and Performance of a Low-Cost Wrought Magnesium Alloy: ZAXEM11100

Thomas Avey, Jiashi Miao, Joshua Caris, Anil K. Sachdev, and Alan Luo

Abstract

Lightweight components improve the fuel efficiency of internal combustion vehicles and contribute to extending the driving range of electrified vehicles. Many Mg alloys have been developed over the years to meet these demands; however, low formability at room temperature, corrosion, and high cost have inhibited widespread adoption in the automotive industry. The new alloy Mg-1Zn-1Al-0.5Ca-0.2Ce-0.4Mn (ZAXEM 11100, all in weight %) has shown excellent post-rolling formability with an Ericksen Index of 7.8 mm and a post-T6 yield stress of 270 MPa in lab scale sheet samples. In this work, a processing map based on Gleeble thermomechanical tests has been developed for the new alloy. This processing map provided important guidance to a production scale extrusion. This work details the mechanical performance of ZAXEM11100 as an extrusion alloy.

Keywords

Extrusion • Magnesium alloys • Processing map

T. Avey · J. Miao · A. Luo (✉)
Department of Materials Science and Engineering, The Ohio State University, Columbus, OH 43210, USA
e-mail: luo.445@osu.edu

J. Caris
Terves LLC, Euclid, OH 44117, USA

A. K. Sachdev
General Motors Global Research and Development, Warren, MI 48092, USA

A. Luo
Department of Integrated Systems Engineering, The Ohio State University, Columbus, OH 43210, USA

Introduction

Vehicle lightweighting has been a strong motivator for magnesium alloy development as a lightweight material [1, 2]. With increasing demand and production of electric vehicles, where large capacity and heavy batteries are needed, the benefit of removing weight from other areas of the vehicle has become a greater initiative [3]. Magnesium (Mg) and its alloys have been formed successfully into automotive parts by both high-pressure die casting (HPDC) and wrought processing [1, 4]. A major hurdle toward greater use of Mg has been its low formability at room temperature, largely due to strong c-axis texture after deformation [4, 5]. This has limited the use of wrought Mg to only higher temperature production methods like rolling, forging, and extrusion [6]. The increase in working temperature can improve formability but will also increase processing time, cost, and complexity compared to steel and aluminum alloys. [1]

Currently, there are very few commercial wrought Mg alloy compositions, compared to steel and Al alloys, used for structural applications [4, 5, 7]. The most common are the Al- and Zn-containing alloys (AZ61, AZ80, AZ31), Zn- and Zr-containing alloys (ZK60, ZK31), and Al- and Mn-containing alloys (AM30). Extensive research has been conducted into modifications to these alloys with rare earths, Li, and/or Ca [8–12]. These additions have shown improvement in room temperature ductility and formability [4, 12]. This is due to the ability of these elements to modify texture of the deformed microstructure, the CRSS of slip systems, and the occurrence and frequency of twinning.

Shi et al. [5], in cooperation with the United States Automotive Materials Partnership (USAMP), developed Mg-1Zn-1Al-0.5Ca-0.4Mn-0.2Ce alloy (all in weight %, hereafter referred to as ZAXEM11100) that possessed high formability in post-rolled T4 condition and excellent tensile properties after T6. The improved formability was credited to texture modifications (Ca and Ce in solid solution) and the

S. Barela et al. (eds.), *Magnesium Technology 2023*, The Minerals, Metals & Materials Series,
https://doi.org/10.1007/978-3-031-22645-8_35

improved strength post-heat treatment to clustering of solute atoms.

Forming at elevated temperatures makes possible deformation modes that are not readily activated at room temperature. In wrought Mg-Zn-Al alloys, deformed at elevated temperatures, Dynamic Recovery (DRV), Dynamic Recrystallization (DRX), and Dynamic Precipitation (DP) have been reported each with different and interdependent effects on flow behavior [13, 14]. The processing map was developed to display the dominate microstructure mechanisms that are acting at a given process conditions (temperature and strain rate) [15–17].

The current paper extends the work on this alloy begun by Shi et al. by observing the compressive stress vs. strain behavior under various temperatures and strain rates to produce a processing map. The extrudability and scalability of this alloy is studied with large-scale extrusion testing.

Numerical Methods

Processing Map Construction

A processing map is a well-studied tool for hot working processes, which can be used to find the optimal temperature and strain rates for deformation [18]. The maps are constructed with two parts: an efficiency parameter and a stability criterion [19–21]. Most published processing maps make use of the Prasad method [22] due in part to the computational simplicity as well as the large body of literature to support its reliability. A brief derivation of the important equations is given as well as a discussion as to important assumptions made in the derivation that must be satisfied. The assumptions of the processing maps are that the work piece acts as a power dissipator, where the total applied power (P) is dissipated by plastic work/heat and microstructural change, through irreversible processes. This process is expressed as:

$$P = \sigma \cdot \dot{\varepsilon} = \int_0^{\dot{\varepsilon}} \sigma * d\dot{\varepsilon} + \int_0^{\sigma} \dot{\varepsilon} * d\sigma \tag{1}$$

which can also be written more simply as

$$P = G + J \tag{2}$$

with G being the dissipator content and represents the power dissipated by plastic work (heat) and J being the dissipator co-content which represents the power dissipated by dynamic microstructural and metallurgical processes. J is the term of interest and can be evaluated as

$$J = \int_0^{\sigma} \dot{\varepsilon} * d\sigma = \frac{\sigma * \dot{\varepsilon} * m}{m+1} \tag{3}$$

under the assumption of a power-law stress–strain rate relationship ($\sigma = A\dot{\varepsilon}^m$). The simplification in Eq. 3 is only valid if strain rate ($\dot{\varepsilon}$) and the strain rate sensitivity parameter (m) are independent. This independence can be verified by plotting $\ln(\sigma)$ versus $\ln(\dot{\varepsilon})$. If $\dot{\varepsilon}$ and m are independent, the resulting curve will be linear and the efficiency that power is dissipated by microstructural processes, η, is defined by Prasad as the ratio

$$\eta = \frac{J}{J_{max}} = \frac{2m}{m+1} \tag{4}$$

where J_{max} is the value of J at $m = 1$ or where the workpiece acts as a liner dissipator. However, if $\dot{\varepsilon}$ and m are not independent, both Eqs. 3 and 4 are invalid. To compute processing maps in this case, Murty and Rao et al. developed a more general solution that does not rely on the power-law relationship [23]. Equation 4 was modified to be

$$\eta = \frac{J}{J_{max}} = \frac{(P-G)}{J_{max}} = 2\left[1 - \frac{1}{\sigma * \dot{\varepsilon}} \int_0^{\dot{\varepsilon}} \sigma d\dot{\varepsilon}\right] \tag{5}$$

Equation 5 requires an integration of G beginning at a strain rate of zero which is not done experimentally. To mediate this, Murty and Rao separated the integral in Eq. 5 so that a power-law relationship is assumed near a strain rate of zero via Eq. 6 and substituted into Eq. 5.

$$G_o = \frac{1}{m+1} \sigma \dot{\varepsilon} \tag{6}$$

$$\begin{aligned} \eta &= 2\left[1 - \frac{1}{\sigma * \dot{\varepsilon}} \int_0^{\dot{\varepsilon}} \sigma d\dot{\varepsilon}\right] \\ &= 2\left[1 - \frac{1}{\sigma * \dot{\varepsilon}} \left(\int_0^{\dot{\varepsilon}_{min}} \sigma d\dot{\varepsilon} + \int_{\dot{\varepsilon}_{min}}^{\dot{\varepsilon}} \sigma d\dot{\varepsilon}\right)\right] \\ &= 2\left[1 - \frac{1}{\sigma * \dot{\varepsilon}} \left(G_0 + \int_{\dot{\varepsilon}_{min}}^{\dot{\varepsilon}} \sigma d\dot{\varepsilon}\right)\right] \end{aligned} \tag{7}$$

There are many proposed equations to predict instability for the construction of processing maps [24]. Kumar [19] and Prasad [20] built on the instability criteria outlined by Ziegler [25]

$$\frac{dD}{d\dot{\varepsilon}} > \frac{D}{\dot{\varepsilon}} \tag{8}$$

where $D(\dot{\varepsilon})$ is the materials dissipative function which is substituted for J as the metallurgical dissipative function is of interest. The full derivation can be found in the cited literature. This substitution in a power-law stress and strain rate relationship yields the inequality

$$\xi(\dot{\varepsilon}) = \frac{d\ln\left(\frac{m}{m+1}\right)}{d\ln\dot{\varepsilon}} + m > 0 \tag{9}$$

wherever this inequality is not satisfied, instable flow is predicted to occur. When the power law relationship is not assumed, the stability criteria is seen in Eq. 10 as derived by Murty and Rao [24].

$$2m > \eta \tag{10}$$

Materials and Methodology

The material used for this study was gravity cast ZAXEM11100 alloy billets (22.86 cm diameter by 76.2 cm length) provided by Terves LLC. The composition was verified with spark optical emission spectroscopy (Spark-OES) and is included in Table 1. A test billet was subjected to a multistage heat treatment, developed in a previous work [5], to sequentially dissolve all secondary phases and improve processability.

Compression cylindrical samples were machined by electronic discharge machining (EDM) to the size of 10 mm diameter by 15 mm long. Samples were taken near the outer edge of the heat-treated billet slice. Uniaxial compression tests at fixed temperatures and true strain rates were performed on a Gleeble 3800 thermomechanical testing frame. The test temperatures were from 325 to 425 °C in 25 °C increments and strain rates in one-decade increments from 10^{-3} to 10^{0} s^{-1}. Strain was measured by using the displacement of the Gleeble arm. All compression tests were conducted to a true strain of 1 mm/mm. Samples were resistively heated at a rate of 5 °C/s until the desired temperature was reached. The temperature was measured with a thermocouple that was impact welded to the center of the sample. All sample temperatures remained within ±3 °C throughout testing. After testing, samples were quenched into room temperature water within 5 s of test termination to prevent post-deformation microstructure evolution.

Large-scale extrusions were done on an Extral 3800 MT extrusion press. The billets were skinned with a lathe to remove scaling before preheating to 390 °C. A 0.4″ × 5.5″ (1 × 13.9 cm) plate die with at extrusion ratio of 28.9 was used. Feed rate was varied to change the average strain rate, as calculated through slab analysis, of the extrusion from 0.04 to 0.15 s^{-1}.

Results and Discussion

Figure 1 shows the flow behavior of ZAXEM11100 alloy, following an expected trend of yield stress and flow stress decreasing with increasing temperature and increasing with increasing strain rate. The flow stress data was first processed with a custom-built MATLAB program using a cubic spline fit to extrapolate the experimental data to intermediate strain rates and temperatures. After this extrapolation, a ln(σ) versus ln($\dot{\varepsilon}$) plot, Fig. 2, was constructed, and ZAXEM11100 was determined to not follow a linear relationship, and, therefore, the Murty and Rao method for processing map construction (Eq. 7) and the instability relationship from Eq. 10 will be used in this work.

The flow stress data at 20, 40, 60, and 80% strains was used to construct the processing maps seen in Fig. 3. Two domains and one instability region were identified by conventional means. Domain I has a maximum efficiency 0.95–1, exists at low strain rate (10^{-3} s^{-1}), and shifts from 350 to 375 °C as the strain increases. The peak efficacy of Domain II cannot be determined because the bulk of this domain is at strain rates higher than the tested rates. The bulk of the processing map is dominated by the instability region. This region exists between strain rates of $10^{-2.7}$ s^{-1} to 10^{-1} s^{-1} at all temperatures investigated with little movement seen in location the instability region at different strains.

Generally, curves at low strain rates (10^{-3}–10^{-2}) and high temperatures (400–425 °C) show ideal plastic flow behavior while curves at 10^{0} strain rate and 325–375 °C show strain softening that is evidence of dynamic microstructural changes. The early peak in flow stress followed by softening seen in 10^{-2} s^{-1}, 325 °C condition and the 10^{0} s^{-1}, 425 °C both show strong evidence for significant DDRX [26] (Fig. 1).

The power of a processing map is its ability to identify regions ideal for extensive deformation processing. Conventional wisdom is to select strain rates and temperatures that result in a high degree of DRX. In Mg alloys, this is even more important due to the reduced *c*-axis texture and grain size of heavily recrystallized microstructures. A peak efficiency between 30 and 50%, depending on the material's stacking fault energy, is a good indication for DRX being the dominating microstructural process [22]. Domain II could be an option as the efficacy seen in the range investigated is in the 30–50% range that is typical of good dynamic recrystallization [18].

Table 1 OES measured composition of ZAXEM11100 billet

Element	Zn	Al	Ca	Ce	Mn	Other elements	Mg
Composition (wt.%)	1.38	1.25	0.52	0.2	0.46	0.09	96.1

Fig. 1 True stress vs strain plots of ZAXEM 11100 at strain rates of **a** 10^{-3} s^{-1}, **b** 10^{-2} s^{-1}, **c** 10^{-1} s^{-1}, and **d** 10^{0} s^{-1}

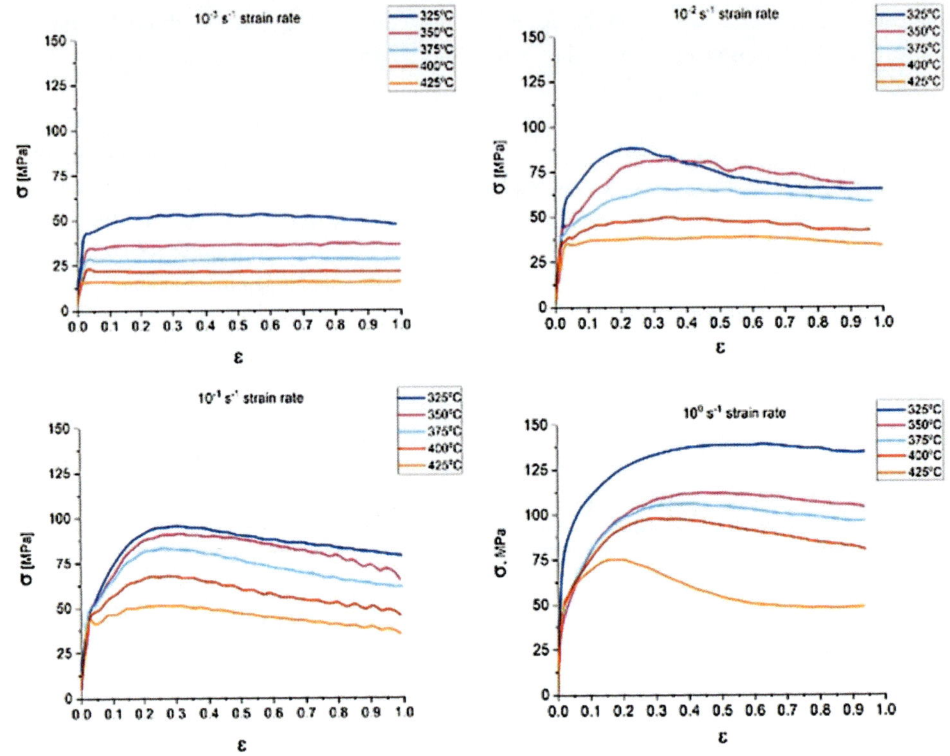

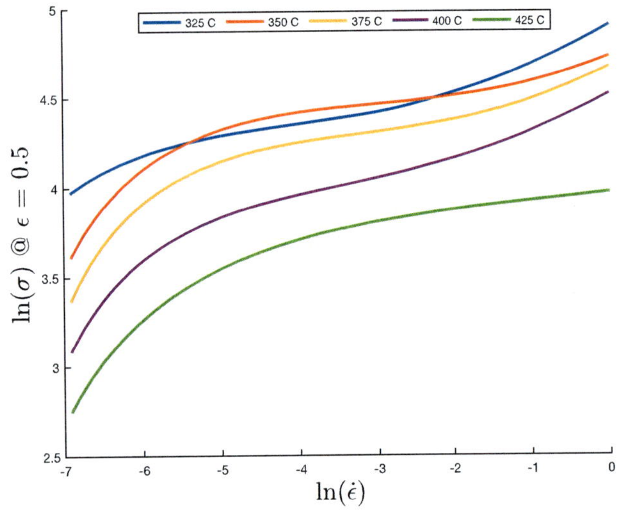

Fig. 2 ln(σ) versus ln($\dot{\varepsilon}$) plot of ZAXEM11100 at a fixed true strain of 0.5

Production Scale Extrusions

A pilot study of the performance of this alloy under industrial scale extrusion was performed on the same billet with the same condition used in the production of the processing map. The first trial was done with a billet temperature of 390 °C and a speed of 1″/min (25.4 mm/min). There was no observed macroscopic cracking, Fig. 4a. A second trial was done at the same billet temperature but with an elevated extrusion speed of 2″/min (50.8 mm/min). This second trial, Fig. 4b, resulted in extensive hot cracking after extrusion despite the calculated average strain rate, 0.15 s^{-1}, and extrusion temperature falling within the stable region of the processing map except for the initial strains, Fig. 3a.

Through CALPHAD analysis Fig. 5, the solidus of ZAXEM11100 was calculated to be quite low at 425 °C. The measured temperature of the plate, after leaving the extrusion press, fluctuated between 420 and 450 °C which is 60 °C warmer than the billet before extrusion and above the solidus temperature indicating that friction during extrusion resulted in sufficient heating induce incipient melting. A lower initial billet temperature was attempted but the stress required to begin extrusion, breakthrough stress, was too great and extrusion was not possible. In Table 2, the effect of Zn and Ca content on the solidus was investigated. It was found that reduction of each individually was able to raise the solidus to above the exit temperature of the billet at 2″/min.

While ZAXEM11100 has shown good performance in lab scale rolling experiments and some promise in the processing map analysis, the low solidus gives it a very narrow temperature range for successful production scale extrusion. A change in the Zn or Ca content could increase this window and lead to a more extrudable alloy.

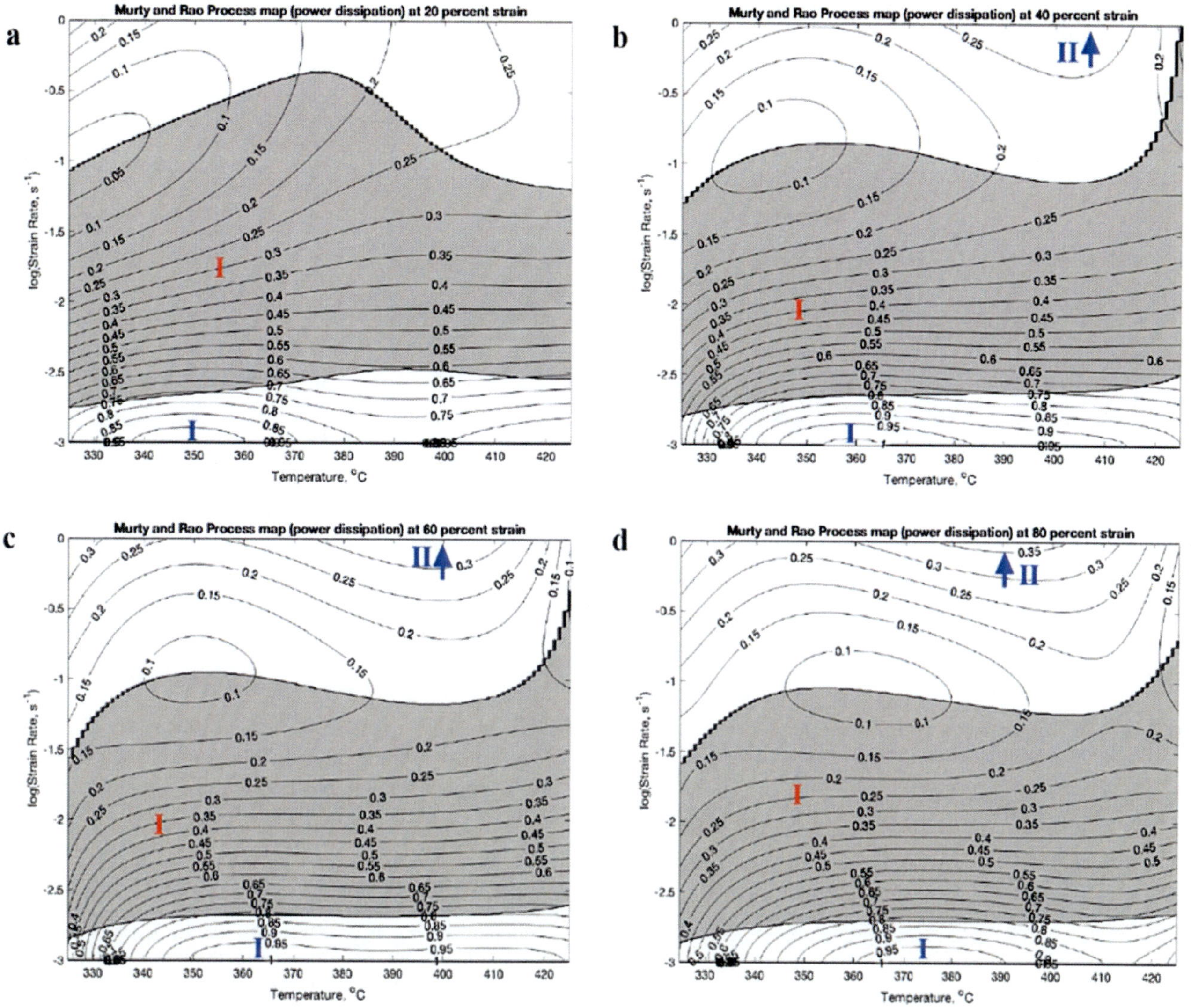

Fig. 3 Processing map Constructed for ZAXEM11100 at **a** 20%, **b** 40%, **c** 60%, and **d** 80% strain with the instability region shaded in gray. Contour lines correspond to lines of equal efficiency

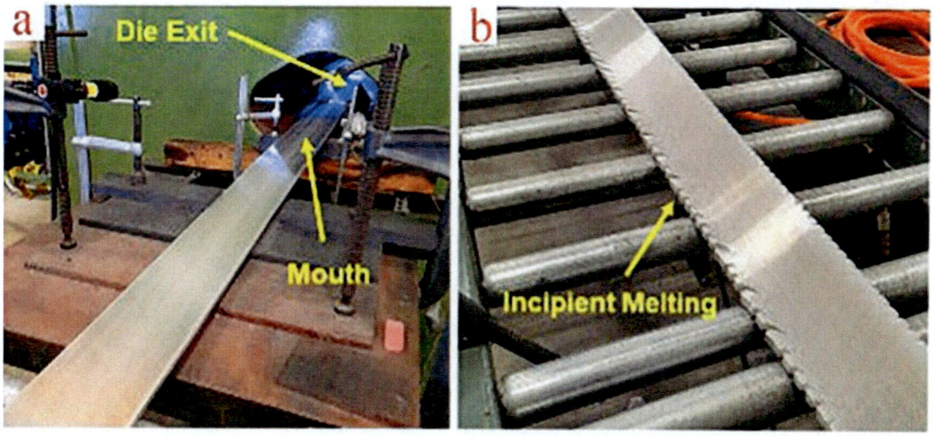

Fig. 4 Images of extruded ZAXEM11100 plate as it left the die with an extrusion speed of **a** 1``/min (25.4 mm/min) and **b** 2'/min (50.8 mm/min)

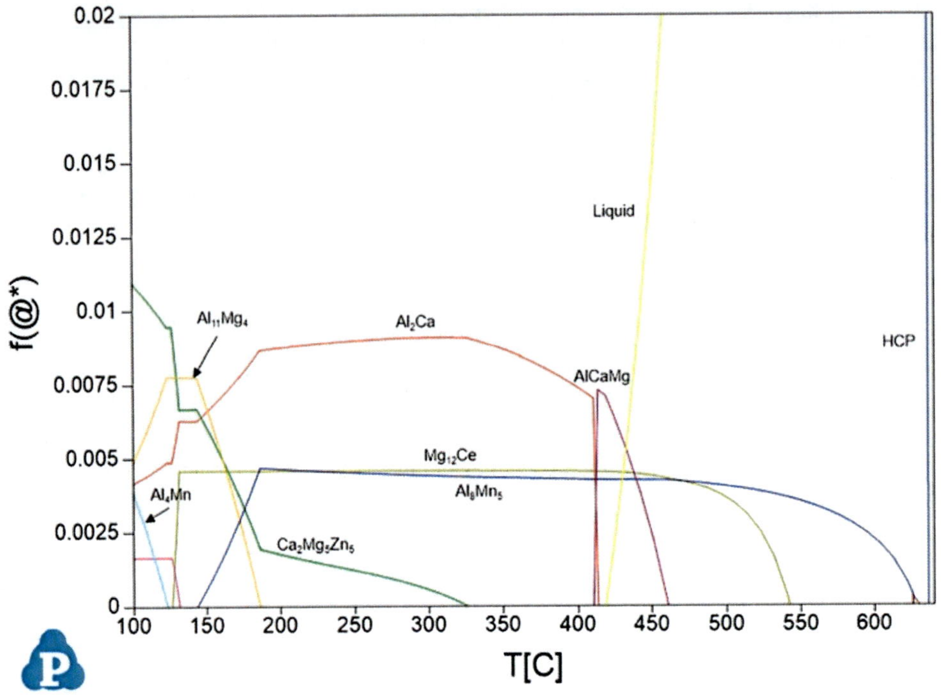

Fig. 5 Equilibrium phase fraction versus temperature plot of ZAXEM11100 billet used in extrusion testing calculated with the PanMg2021 database

Table 2 Effect of Zn and Ca content on the solidus of nominal ZAXEM11100 composition (Mg-1Zn-1Al-0.5Ca-0.2Ce-0.4Mn)

Zn content (wt%)	Solidus temp (°C)	Ca content (wt%)	Solidus temp (°C)
0.7	465	0	570
1	442	0.5	442
1.4	417		

Conclusions

An investigation of the processing feasibility of ZAXEM11100 was conducted through lab scale thermo-mechanical testing and through production scale extrusions.

- ZAXEM11100 does not exhibit a linear strain sensitivity parameter and therefore needs to be modeled with the less conventional but more general solution developed by Murty and Rao [23].
- The resulting processing map shows a wide instability region from 350 to 425 °C and at strain rates between $10^{-2.5}$ and 10^{-1} s^{-1}.
- Production scale extrusion found that extrusion speed was limited by incipient melting resulting from frictional heating and ZAXEM11100's low solidus temperature.
- CALPHAD analysis was done, and the solidus of ZAXEM11100 was shown to be very sensitive to Zn and Ca content.
- The thermal window to successfully extrude ZAXEM11100 limits its usability but a change to the Zn or Ca composition should improve its workability.

Acknowledgements Funding for this work was provided by DOE project award DE-SC0020806. This report was prepared as an account of work sponsored by an agency of the United States government. Neither the United States government nor any agency thereof, nor any of their employees, makes any warranty, express or implied, or assumes any legal liability or responsibility for the accuracy, completeness, or usefulness of any information, apparatus, product, or process disclosed, or represents that its use would not infringe privately owned rights. Reference herein to any specific commercial product, process, or service by trade name, trademark, manufacturer, or otherwise does not necessarily constitute or imply its endorsement, recommendation, or favoring by the United States government or any agency thereof. The views and opinions of authors expressed herein do not necessarily state or reflect those of the United States government or any agency thereof.

References

1. Joost WJ, Krajewski PE. Towards magnesium alloys for high-volume automotive applications. Scr Mater 2017;128:107–12. https://doi.org/10.1016/J.SCRIPTAMAT.2016.07.035.
2. Taub A, De Moor E, Luo A, Matlock DK, Speer JG, Vaidya U. Materials for Automotive Lightweighting. Annu Rev Mater Res 2019;49:327–59. https://doi.org/10.1146/annurev-matsci-070218-010134.
3. Muratori M, Alexander M, Arent D, Bazilian M, Cazzola P, Dede EM, et al. The rise of electric vehicles—2020 status and future expectations. Prog Energy 2021;3:022002. https://doi.org/10.1088/2516-1083/ABE0AD.

4. Luo AA, Shi R, Miao J, Avey T. Review: Magnesium Sheet Alloy Development for Room Temperature Forming. JOM 2021;73:1403–18. https://doi.org/10.1007/s11837-021-04616-y.
5. Shi R, Miao J, Avey T, Luo AA. A new magnesium sheet alloy with high tensile properties and room-temperature formability. Sci Rep 2020;10:1–10. https://doi.org/10.1038/s41598-020-67161-9.
6. Pekguleryuz MO. Current developments in wrought magnesium alloys. Woodhead Publishing Limited; 2012. https://doi.org/10.1533/9780857093844.1.3.
7. Luo AA. Applications: Aerospace, automotive and other structural applications of magnesium. Fundam. Magnes. Alloy Metall. A Vol. Woodhead Publ. Ser. Met. Surf. Eng., Elsevier Inc.; 2013, p. 266–316. https://doi.org/10.1533/9780857097293.266.
8. Bettles CJ, Gibson MA. Current wrought magnesium alloys: Strengths and weaknesses. JOM 2005 575 2005;57:46–9. https://doi.org/10.1007/S11837-005-0095-0.
9. Liu C, Chen X, Chen J, Atrens A, Pan F. The effects of Ca and Mn on the microstructure, texture and mechanical properties of Mg-4 Zn alloy. J Magnes Alloy 2021;9:1084–97. https://doi.org/10.1016/J.JMA.2020.03.012.
10. Meng X, Wu R, Zhang M, Wu L, Cui C. Microstructures and properties of superlight Mg–Li–Al–Zn wrought alloys. J Alloys Compd 2009;486:722–5. https://doi.org/10.1016/J.JALLCOM.2009.07.047.
11. Liu P, Jiang H, Cai Z, Kang Q, Zhang Y. The effect of Y, Ce and Gd on texture, recrystallization and mechanical property of Mg–Zn alloys. J Magnes Alloy 2016;4:188–96. https://doi.org/10.1016/J.JMA.2016.07.001.
12. Mishra RK, Gupta AK, Rao PR, Sachdev AK, Kumar AM, Luo AA. Influence of cerium on the texture and ductility of magnesium extrusions. Scr Mater 2008;59:562–5. https://doi.org/10.1016/J.SCRIPTAMAT.2008.05.019.
13. He YY, Fang G. Characterization of dynamic precipitation behaviors accompanying dynamic recrystallization in an Mg-Al-Zn-RE alloy. J Alloys Compd 2022;901:163745. https://doi.org/10.1016/j.jallcom.2022.163745.
14. Myshlyaev MM, Mcqueen HJ, Mwembela A, Konopleva E. Twinning, dynamic recovery and recrystallization in hot worked MgÁ/ AlÁ/ Zn alloy. Mater Sci Eng A 2002;337:121–33.
15. Ashby MF. A first report on deformation-mechanism maps. Acta Metall 1972;20:887–97. https://doi.org/10.1016/0001-6160(72)90082-X.
16. Fatemi SM, Aliyari S, Miresmaeili SM. Dynamic precipitation and dynamic recrystallization during hot deformation of a solutionized WE43 magnesium alloy. Mater Sci Eng A 2019;762:138076. https://doi.org/10.1016/J.MSEA.2019.138076.
17. Mohamed FA, Langdon TG. Deformation mechanism maps based on grain size. Metall Mater Trans B 1974 511 1974;5:2339–45. https://doi.org/10.1007/BF02644014.
18. Sutton SC, Luo AA. Constitutive behavior and processing maps of a new wrought magnesium alloy ZE20 (Mg-2Zn-0.2Ce). J Magnes Alloy 2020;8:111–26. https://doi.org/10.1016/j.jma.2019.11.007.
19. Kumar AKSK. Criteria for Predicting Metallurgical Instabilities in Processing. Master's Thesis. India Institute of Science, Bangalore, India, 1987.
20. Prasad YVRK, Gegel HL, Doraivelu SM, Malas JC, Morgan JT, Lark KA, et al. Modeling of Dynamic Material Behavior in Hot Deformation: Forging of Ti-6242. n.d.
21. Prasad YVRK. Recent Advances in the Science of Mechanical processing. Indian J Technol 1990;28:435–51.
22. Prasad Y, Rao K, Sasidhar S. Hot working guide: a compendium of processing maps. ASM international; 2015.
23. Murty SVSN, Sarma MS, Rao BN. On the evaluation of efficiency parameters in processing maps. Metall Mater Trans A Phys Metall Mater Sci 1997;28 A:1581–2. https://doi.org/10.1007/s11661-997-0219-y.
24. Narayana Murty SVS, Nageswara Rao B, Kashyap BP. Instability criteria for hot deformation of materials. Int Mater Rev 2000;45:15–26. https://doi.org/10.1179/095066000771048782.
25. Zeigler H. Progress in solid mechanics. New York: Wiley; 1963.
26. Huang K, Logé RE. A review of dynamic recrystallization phenomena in metallic materials. Mater Des 2016;111:548–74. https://doi.org/10.1016/j.matdes.2016.09.012.

Part VI
Poster Session

Effect of Extrusion and Heat Treatment on Microstructure and Mechanical Properties of Mg-1.27Zn-0.75Gd-0.17Zr Alloy

Siqi Yin, Yifan Zhang, Dongting Hou, Guangzong Zhang, and Zhiqiang Zhang

Abstract

Mg-1.27Zn-0.75Gd-0.17Zr alloy (at.%) reinforced by long period stacking ordered (LPSO) structure X-$Mg_{12}GdZn$ was fabricated and then subjected to indirect extrusion and T5 and T6 heat treatments, respectively. Effect of indirect extrusion and heat treatments on microstructure evolution and room temperature tensile mechanical properties were systematically studied and discussed. Results show that the morphology of *X*-phase changes from plate-like to lamellar-like and further to disconnected block-like with the technical process. Large amounts of fine β' phases precipitate in α-Mg matrix of T5-treated alloy. Tensile and yield strength of the as-extruded alloy is 338 and 230 MPa, which is owing to the hard and fragile lamellar-like *X*-phase is easy to start cracks. T5-state alloy exhibits optimal mechanical properties with ultimate tensile strength of 390 MPa, tensile yield strength of 295 MPa and elongation of 11.9%, owing to the β' precipitates strengthening and LPSO *X*-phase strengthening.

Keywords

Mg-Zn-Gd-Zr alloy • Ternary phases • Heat treatment • β' precipitate • Mechanical properties

Introduction

Recently, magnesium alloys have shown the potential to serve as excellent structural and engineering materials that can reduce the weight of vehicle or aerospace components with superior high specific strength and specific stiffness [1, 2]. However, a serious difficulty in further application of magnesium alloys is the poor mechanical properties, especially its low strength and mechanical processing [3].

The strengthening of Mg-based alloys has long been a popular subject for the weight reduction. Mg-Zn-RE alloys have attracted more attention because of the superior mechanical properties. The addition of Zn in Mg-RE alloys has been identified to have significant effect in improving both the strength and ductility [4]. The solid solubility of rare earth element Gd in magnesium alloys reduces dramatically with decreasing temperature, whose equilibrium solid solubility at 821 K and 473 K is 4.53 at.% and 0.61 at.%, respectively, forming an ideal system for precipitation hardening. Thus, magnesium alloys with Gd addition have been extensively studied to develop high strength and good heat-resistance. Anyanwu et al. researched that the as-extruded Mg-17Gd-0.51Zr alloy reaches the ultimate tensile strength of 400 MPa after T5 heat treatment, which was mainly due to the fine precipitates in the matrix during aging [5]. Mg-Zn-Gd alloys with excellent mechanical properties are attributed to solution strengthening, aging strengthening and secondary phases strengthening, especially long period stacking ordered (LPSO) structure X-$Mg_{12}GdZn$ strengthening [6]. Lu et al. [7] investigated that 16 ECAP passed $Mg_{97.1}Zn_1Gd_{1.8}Zr_{0.1}$ (at.%) alloy owns the ultimate tensile strength, yield strength and elongation of

S. Yin · Y. Zhang · D. Hou
School of Mechanical Engineering, Dalian Jiaotong University, Dalian, 116028, China
e-mail: yinsiqi89@163.com

Y. Zhang
e-mail: 1940300452@qq.com

D. Hou
e-mail: 21940300452@qq.com

G. Zhang (✉)
Engineering Research Center of Continuous Extrusion, Ministry of Education, Dalian Jiaotong University, Dalian, 116028, China
e-mail: gzzhang@djtu.edu.cn

Z. Zhang
Key Lab of Electromagnetic Processing of Materials, Ministry of Education, Northeastern University, Shenyang, 110819, China
e-mail: zqzhang@mail.neu.edu.cn

S. Barela et al. (eds.), *Magnesium Technology 2023*, The Minerals, Metals & Materials Series,
https://doi.org/10.1007/978-3-031-22645-8_36

387 MPa, 324 MPa and 23.2%, respectively, which could be attributed to a large number of 14H LPSO *X*-phase.

As a kind of effective strengthening phase, Mg-Zn-Gd alloys with LPSO phase show good possibility to be a type of high-strength magnesium alloys. The accumulated experimental results indicate that the mechanical properties of Mg-Zn-Gd alloys could be tailored by reasonable control of the volume fraction and morphology of the LPSO structures. Simultaneously, Zn and Gd ratio could influence the volume fraction of LPSO *X*-phase and show a range of mechanical properties. More recently, it is believed that the LPSO-containing magnesium alloys demonstrate ordinary performance in as-cast condition, but the mechanical properties are significantly enhanced after conventional plastic deformations such as hot extrusion, equal channel angular pressing (ECAP) process and rolling [8]. Hot extrusion is an extremely effective method to refine the microstructure and phase morphology so as to improve the mechanical properties of magnesium alloys. On the basis of considering technical conditions could also cause the transformation of secondary phase volume fraction and morphology, and the effect and strengthening mechanism have not yet been widely reported.

Based on this background, microstructures and mechanical properties of a Mg-1.27Zn-0.75Gd-0.17Zr alloy during extrusion, and following solution treatment and isothermal aging (T5 and T6 treatments) were investigated in this study, in order to explore the possibility of preparing Mg-Zn-Gd-Zr alloys strengthened via co-existed LPSO phases and precipitates. A trace amount of Zr added into the investigated alloys was aiming at grain refinement.

Materials and Methods

A *Φ*60 mm Mg-1.27Zn-0.75Gd-0.17Zr (at.%) billet was successfully prepared by conventional casting method [9]. Chemical composition of the alloy is presented in Table 1. Samples cut from the ingot are homogenized at 430 °C for 14 h and quenched into cold water. Indirect extrusion is from *Φ*47 mm cylinders to *Φ*12 mm rods with a speed of 5.6 cm s^{-1} in a *Φ*50 mm extruding container at 400 °C. Aging hardening behaviors are investigated at 200 °C for different times between 2 and 117 h. Parts of the extruded rods and tensile specimens are solution-treated at 430 °C for 8 h, among which a fraction of them is followed by aging at 200 °C to achieve T6-state. T5 processing is performed by aging of the extruded samples directly at 200 °C. Finally, the annealed samples are dipped into the water for quenching.

Tensile tests of the specimens are performed in triplicate at a crosshead speed of 1 mm/min, according to the relevant standard of using cylindrical specimen with the dimension of 25 mm gauge length and 6 mm diameter. All tensile tests are carried out on a Shimadzu AG-X (10 kN) machine at room temperature. Image analysis technique using at least ten areas is applied to estimate the volume fraction of second phase. Vickers hardness test is carried out by 3 kg load for 15 s. Linear intercept method is used to determine the grain size of the alloys. Microstructures are observed by an optical microscope (OM), a transmission electron microscope (TEM; JEM-ARM200F) and a scanning electron microscope (SEM; Zeiss Ultra 55, Germany) with an energy dispersive spectroscope analyzer (EDS). The acceleration voltage of TEM is 200 kV, the corresponding bright-field and high-resolution images are characterized. Similarly, the secondary electron images of SEM are detected, 500–5000 magnifications are used, and the acceleration voltage is 15 kV. Results of diffraction peaks are manipulated through the normalized process.

Results and Discussion

Age hardening behaviors: The age-hardening values with different aging time of the sample aged at 200 °C is shown in Fig. 1. Originally, the hardness value increases slowly and then increases rapidly after 8 h aging. It reaches the peak at 16 h and then decreases gradually. Subsequently, the hardness adopts a saw-tooth-like model with the prolonged aging time and then turns into a period of fluctuation. In order to correlate the hardness with the microstructures, peak hardness of 16 h will be investigated further.

Microstructure evolution: Fig. 2 shows the XRD patterns of the Mg-1.27Zn-0.75Gd-0.17Zr alloy at different conditions. It reveals that as-cast Mg-1.27Zn-0.75Gd-0.17Zr alloy is composed of α-Mg solid solution, *W*-phase ($Mg_3Zn_3Gd_2$), and *X*-phase ($Mg_{12}GdZn$). After the indirect extrusion and the subsequent heat treatments, second phases composition remain the same but diffraction peaks intensity of *X*-phase is stronger than that in as-cast condition, and diffraction peaks intensity of *W*-phase is weaker than the as-cast alloy.

Figure 3 manifests the microstructures of the longitudinal section for the Mg-1.27Zn-0.75Gd-0.17Zr alloy. After

Table 1 Chemical analysis constitution of the experimental alloy

Alloy	Chemical analysis constitution (at. %)			
	Zn	Gd	Zr	Mg
Mg-1.27Zn-0.75Gd-0.17Zr	1.268	0.753	0.171	Bal

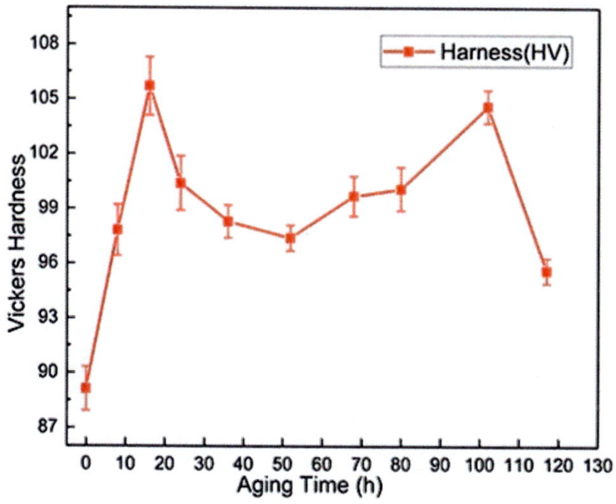

Fig. 1 Age hardening values of experimental alloy

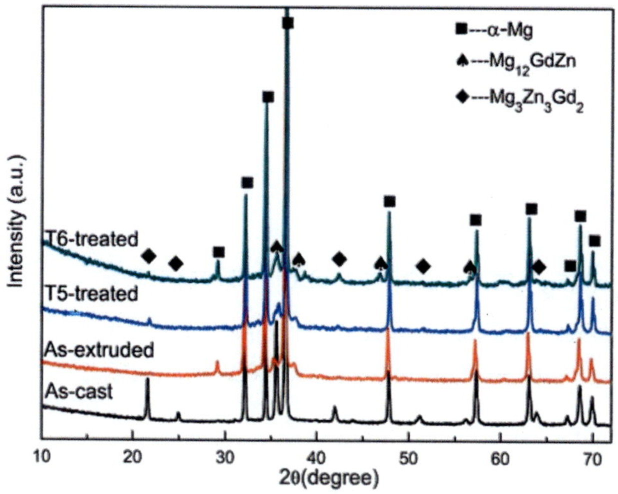

Fig. 2 XRD patterns of the Mg-1.27Zn-0.75Gd-0.17Zr alloy

extrusion and heat treatments, the second phases are cracked and zonal distributing along the extruding direction. Meanwhile, dynamic recrystallization (DRX) occurs during the extrusion process and large amount of dynamic recrystallized (DRXed) grains appear. Average size of the dynamic recrystallization grains ranges from 8 ~ 12 to 10 ~ 15 μm via T5-treated procedure, and it reaches about 18 μm after T6 treatment. This phenomenon illustrates that heat treatment process could coarsen the grains. EDS result of point A suggests that the chemical composition of the block-like phase is Mg-9.17 at. % Zn-6.49 at. % Gd, suggesting it is W-$Mg_3Zn_3Gd_2$ phase. The lamellar-like structure of point B in Fig. 2a contains Mg (88.47 at. %), Zn (5.13 at. %) and Gd (6.36 at. %), which is close to X-$Mg_{12}GdZn$ phase. Volume fraction of the second phases in matrix increase with the process of T5 and T6 treatments. Meanwhile, *X*-phase morphology develops into bulk of thick plates throughout the matrix (Fig. 3e).

The TEM observation and corresponding SAED patterns for the as-annealed specimens are shown in Fig. 4. The 14-layer period-stacking structure *X*-phase is always existed in as-extruded and as-annealed conditions as shown in Fig. 4b and c. Figure 4a shows the TEM bright-field image of the alloy aged at 200 °C for 16 h. A high density of fine spherical precipitates with black contrast could be observed in the grains. These particles are distributing in the matrix uniformly and compactly. According to the corresponding selected area electron diffraction (SAED) pattern given in Fig. 4d, extra diffraction spots of the precipitates are observed at the 1/2 {1–210} and 1/2 {−1012} positions, indicating these precipitates are β' phases [10]. Figure 4b shows the image of the alloy at T6 condition. A little bit of small β'ellipsoidal particles emerges in the matrix.

The mechanical properties of the alloy in as-extruded condition, T5-state and T6-state are demonstrated in Fig. 5.

Fig. 3 Microstructures of the Mg-1.27Zn-0.75Gd-0.17Zr alloy: **a** Optical microstructure and **b** SEM image of the as-extruded alloy; **c** Optical microstructure and **d** SEM image of the T5-treated alloy; **e** Optical microstructure and **f** SEM image of the T6-treated alloy

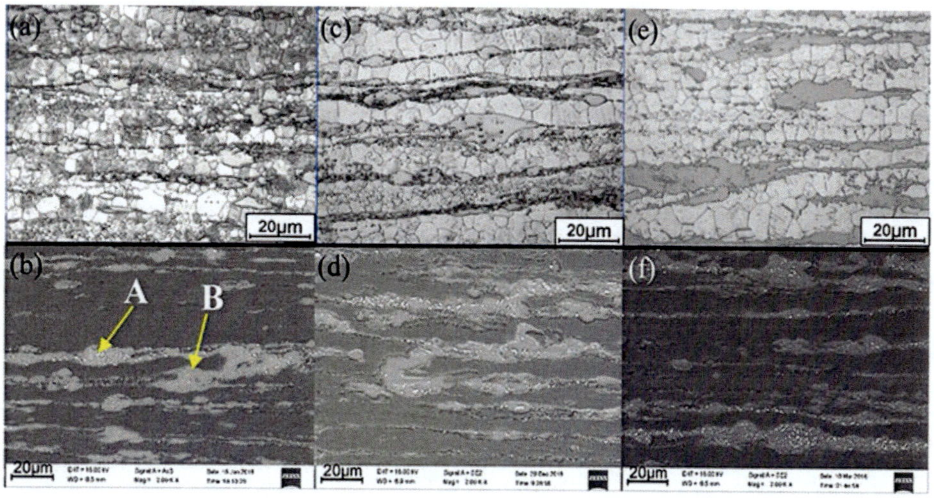

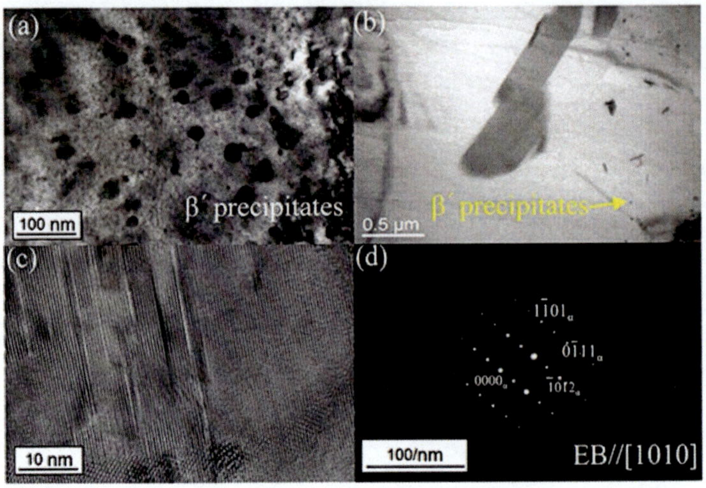

Fig. 4 TEM images and corresponding SAED patterns of the as-annealed experimental alloy: **a** T5-state and **b** T6-state Mg-1.27Zn-0.75Gd-0.17Zr alloy; **c** HRTEM image of T6-state Mg-1.27Zn-0.75Gd-0.17Zr alloy and **d** corresponding SAED pattern of (**a**)

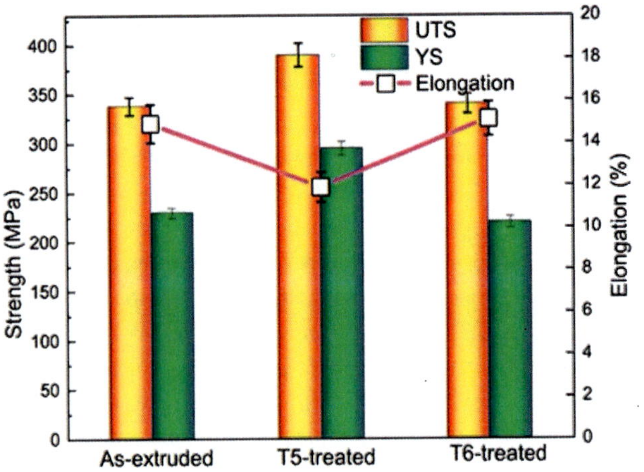

Fig. 5 Tensile properties of the experimental alloy at different conditions

With the peak-aging treatment at 200 °C for 16 h, the ultimate tensile strength is improved from 338 to 390 MPa. However, the elongation of the alloy is decreased from 14.9 to 11.9%. Nevertheless, ultimate tensile strength decreases to 340 MPa and the elongation slightly increases to 15.1% after the solution and aging treatment (T6). Generally speaking, volume fraction of second phases increases gradually than the extruded condition. Meanwhile, *X*-phase morphology develops into bulk of thick plates throughout the matrix (Fig. 3e). Although the aging treatment at 200 °C does not affect the volume fraction of LPSO *X*-phase in the alloy, it has a significant effect on the formation of RE precipitates. With prolonged aging at 200 °C, large amount of RE precipitates (β' phases) is formed in α-Mg matrix of the alloy and its volume fraction also increased gradually. The large amount of fine, dense and uniformly dispersed β′ phases precipitate in the T5-treated alloy contribute to the high strength and good thermal stability of the alloy. The precipitates have great ability to retard dislocation motion on the basal planes [11]. The β' precipitates form on the prismatic planes and the LPSO structures form on the basal planes, interconnecting into a network within the alloy and thus leading to the unique mechanical properties of the alloy.

To sum up, aging (T5) treatment is the effective measure to strengthen the LPSO *X*-phase containing alloys. The worse mechanical properties of T6-staged alloy may be attributed to the dissevered effect of plate-shaped *X*-phase and the relatively large DRXed grains. In the meanwhile, strengthening precipitate can easily change the semi-coherent crystal lattice with high solid solution temperature, so as to decrease the strengthening effect.

Conclusion

1. Mg-1.27Zn-0.75Gd-0.17Zr (at.%) alloy consists of α-Mg, *X*-phase and *W*-phase. Volume fraction of secondary phases increase through heat treatments, but the second phase species remain unchanged.

2. Large amounts of fine β′ phases precipitate in α-Mg matrix of T5-treated alloy. The alloy owns the superior ultimate and yield strength of 390 and 295 MPa at T5-state, respectively, as well as the elongation of 11.9%.

3. Little β'ellipsoidal particles emerge in the matrix after T6 treatment. Through T6 treatment, the ultimate and yield strength is 340 and 220 MPa, the elongation is 15.1%.

Acknowledgements This research was financially supported by "Double Tops" Construction Program of Liaoning Province [grant nos. 032077 and 032089] and National Natural Science Foundation of Liaoning Province [grant nos. 2022-BS-261 and 2022-BS-262].

References

1. Joost WJ, Krajewski PE (2017) Towards magnesium alloys for high-volume automotive applications. Scr. Mater. 128:107-112.
2. Yamasaki M, Hashimoto K, Hagihara K, Kawamura Y (2011) Effect of multimodal microstructure evolution on mechanical properties of Mg-Zn-Y extruded alloy. Acta. Mater. 59:3646-3658.
3. Hagihara K, Kinoshita A, Sugino Y, Yamasaki M, Kawamura Y, Yasuda HY, Umakoshi Y (2010) Ultra high-strength Mg-Gd-Y-Zn-Zr alloy sheets processed by large-strain hot rolling and ageing. Acta. Mater. 58:6282-6293.
4. Matsuda M, Ii S, Kawamura Y, Ikuhara Y, Nishida M (2004) Enhanced age hardening response and creep resistance of Mg-Gd alloys containing Zn. Mater. Sci. Eng. A. 386:447-452.
5. Anyanwu IA, Kamado S, Kojima Y (2011) Creep properties of Mg-Gd-Y-Zr alloys. Mater. Trans. 42:1212-1218.
6. Zhang XB, Wang Q, Chen F, Wu YJ, Wang ZZ, Wang Q (2015) Relation between LPSO structure and biocorrosion behavior of biodegradable GZ51K alloy. Mater. Lett. 138:132-135.
7. Lu F, Ma A, Jiang J, Yang D, Song D, Yuan Y, Chen J (2014) Effect of multi-pass equal channel angula rpressing on microstructure and mechanical properties of $Mg_{97.1}Zn_1Gd_{1.8}Zr_{0.1}$ alloy. Mater. Sci. Eng. A. 594:330-333.
8. Wang JF, Song PF, Gao S, Huang XF, Shi ZZ, Pan FS (2011) Effects of Zn on the microstructure, mechanical properties, and damping capacity of Mg-Zn-Y-Zr alloys. Mate. Sci. Eng. A. 528:5914-5920.
9. Yin SQ, Duan WC, Liu WH, Wu L, Yu J M, Zhao ZL, Liu M, Wang P, Cui JZ, Zhang ZQ (2020) Influence of specific second phases on corrosion behaviors of Mg-Zn-Gd-Zr alloys. Corros. Sci. 166:108419.
10. Liu XB, Chen RS, Han EH (2008) Effects of ageing treatment on microstructures and properties of Mg-Gd-Y-Zr alloys with and without Zn additions. J. Alloys Compd. 465:232-238.
11. Nie JF (2003) Effects of precipitate shape and orientation on dispersion strengthening in magnesium alloys. Scr. Mater. 48:1009-1015.

Effect of Grain Size on Bio-corrosion Properties of AZ31 Magnesium Alloy

Wenli Zhao and Qizhen Li

Abstract

As a type of biodegradable material, AZ31 magnesium alloy has attracted much attention in the bio-implanting field. Depending on the actual applications, it is often necessary to modify the microstructure of materials to achieve different properties. It is critical to understand the relationship between microstructure and corrosion property of AZ31 magnesium alloy. Grain size is the microstructural feature studied in this work. Heat treatment was employed to obtain the samples with different grain sizes. Bio-corrosion test was conducted for the samples through being immersed in simulated body fluid for various time intervals. Scanning electron microscope was employed for observing morphology changes due to corrosion testing. The results show that the corrosion rates increased with the immersion duration for the samples, the samples with larger grain size exhibited better corrosion resistance, and the corrosion mechanism was uniformly pitting corrosion.

Keywords

Magnesium alloy • Grain size • Corrosion • Pitting

Introduction

As the eighth abundant elements in the earth crust and due to its high strength to density ratio, magnesium and its alloys are playing a pivotal role in the fields of automotive industry, aeronautics industry and clinic applications. Specifically for clinic applications, there are many kinds of materials that can be used as biomaterials such as stainless steels, titanium alloys, magnesium alloys, ceramics as well as polymers. However, magnesium alloys have critical advantages among all these choices as the material candidates for bone implants [1–4], orthopedic applications [5] as well as vascular stents [6]. AZ31 magnesium alloy was intensively studied for the bio-applications. Bertuola et. al studied the corrosion protection capability of a polymeric coating from a phytocompound for AZ31 alloy to constrain bacterial adhesion [7]. Wang et. al studied the cellular response of AZ31 magnesium alloy stent in artery [8]. Yadav et. al studied the surface properties and biodegradation kinetics for AZ31 magnesium alloys with bioceramic hydroxyapatite coating [9]. Although mechanical properties of magnesium and its alloys are important [10, 11], their corrosion behavior is also crucial. It is known that the microstructure of materials affect the corrosion behavior. This study investigated the effect of grain size on the corrosion behavior of AZ31 magnesium alloy including weight loss, corrosion rate, and surface morphology.

Experimental

The cylindrical samples were prepared from AZ31 magnesium alloy bar with the chemical composition of ~3% aluminum, ~1% zinc, and ~96% magnesium. The samples were with a diameter of 12 mm and a height of 10 mm. To achieve different grain sizes, one batch of the samples were heat-treated at 200 °C for 72 h. Both as-received and heat-treated samples were polished using P600 sandpapers to ensure the same surface smoothness. Optical microscope was employed to characterize grain sizes of the samples before and after the heat treatment. SBF (simulated body fluid) solution [12] was utilized for the immersion testing of the samples. For the immersion testing, four specimens were prepared at every different immersion durations. The immersion durations used in this study are 0.5 day, 1 day, 7 days, 14 days, and 28 days. The testing was performed at

W. Zhao · Q. Li (✉)
School of Mechanical and Materials Engineering, Washington State University, Pullman, WA, USA
e-mail: qizhen.li@wsu.edu

S. Barela et al. (eds.), *Magnesium Technology 2023*, The Minerals, Metals & Materials Series,
https://doi.org/10.1007/978-3-031-22645-8_37

36.5 ± 1.5 °C and pH in the range of 6.3–7.8. The ratio of SBF volume to the surface area of a sample should be 30 ml/cm^2. SBF was renewed every 2–3 days to ensure that pH remained at physiological values, which can simulate the human fluid environment optimally. After the immersion testing, the samples were washed by ethanol and dried by warm flowing air. The masses of the samples were measured before and after the immersion testing. Scanning electron microscopy (SEM) was utilized to study the morphology of the sample surface after being corroded in SBF.

Results and Discussion

Figure 1 reports the grain size distribution histograms for both as-received AZ31 magnesium alloy and AZ31 magnesium alloy after heat treatment at 200 °C for 72 h. The average grain sizes are about 7.43 ± 4.79 μm and 14.69 ± 12.28 μm for the as-received and heat-treated samples, respectively. The average grain size for the heat-treated sample is about twice of that for the as-received sample. The heat treatment process resulted in the significant increase of the grain size. However, in terms of dispersion, grain size distributions before and after heat treatment remained almost the same.

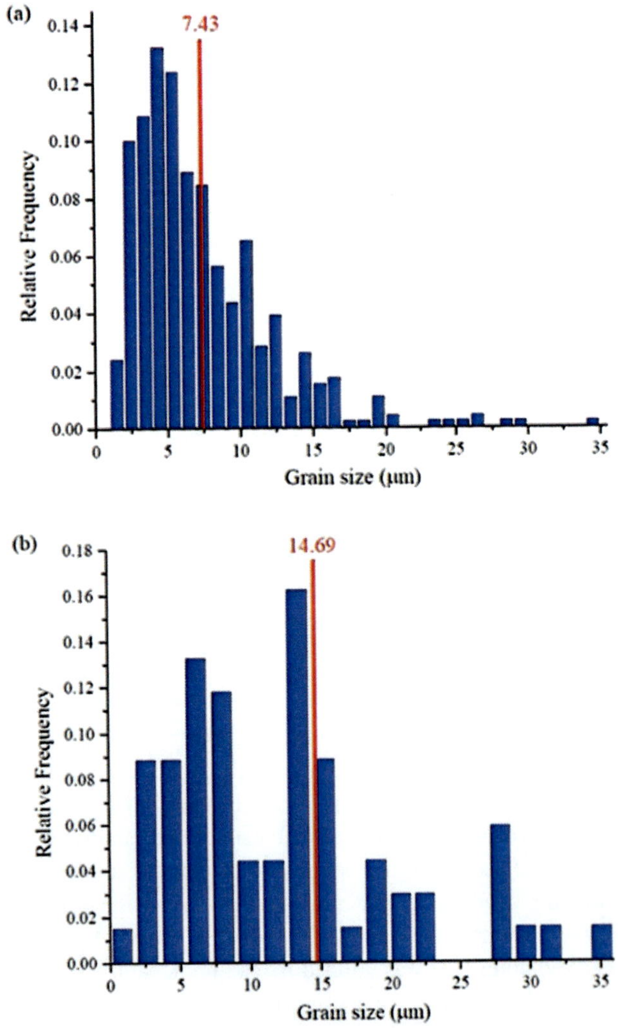

Fig. 1 Grain size distribution histogram for **a** as-received AZ31 magnesium alloy and **b** AZ31 magnesium alloy after heat treatment at 200 °C for 72 h

During the whole SBF immersion tests, the pH value was strictly monitored. The SBF solution was changed every 48 h to keep a normal physiological environment. The variation of pH values during SBF testing is reported in Fig. 2 for both as-received and heat-treated samples. The data show that the pH value increased with the immersion durations for both types of samples. The reason for the pH value change is due to the evolution of hydrogen as well as magnesium matrix being anode during the process. The overall chemical reaction is as follows:

$$2\,Mg + 2\,H^+ + 2\,H_2O \rightarrow 2\,Mg^{2+} + 2\,OH^- + 2\,H_2(gas) \quad (1)$$

As a result, the pH value increased at the expense of dissolution of the Mg matrix. The more significant the pH value changes, the more corroded the sample would be. For both as-received and heat-treated samples, the change of pH values became less with the increase of immersion durations. The comparison between Fig. 2b indicates that the heat-treated sample would be more corrosion resistant than the as-received sample, since the highest pH value for the heat-treated sample is less than that for the as-received sample.

The weight loss was measured for both as-received and heat-treated samples after the immersion testing as shown in Fig. 3. The weight loss increased with the increase of immersion time for both samples. Furthermore, the rate of dissolution for the samples increased with the increase of immersion time. The weight loss for the as-received sample is higher than that for the heat-treated samples for the studied immersion durations. This may be related to the different grain sizes for the two types of samples. The heat-treated sample has larger average grain size, and thus less grain boundary area compared to the as-received sample. Generally, grain boundaries are prone to be attacked in the solution and the increase of grain boundary area can lead to worse corrosion resistant.

The corrosion rate could be calculated using the mass information of the dried samples through the following equation [13],

$$CR = 8.76 \times 10^4\ W/(ATr) \quad (2)$$

where CR is the corrosion rate, W is the mass loss (g), A is the original surface area of each sample (cm^2), T is the

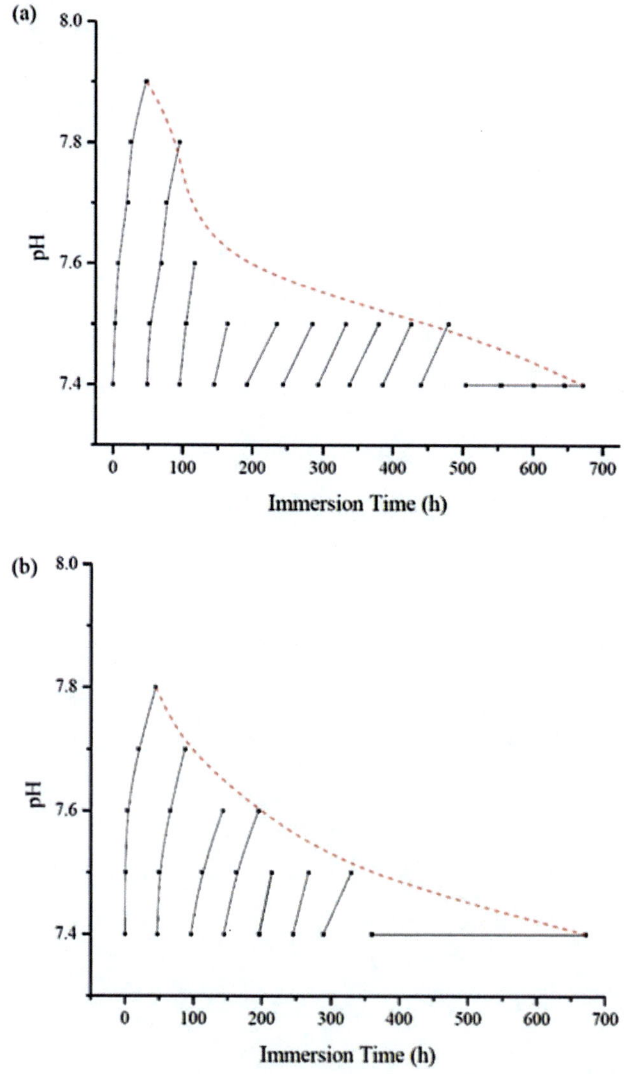

Fig. 2 Variation of pH value after different immersion durations for **a** as-received AZ31 magnesium alloy and **b** AZ31 magnesium alloy after heat treatment at 200 °C for 72 h

immersion time (h), and r is the sample density (g/cm^3). For both as-received and heat-treated samples, a similar trend was followed that the corrosion rate increased significantly with the increase of the immersion duration in the SBF solution. For immersion time from 0 to 72 h, the corrosion rate was increasing with the same rate; and for samples exposed in SBF solution from 72 to 168 h, the rate of corrosion increased dramatically; as to further immersion up to 336 h, samples remained the same corrosion rate as the corrosion rate calculated at 168 h, which may mean a balance between the formability of the passive protection film on the sample's surface and the dissolution of that film accompanied by the attacking of Cl^- ions in the solution; and afterward, it was increased significantly for longer explosion in the SBF solution (Fig. 4).

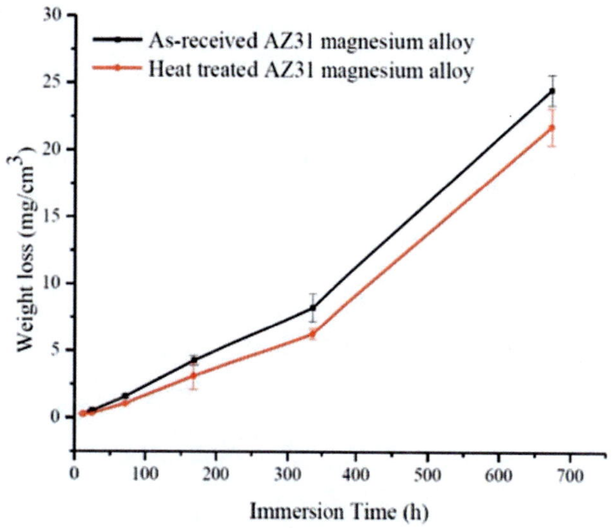

Fig. 3 Weight loss for both as-received and heat-treated AZ31 magnesium alloy under different immersion durations

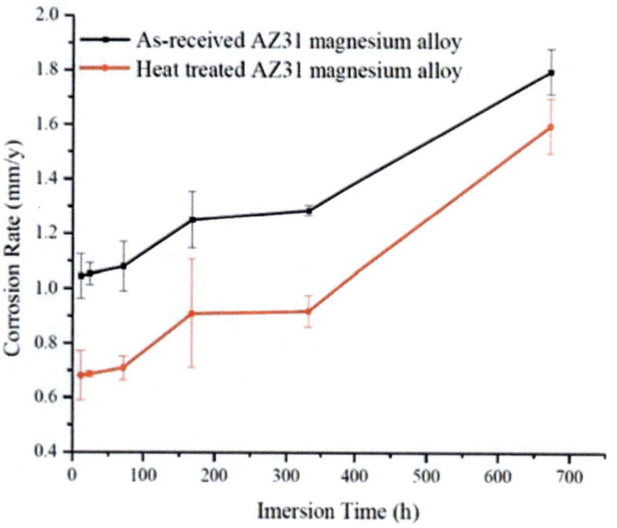

Fig. 4 Corrosion rate for both as-received and heat-treated AZ31 magnesium alloy under different immersion durations

Fig. 5 AZ31 magnesium alloy before corrosion

Figure 5 shows the image of the as-received sample before experiencing corrosion as a reference. Figures 6 and 7 exhibited the SEM images of surface morphology

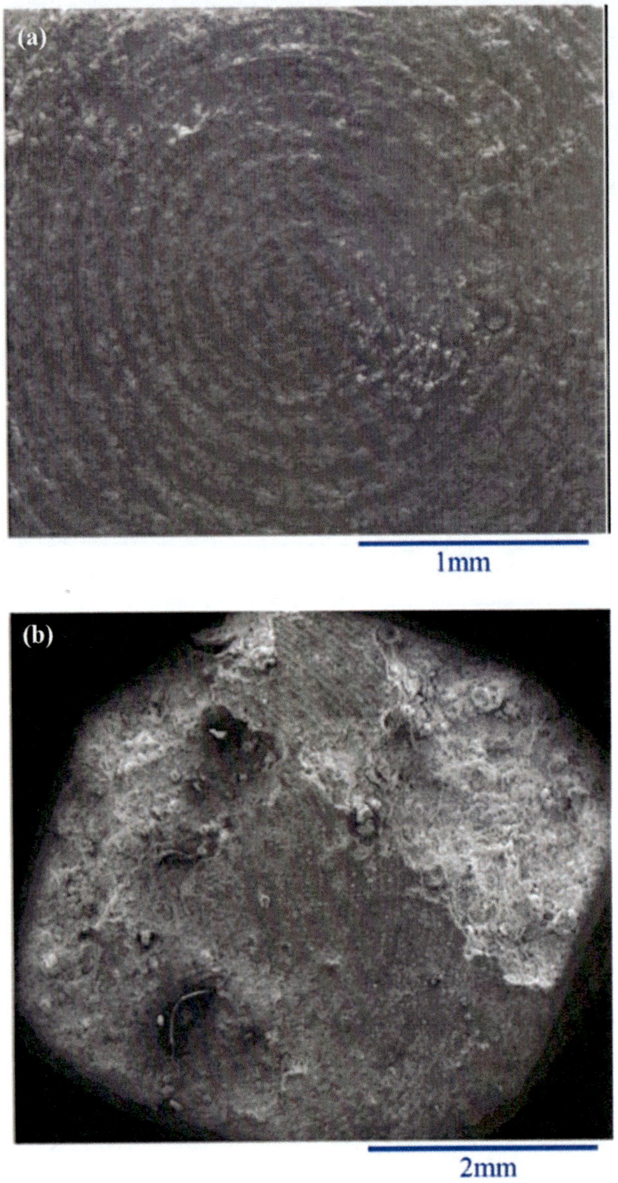

Fig. 6 Surface morphology of the as-received AZ31 magnesium alloy samples after the immersion in the SBF solution for **a** 12 h and **b** 672 h

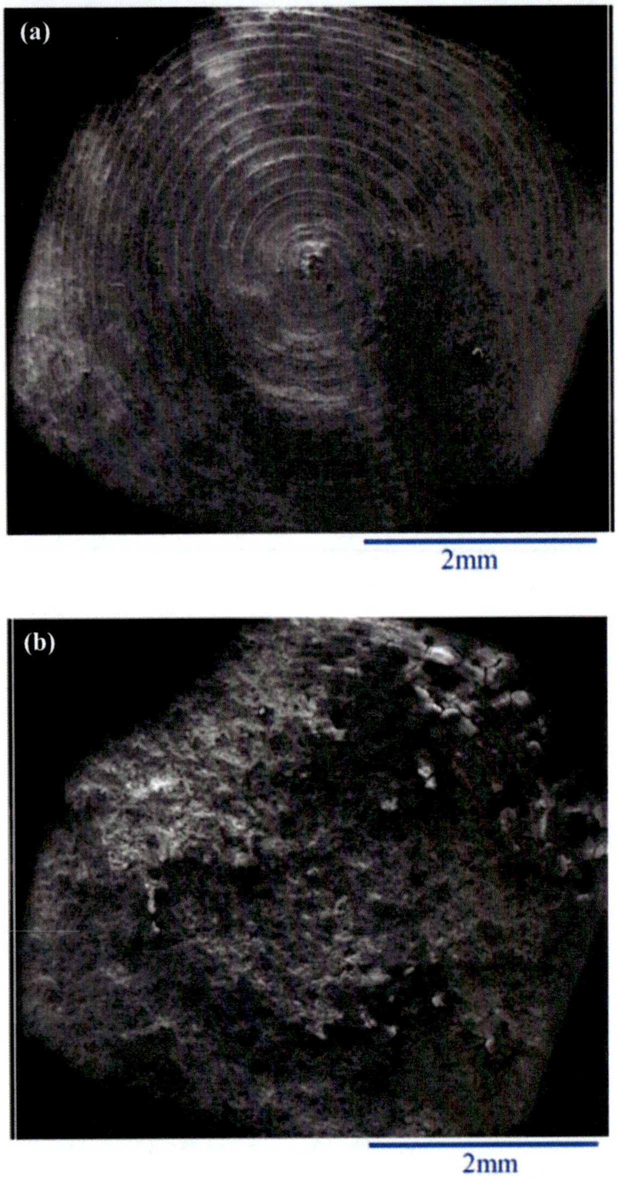

Fig. 7 Surface morphology of the heat-treated AZ31 magnesium alloy samples after the immersion in the SBF solution for **a** 12 h and **b** 672 h

variation when exposed to SBF solution for different time periods, for both as-received and heat-treated samples. The figures show the micrographs for the immersion durations of 12 and 672 h as the representing conditions. For both types of samples, the variation of surface morphology followed the similar trend with the increase of the immersion duration. At the initial stage, after being exposed to humid environment for up to 24 h, the outline of the surface was complete and no pitting was observed. This observation indicated a process of formation of a noble film on the surface of the samples, which was in good agreement with many other studies in pure magnesium and magnesium alloys [14–16]. This also agrees with the increase of pH value shown in Fig. 2.

As being exposed in the SBF solution for up to 168 h, pitting obviously appeared on the surface, which indicated the breaking down of the film layer mainly composed of $Mg(OH)_2$ at some locations. When the immersion duration is longer than 168 h, a compact protective layer was no longer existing and localized corrosion started to occur. But at some other locations on the surface, the intact portion of the layer was still acting as a protection.

Conclusion

This study investigated the effect of grain size on the corrosion behavior of AZ31 magnesium alloy including pH value variation, weight loss, corrosion rate, and surface morphology. Heat treatment was used to increase the average grain size. The following insights can be drawn from the results.

- The pH value increased with the immersion durations for both as-received and heat-treated samples.
- For both as-received and heat-treated samples, the change of pH values became less with the increase of immersion durations.
- The weight loss and corrosion rate for the as-received sample is higher than that for the heat-treated samples.
- The heat-treated sample would be more corrosion resistant than the as-received sample.
- As being exposed in the SBF solution for a duration of at least 168 h, pitting became the corrosion mechanism.

References

1. Witte, F., V. Kaese, H. Haferkamp, E. Switzer, A. Meyer-Lindenberg, C.J. Wirth, and H. Windhagen, In vivo corrosion of four magnesium alloys and the associated bone response. Biomaterials, 2005. 26(17): p. 3557–63.
2. Zhang, E., L. Xu, G. Yu, F. Pan, and K. Yang, In vivo evaluation of biodegradable magnesium alloy bone implant in the first 6 months implantation. Journal of biomedical materials research Part A, 2009. 90(3): p. 882-893.
3. Zhang, E. and L. Yang, Microstructure, mechanical properties and bio-corrosion properties of Mg–Zn–Mn–Ca alloy for biomedical application. Materials Science and Engineering: A, 2008. 497(1–2): p. 111–118.
4. Staiger, M.P., A.M. Pietak, J. Huadmai, and G. Dias, Magnesium and its alloys as orthopedic biomaterials: a review. Biomaterials, 2006. 27(9): p. 1728–1734.
5. Zeng, R., W. Dietzel, F. Witte, N. Hort, and C. Blawert, Progress and challenge for magnesium alloys as biomaterials. Advanced Engineering Materials, 2008. 10(8).
6. Heublein, B., R. Rohde, V. Kaese, M. Niemeyer, W. Hartung, and A. Haverich, Biocorrosion of magnesium alloys: a new principle in cardiovascular implant technology? Heart, 2003. 89(6): p. 651–656.
7. M. Bertuola, A. Miñán, C.A. Grillo, M.C. Cortizo, M.A. Fernández Lorenzo de Mele, Corrosion protection of AZ31 alloy and constrained bacterial adhesion mediated by a polymeric coating obtained from a phytocompound, Colloids and Surfaces B: Biointerfaces, 2018, 172: p. 187–196.
8. N. Wang, K. Li, Y. Zou, and J. Wang, Cellular response of biodegradable AZ31 magnesium alloy stent in artery, Journal of Physics: Conference Series, 2021: 1732, 012114.
9. V.S. Yadav, A. Kumar, A. Das, D. Pamu, L.M. Pandey, M.R. Sankar, Degradation kinetics and surface properties of bioceramic hydroxyapatite coated AZ31 magnesium alloys for biomedical applications, Materials Letters, 2020: 270, 127732.
10. Q. Li, Carbon nanotube reinforced porous magnesium composite: 3D nondestructive microstructure characterization using x-ray micro-computed tomography, Materials Letters 133 (2014) 83–86.
11. H. Xu, Q. Li, Deformation mechanisms and mechanical properties of porous magnesium/carbon nanofiber composites with different porosities, Journal of Materials Science 53(20) (2018) 14375–14385.
12. Kokubo, T. and H. Takadama, How useful is SBF in predicting in vivo bone bioactivity? Biomaterials, 2006. 27(15): p. 2907–2915.
13. Wang, Y., M. Wei, J. Gao, J. Hu, and Y. Zhang, Corrosion process of pure magnesium in simulated body fluid. Materials Letters, 2008. 62(14): p. 2181–2184.
14. Song, Y., D. Shan, R. Chen, F. Zhang, and E.-H. Han, Biodegradable behaviors of AZ31 magnesium alloy in simulated body fluid. Materials Science and Engineering: C, 2009. 29(3): p. 1039–1045.
15. Zhang, X., G. Yuan, J. Niu, P. Fu, and W. Ding, Microstructure, mechanical properties, biocorrosion behavior, and cytotoxicity of as-extruded Mg–Nd–Zn–Zr alloy with different extrusion ratios. Journal of the mechanical behavior of biomedical materials, 2012. 9: p. 153–162.
16. Zhang, X., G. Yuan, L. Mao, J. Niu, P. Fu, and W. Ding, Effects of extrusion and heat treatment on the mechanical properties and biocorrosion behaviors of a Mg–Nd–Zn–Zr alloy. Journal of the mechanical behavior of biomedical materials, 2012. 7: p. 77–86.

Hot Compression Processing Map and Microstructure Evolution of a Mg–Sn–Al–Zn–Mn Alloy

Wei Sun, Yangchao Deng, Hongyi Zhan, and Guang Zeng

Abstract

In the present study, the microstructural evolution of solid-solution treated Mg–4Sn–3Al–1Zn–0.3Mn (TAZM) was explored through hot compression tests in the temperature range of 300–460 °C and the strain rate range of 0.001–1 s^{-1}. Dynamic recrystallization (DRX) behaviors induced by hot compression and the effects of Sn alloying were studied, using analytical electron scanning microscopy. The correlation between DRX of α-Mg grains and dynamic precipitation of Mg_2Sn during hot deformation was revealed. The influences of Mg_2Sn dynamic precipitates over plastic deformation of α-Mg grains, texture evolution, and cracking were also discussed.

Keywords

Mg–Sn alloy • Dynamic recrystallization • Dynamic precipitation

Introduction

As the lightest structural metal, magnesium alloys are considered as a promising alternative for the application of automobile lightweight, due to their excellent properties of low density, high specific strength, and specific stiffness [1–3]. Magnesium alloys normally yield insufficient formability under room temperature [4, 5]. It is well known that the processing parameters have a great impact on plastic deformation mechanisms and therefore the formability of magnesium alloys [6–8]. Another reason is that the processing parameters such as temperature, strain, and strain rates are also able to effectively control precipitation behavior by regulating the diffusion of solute elements [9–13]. Attention has been drawn to Mg-Sn-based alloys due to their superior precipitation strengthening effect introduced by Sn alloying [13–28]. Mg_2Sn acts as the main precipitate in Mg–Sn system and shows excellent thermal stability, leading to a good combination of strength and formability [13]. Past work has shown that dynamic precipitates of Mg_2Sn during hot working/deformation improved the mechanical performance of magnesium alloys [17, 29–31]. Zhang et al. [32] reported a novel lean Mg–Sn–Ca–Mn wrought alloy with yield strength of 450 MPa and elongation of 5%. Sasaki et al. [30] also developed a Mg–6.6Sn–5.9Zn–2.0Al–0.2Mn (TZAM6620) (wt.%) extruded alloy with a combination of strength (370 MPa) and elongation (14%) via double aging.

Numerous research focuses on how to reduce the activation of non-basal slip systems in magnesium alloys [33]. It is well established that the addition of alloying elements, such as a few weight percentages of Sn alloying element, could weaken basal texture and promote the activation of non-basal slip systems, by altering the stacking fault energy (SFE) [34–36]. Muzyk et al. [37] suggested that the addition of Sn could decrease the energy barrier for partial dislocations and stacking fault formation significantly compared with the other alloying elements. Consequently, Mg_2Sn precipitates can influence the activation of slip systems [12, 38, 39]. It is widely accepted that dynamic recrystallization (DRX) behavior significantly influences the microstructure and mechanical properties of Mg alloy [40–44]. Mg_2Sn precipitates are generally believed to increase the number of the DRXed grains via Particle Stimulated Nucleation (PSN) mechanism [44–47]. Zhao et al. [48] reported that significant precipitation occurred in Mg-4Sn after hot extrusion, and the average DRXed grain size decreased from 15.6 um to 3.6 μm with increasing the content of Sn from 1.3 to 4.7 wt.%. Alternatively, the morphology of Mg_2Sn

W. Sun · Y. Deng · G. Zeng (✉)
School of Materials Science and Engineering, Central South University, Changsha, 410083, China
e-mail: g.zeng@csu.edu.cn

H. Zhan
China Science Lab, General Motors Global Research and Development, Shanghai, 201206, China

S. Barela et al. (eds.), *Magnesium Technology 2023*, The Minerals, Metals & Materials Series,
https://doi.org/10.1007/978-3-031-22645-8_38

precipitated phase is also a function of processing parameters. Kabir et al. [16] found that there were more dynamic precipitates at low temperature (250 °C) and low strain rate (0.001 s^{-1}) during hot deformation in Mg–3Al–3Sn (wt.%). Most precipitates were present mostly in the recrystallized area on the recrystallized grain boundaries. It is of great interest to explore the mutual influences of Mg_2Sn dynamic precipitation and DRX behavior during deformation at elevated temperatures.

In the present work, we constructed hot compression processing map of a Mg–4Sn–3Al–1Zn–0.6Mn (TAZM4310) alloy with −0.9 true strain, in the temperature range of 340–460 °C and strain rate range of 10^{-3}–1 s^{-1}. We focus on microstructure evolution, especially Mg_2Sn dynamic precipitates and DRX behaviors of α-Mg grains, in Mg–Al–Zn–Mn alloys during hot compression.

Experimental

Mg–4Sn–3Al–1Zn–0.6Mn (wt.%) (TAZM4310) alloy with the composition given in 3D X-ray Microscopy (XRM) (ZEISS Xradia 620 Versa, Carl Zeiss X-ray Microscopy Inc., USA) was performed on ø 3 × 12 mm samples of both alloys. The samples were imaged at 80 kV with a resolution of 0.7 μm·$voxel^{-1}$ and an exposure time of 3.5 s per projection. A total of 2400 projections were collected per each tomography scan. 3D reconstructions were performed using the reconstruction software (XMReconstructor, Carl Zeiss X-ray Microscopy Inc., USA), which is based on the filtered back-projection-based algorithm. Reconstructed datasets were segmented and quantitatively analyzed using ImageJ (US NIH, USA) and Avizo 9.2 (Thermo Fisher Scientific, USA).

Table 1 was prepared from pure Mg, Al, Zn, Sn (>99.9 wt.%), and Mg-1.83wt.%Mn master alloy, by melting in an electronic resistance furnace at 700 °C under the protection of cover gas with 99.5% CO_2 and 0.5% SF_6 mixture. The alloying melt was held at 700 °C for 30 min to homogenize, followed by cooling down to 680 °C and then cast into the permanent mold which had coated and preheated to 250 °C. Ingots with the dimensions of Φ95 mm × 500 mm were cast by pouring the melt into the mold.

Cylindrical samples (10 mm in diameter and 12 mm in length) for compression tests were homogenized alloy at 420 °C for 24 h. Uniaxial hot compression tests were conducted on a Gleeble-3180 thermal–mechanical simulation testing system. Nickel-based lubricants and graphite sheets were used on both ends of the sample to reduce friction. Before the tests, two thermocouples were welded on the surface of the sample for temperature control. Then, the sample was heated at a heating rate of 5 °C/s to 350 °C and held for 120 s to ensure uniform temperature distribution. For hot compression, the strain rates (true strain rate) were 0.001, 0.01, 0.1, and 1 s^{-1}, the deformation temperatures were 300, 340, 380, 420, and 460 °C, and the strain was constantly kept to −0.9.

The hot compressed microstructure was characterized using an optical microscope (OM, Olympus DP72, UK) and a scanning electron microscope (SEM, TESCAN MIRA3, Czech) equipped with an electron backscatter diffraction (EBSD, Oxford Instrument C-Nano, UK) detectors. The samples were cross-sectioned transversely along the centerline and then ground and polished following standard procedures. High-angle annular dark-field images (HAADF-STEM) and energy dispersive spectrometer maps (EDS) were obtained using a transmission electron microscope (TEM, FEI Titan G2 60–300, Thermo Fisher Scientific, USA). Samples for OM and SEM observations were etched by a solution of 5 wt.% picric acid, 5 vol.% acetic acid, and 10 vol.% deionized water-alcohol for 10 s. Electrochemical polishing was performed on polished samples for EBSD characterization, using a solution of 4 vol.% perchlorate-alcohol at −40 °C for 90 s. EBSD datasets were processed using AZtecCrystal 2.1 software (Oxford Instruments, UK) and MATLAB ™ R2021b (MathWorks, USA) with MTEX 5.8.3 toolbox [49, 50].

3D X-ray Microscopy (XRM) (ZEISS Xradia 620 Versa, Carl Zeiss X-ray Microscopy Inc., USA) was performed on ø3 × 12 mm samples of both alloys. The samples were imaged at 80 kV with a resolution of 0.7 μm·$voxel^{-1}$ and an exposure time of 3.5 s per projection. A total of 2400 projections were collected per each tomography scan. 3D reconstructions were performed using the reconstruction software (XMReconstructor, Carl Zeiss X-ray Microscopy Inc., USA), which is based on the filtered back-projection-based algorithm. Reconstructed datasets were segmented and quantitatively analyzed using ImageJ (US NIH, USA) and Avizo 9.2 (Thermo Fisher Scientific, USA).

Table 1 Alloy composition as determined by inductively coupled plasma emission spectrometer (ICP-AES)

Alloy	Compositions (wt.%)								
	Mg	Sn	Al	Zn	Mn	Fe	Cu	Si	Ni
TAZM4310	Bal	3.92	2.86	1.32	0.59	0.002	0.005	0.008	<0.001

Results and Discussion

The true stress–strain curves of the TAZM4310 alloy under various deformation temperatures and strain rates are shown in Fig. 1a. All the samples were strained to −0.9 via uniaxial hot compression. It is observed that the flow stress rapidly increases to a peak value and then gradually decreases to a relatively steady state. It can be found that the flow stress curve normally experienced three stages: stage I (work hardening stage) approximately in the strain of 0 to −0.2, stage II (softening stage) approximately in the strain of 0 to −0.2 to −0.5, and stage III (steady stage) approximately in the strain range of −0.5 to −0.9. The flow stress levels decreased with the rise of deformation temperature, and also the reduction of strain rates, in the form of Zener–Hollomon parameter. It is also noted that the hardening rate gradually decreases with the compression until it approaches 0 at a strain of −0.2, as the dynamic recovery (DRV) and DRX began.'

The processing maps of hot deformation containing a power dissipation map and an instability map are usually used to evaluate the machinability of alloys. Power dissipation efficiency (η) and instability criterion (ξ) are calculated by the following Eqs. (1–3) [51]:

$$m = \left(\frac{\partial \ln\sigma}{\partial \ln\dot{\varepsilon}}\right)_{\varepsilon,T} \quad (1)$$

$$\eta = \frac{2m}{m+1} \quad (2)$$

$$\xi = 2m - \eta \quad (3)$$

Processing maps of the alloys were drawn in Fig. 2. It is shown that η increases with decreasing strain rate and the rise of temperature. Similarly, instability regions tend to yield lower η values. The efficiency indicator increases, implying DRX becomes more dominant during deformation, and this also represents more suitable temperature and strain rates. From the processing map, TAZM4310 alloys have a wide area of window for processing parameters with good formability, as indicated by the right bottom corner of the maps.

Optical micrographs in Fig. 3 indicate that macro-SBs (as indicated by yellow arrows) developed as hot compressed to the strain of −0.9. The overall microstructure shows that the inhomogeneous deformation is more evident under the lower compression temperature and slower strain rate. Cracks have been clearly observed at 340 °C at both 0.01 and 1 s^{-1} strain rates. Few twinning signatures were observed. It can be clearly seen that both alloys exhibit a typical bimodal structure, with fine and equiaxial recrystallized (DRXed) grains and coarse unrecrystallized (unDRX) grains elongated along the compression direction. Moreover, the nano-size Mg_2Sn dynamic precipitates presented in the samples were found to be aligned along the DRXed region in TAZM4310 alloys at the temperature of 340 °C with the strain rate of 0.01 s^{-1} in Fig. 4. Substantial dynamic precipitation occurred in TAZM4310 alloy, which is more obvious at 340 °C as shown Back-Scatter Electron (BSE) images. In contrast, there were few precipitates of particles in hot compressed microstructure under 420 °C. It clearly shows the formation of numerous Mg_2Sn as compared to hot compressed AZ31 alloy, when the sample is strained to −0.9.

Figure 5 shows the EBSD IPF mapping, view along axis direction (AD), revealing the grain morphology and orientation of hot deformed microstructure, with a strain of −0.9. Region of interests (ROIs) for EBSD characterization were all selected from the center of the tested samples. Grain reconstruction for EBSD datasets was performed, based on the threshold misorientation angle of 5° between neighbour measurements. Within the shear band, a 'necklace structure' consisting of refined DRXed grains was formed around the deformed coarse grains as shown in Fig. 5a. The crystal orientation of DRXed grains was randomized away from

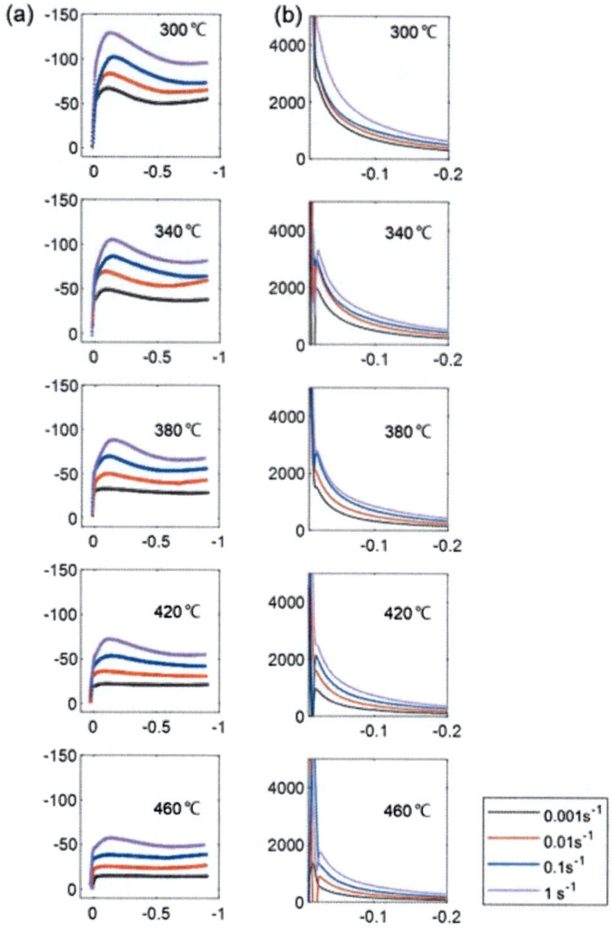

Fig. 1 **a** Flow curves **b** plots of hardening rate versus strain of TAZM, deformed at various temperatures and strain rates

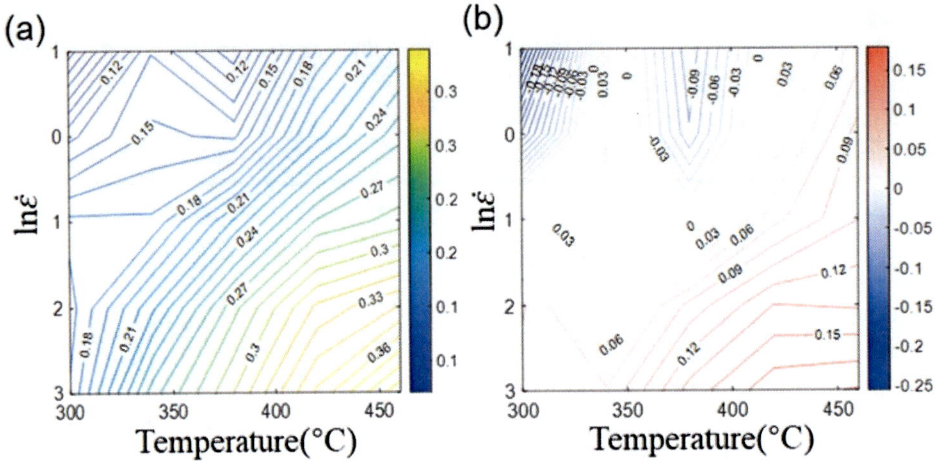

Fig. 2 Hot processing map of the TAZM4310 alloy: **a** power dissipation efficiency (η) and **b** instability criterion (ξ)

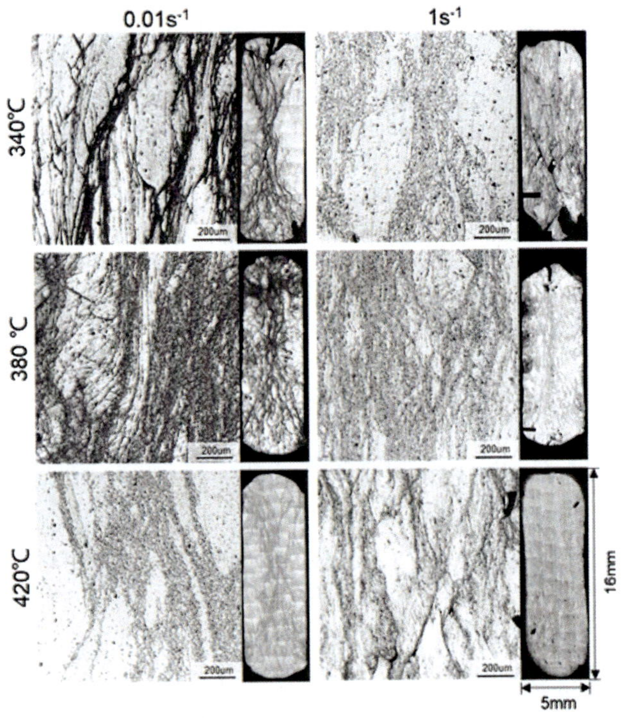

Fig. 3 Optical images of microstructure in TAZM4310 hot compressed at various temperatures and strain rates

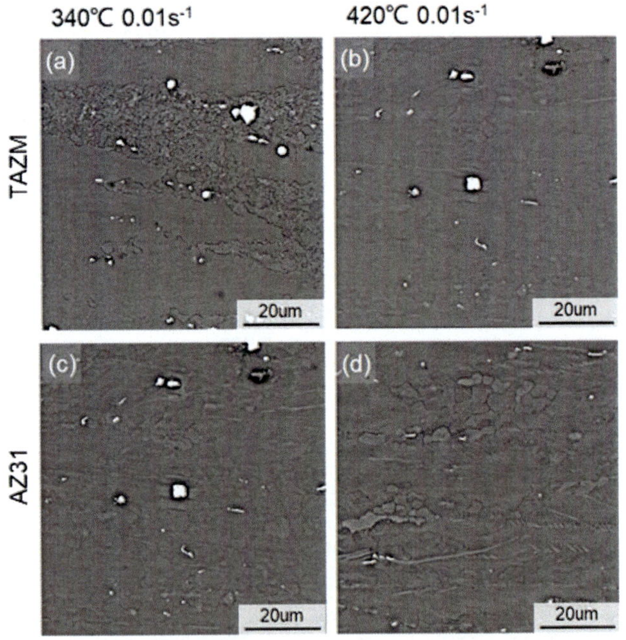

Fig. 4 Dynamic precipitation of Mg_2Sn as **a**, **b** TAZM4310 and **c**, **d** AZ31 being hot compressed to the strain of −0.9

basal-textured deformed grains, as subjected to hot compression in AZ31 alloy. Many twins have been identified in corresponding grain boundary maps. Most of them are tensile twins, mainly distributed in the deformed grain area.

Figure 6 confirms the texture evolution of hot compressed microstructure as influenced by processing parameters derived from Fig. 5. Deformed grains developed a basal-textured microstructure for two alloys, while no obvious texture weakening was observed for the TAZM4310 alloy via DRX. From Fig. 7, it can be seen that the mean size of DRXed grains in TAZM4310 alloy is larger than AZ31 DRXed grains being strained under the same condition.

As shown in Fig. 8, three-dimensional(3D) morphology of hot compressed microstructure was revealed by micro-CT. The intermetallic particles and cracks were separately segmented from 2D slices. It clearly shows that there is no obvious variation in the morphology of intermetallic particles and their size distribution for TAZM4130 and AZ31 alloys. Therefore, the improvement in hot processing performance in TAZM4310 is mainly related to the Sn alloying element. Combining optical micrograph of Fig. 4 with micro-CT results in Fig. 8, it indicates that the macro

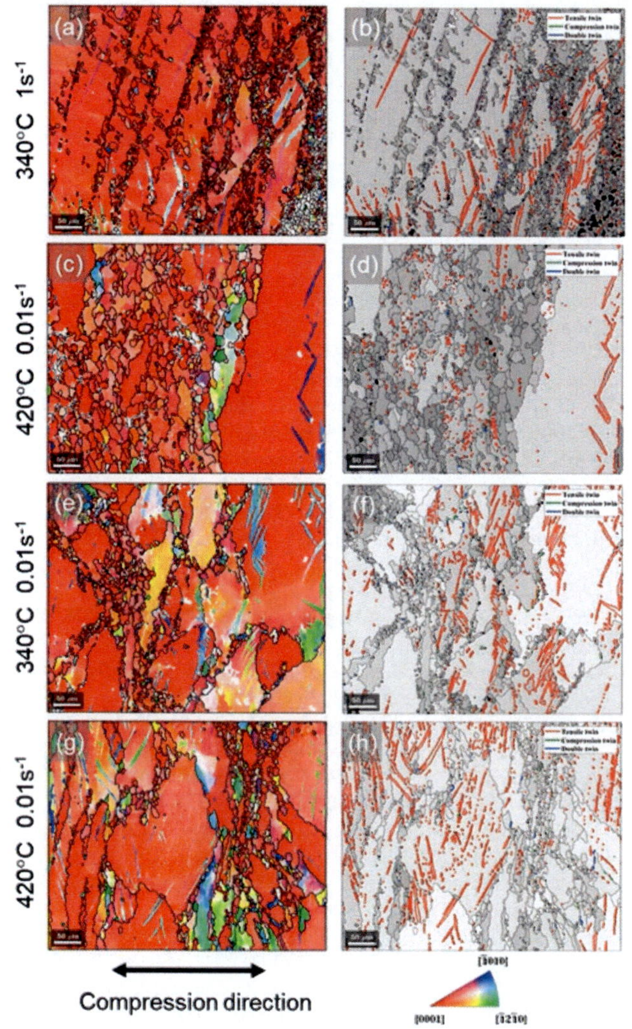

Fig. 5 **a**, **c**, **e**, **g** EBSD IPF-AD map and **b**, **d**, **f**, **h** corresponding grain boundary map of −0.9 strained microstructure for **a–d** TAZM4310 and **e–h** AZ31 alloys

shear bands formed by plastic deformation and dynamic recrystallization were often located along the interface between DRXed grain regions and coarse deformed grains, which is also consistent with our past work on the microstructural analysis of hot compressed Mg–3Zn–Zr alloy [52]. The macro-crack and voids in TAZM4310 alloys were observed in severely hot deformed samples. It may be closely correlated with the growth of the intergranular micro-cracks. In future work, it is of great importance to explore the correlation between dynamically precipitated particles of fine size and these crack zones.

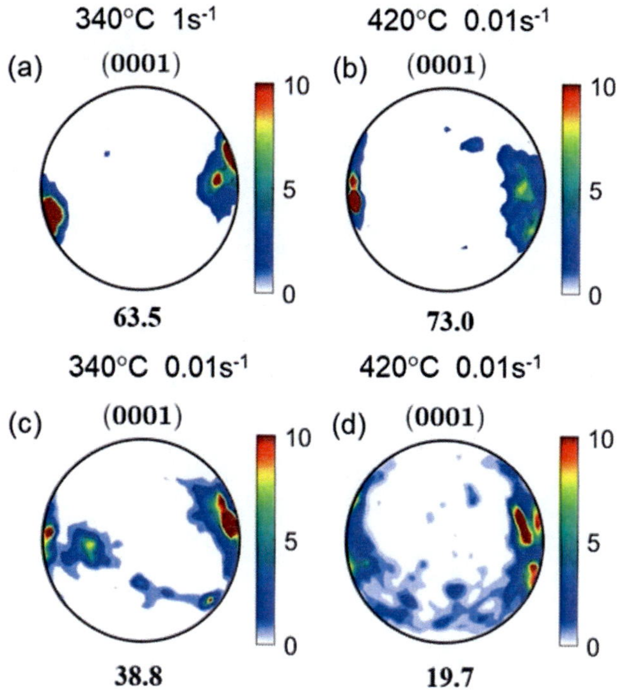

Fig. 6 Pole figures of grains (both deformed and DRXed grains) **a–b** TAZM4310 and **c–d** AZ31 alloys

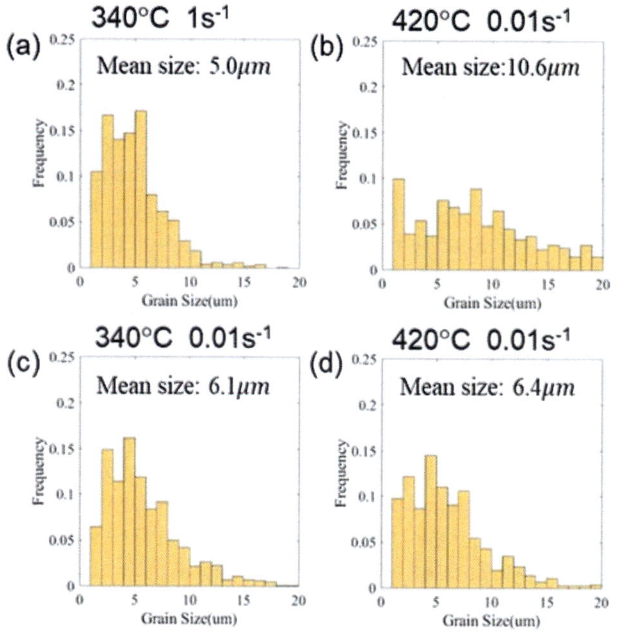

Fig. 7 Size distribution of DRXed grains in **a–b** TAZM4310 and **c–d** AZ31 alloys

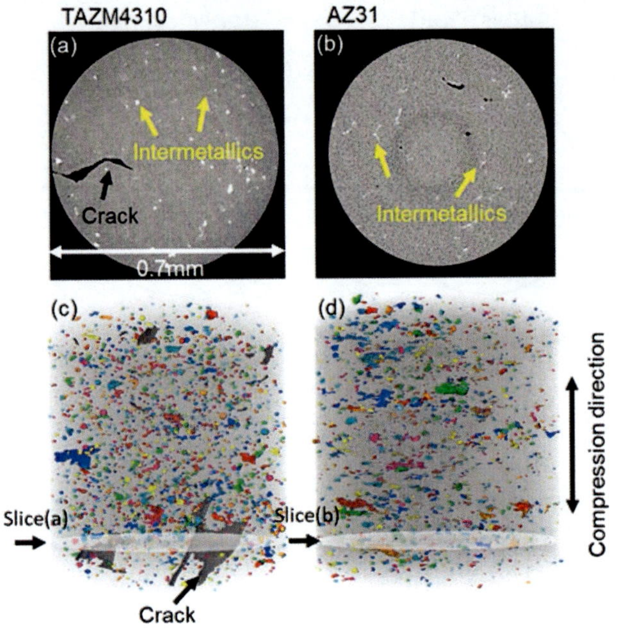

Fig. 8 **a**, **b** Reconstructed slices of hot compressed microstructure using micro-CT. **c**, **d** rendered 3D morphology of intermetallic particles colored by individual particles and cracks in grey. **a**, **c** TAZM4310 alloys and **b**, **d** AZ31 alloys

Conclusions

This paper studied the deformation and DRX behavior of TAZM4310 and AZ31 magnesium alloys in hot compression at the temperature of 340–420 °C and in the range of 0.01–1 s^{-1} strain rates. Under various temperature and strain rates, softening occurred as the strained to −0.1, as the DRX behaviour become dominant as the compression progressed. As seen from the established hot processing map for TAZM4310, there is a wide window of suitable processing parameters, indicating good formability. Hot compressed microstructure exhibits a typical bimodal morphology, with fine and equiaxial recrystallized grains (DRX grains), and coarse unrecrystallized grains elongated along the compression direction (unDRX grains). Nano-size Mg_2Sn dynamic precipitates presented in the samples were found to be aligned along the DRXed region in TAZM4310 alloys. In future work, it is of great importance to explore the correlation between dynamically precipitated particles of fine size and these crack zones.

References

1. V.M. Miller and T.M. Pollock, *Metallurgical and Materials Transactions A: Physical Metallurgy and Materials Science*, 47, 1854–1864 (2016).
2. T.M. Pollock, *Science*, 328, 986–987 (2010).
3. Z. Wu, R. Ahmad, B. Yin, S. Sandlöbes and W.A. Curtin, *Science*, 359, 447–452 (2018).
4. B. Lin, H. Zhang, Y. Meng, L. Wang, J. Fan, S. Zhang and H. J. Roven, *Materials Science and Engineering: A*, 847, 143338 (2022).
5. J. Shao, Z. Chen, T. Chen and C. Liu, *Metallurgical and Materials Transactions A: Physical Metallurgy and Materials Science*, 51, 1911–1923 (2020).
6. R. Zheng, T. Bhattacharjee, S. Gao, W. Gong, A. Shibata, T. Sasaki, K. Hono and N. Tsuji, *Scientific Reports*, 9, 11702 (2019).
7. Y.Z. Wu, H.G. Yan, S.Q. Zhu, J.H. Chen, A.M. Liu and X.L. Liu, *Transactions of Nonferrous Metals Society of China (English Edition)*, 24, 930–939 (2014).
8. C.Y. Wang, X.J. Wang, H. Chang, K. Wu and M.Y. Zheng, *Materials Science and Engineering: A*, 464, 52–58 (2007).
9. Z.-Z. Shi, H.-T. Chen, K. Zhang, F.-Z. Dai and X.-F. Liu, *Journal of Magnesium and Alloys*, 9, 416–431 (2021).
10. S.W. Xu, N. Matsumoto, S. Kamado, T. Honma and Y. Kojima, *Materials Science and Engineering A*, 523, 47–52 (2009).
11. F. Liu, R. Xin, M.X. Zhang, M.T. Pérez-Prado and Q. Liu, *Acta Materialia*, 195, 263–273 (2020).
12. J.J. Bhattacharyya, F. Wang, N. Stanford and S.R. Agnew, *Acta Materialia*, 146, 55–62 (2018).
13. W. Cheng, Y. Bai, L. Wang, H. Wang, L. Bian and H. Yu, *Materials*, 10, 822 (2017).
14. Q. Zhang, Q. Li, X. Chen, J. Bao and Z. Chen, *Materials Science and Engineering: A*, 826, 142026 (2021).
15. B.-C. Suh, J.H. Kim, J.H. Bae, J.H. Hwang, M.-S. Shim and N. J. Kim, *Acta Materialia*, 124, 268–279 (2017).
16. A.S.H. Kabir, M. Sanjari, J. Su, I.-H. Jung and S. Yue, *Materials Science and Engineering: A*, 616, 252–259 (2014).
17. H.-t. Son, J.-b. Lee, H.-g. Jeong and T.J. Konno, *Materials Letters*, 65, 1966–1969 (2011).
18. S. Wei, T. Zhu, M. Hodgson and W. Gao, *Materials Science and Engineering: A*, 585, 139–148 (2013).
19. Y. Luo, Y.a. Chen, L. Ran, X. Pang and F. Pan, *Materials Science and Engineering: A*, 815, 141307 (2021).
20. Y. Deng, G. Zeng, J. Xian, H. Zhan, C. Liu and C.M. Gourlay, *Materials Characterization*, 186, 111807 (2022).
21. J. Ye, X. Chen, Z. Luo, J. Li, Y. Yuan, J. Tan and F. Pan, *Advanced Engineering Materials*, 23, 2100166 (2021).
22. H.J. Kim, S.-C. Jin, J.-G. Jung and S.H. Park, *Journal of Materials Science & Technology*, 71, 87–97 (2021).
23. R. Radha and D. Sreekanth, *Journal of Magnesium and Alloys*, 8, 452–460 (2020).
24. K. Suresh, K.P. Rao, Y.V.R.K. Prasad, N. Hort and K.U. Kainer, *Transactions of Nonferrous Metals Society of China*, 23, 3604–3610 (2013).
25. W.L. Cheng, S.S. Park, B.S. You and B.H. Koo, *Materials Science and Engineering: A*, 527, 4650–4653 (2010).
26. T. Zhou, Q. Zhang, Q. Li, L. Wang, Q. Li and D. Liu, *Journal of Alloys and Compounds*, 889, 161653 (2021).
27. T.T. Sasaki, F.R. Elsayed, T. Nakata, T. Ohkubo, S. Kamado and K. Hono, *Acta Materialia*, 99, 176–186 (2015).
28. A. Dev, N. Naskar, N. Kumar, A. Jena and M. Paliwal, *Journal of Magnesium and Alloys*, 7, 725–737 (2019).
29. Sasaki, TT Oh-Ishi K, Ohkubo T, Hono K (2006) Scripta Materialia 55:251–254
30. T.T. Sasaki, K. Yamamoto, T. Honma, S. Kamado and K. Hono, *Scripta Materialia*, 59, 1111–1114 (2008).
31. W. Cheng, Y. Bai, S. Ma, L. Wang, H. Wang and H. Yu, *Journal of Materials Science & Technology*, 35, 1198–1209 (2019).
32. A. Zhang, R. Kang, L. Wu, H. Pan, H. Xie, Q. Huang, Y. Liu, Z. Ai, L. Ma, Y. Ren and G. Qin, *Materials Science and Engineering: A*, 754, 269–274 (2019).

33. H. Sun, Z. Ding, D. Zhang, H. Zhou, S. Li, E.J. Lavernia, Y. Zhu and W. Liu, *Materialia*, 7, 100352 (2019).
34. M. Yuasa, N. Miyazawa, M. Hayashi, M. Mabuchi and Y. Chino, *Acta Materialia*, 83, 294–303 (2015).
35. S. Sandlöbes, M. Friák, S. Zaefferer, A. Dick, S. Yi, D. Letzig, Z. Pei, L.F. Zhu, J. Neugebauer and D. Raabe, *Acta Materialia*, 60, 3011–3021 (2012).
36. J. Zhang, Y. Dou, G. Liu and Z. Guo, *Computational Materials Science*, 79, 564–569 (2013).
37. M. Muzyk, Z. Pakiela and K.J. Kurzydlowski, *Scripta Materialia*, 66, 219–222 (2012).
38. R. Ni, S.J. Ma, L.J. Long, J. Zheng, H. Zhou, Q.D. Wang and D.D. Yin, *Materials Science and Engineering: A*, 804, 140738 (2021).
39. F. Wang, J.J. Bhattacharyya and S.R. Agnew, *Materials Science and Engineering: A*, 666, 114–122 (2016).
40. S.M. Fatemi-Varzaneh, A. Zarei-Hanzaki and H. Beladi, *Materials Science and Engineering: A*, 456, 52–57 (2007).
41. R. Kaibyshev, *Advances in Wrought Magnesium Alloys*, 186–225 (2012).
42. T. Sakai, A. Belyakov, R. Kaibyshev, H. Miura and J.J. Jonas, *Progress in Materials Science*, 60, 130–207 (2014).
43. J. Zhang, C. Xie, B.W. Zhu, X. Liu, X.F. Wang, T.F. Ma, W.F. Peng and X.D. Shu, *Materials Research Express*, 5, 056532 (2018).
44. J. Zou, J. Chen, H. Yan, W. Xia, B. Su, Y. Lei and Q. Wu, *Materials Science and Engineering: A*, 735, 49–60 (2018).
45. S. Dai, Z. Bian, W. Wu, J. Tao, L. Cai, M. Wang, C. Xia and H. Wang, *Materials Science and Engineering: A*, 792, 139838 (2020).
46. Z.Y. Zhao, R.G. Guan, Y.F. Shen and P.K. Bai, *Journal of Materials Science & Technology*, 91, 251–261 (2021).
47. L. Jiang, W. Huang, D. Zhang, F. Guo, H. Xue, J. Xu and F. Pan, *Journal of Alloys and Compounds*, 727, 205–214 (2017).
48. C. Zhao, X. Chen, F. Pan, S. Gao, D. Zhao and X. Liu, *Materials Science and Engineering: A*, 713, 244–252 (2018).
49. F. Bachmann, R. Hielscher and H. Schaeben, (Trans Tech Publ: 2010), pp 63–68.
50. G. Rafailov, E.a.N. Caspi, R. Hielscher, E. Tiferet, R. Schneck and S.C. Vogel, *Journal of Applied Crystallography*, 53, 540–548 (2020).
51. Y.V.R.K. Prasad and K.P. Rao, *Materials Science and Engineering: A*, 487, 316–327 (2008).
52. J. You, Y. Huang, C. Liu, H. Zhan, L. Huang and G. Zeng, *Materials*, 13, 2348 (2020).

$Mg_3V_2O_8$: A Promising Cathode Material for Aqueous Mg-ion Battery

Xiu-Fen Ma, Hong-Yi Li, Daibo Gao, Weiwei Ren, Jiang Diao, Bing Xie, Guangsheng Huang, Jingfeng Wang, and Fushang Pan

Abstract

The aqueous Mg-ion batteries (RMBs) have attracted lots of attention due to the high safety characteristics, low cost, and similar electrochemical characteristics to lithium. Magnesium orthovanadate ($Mg_3V_2O_8$, MVO) has stable three-dimensional framework and large ion channels and is expected to be a high-performance energy storage electrode material. Herein, low-cost $Mg_3V_2O_8$ cathode for RMBs is prepared by high-temperature calcination with the aqueous Mg^{2+} electrolyte. As a result, the MVO not only exhibits excellent electrochemical rate capability (the reversible discharge specific capacity is 143 mAh g^{-1} at the current density of 0.05 A g^{-1}, and it still remains 61 mAh g^{-1} at a large current density of 4 A g^{-1}), but also expresses good cycling performance (81% capacity retention after 10,000 cycles at the current density of 3 A g^{-1}). This work proves that the MVO is an exceptional candidate for MIBs.

Keywords

Magnesium orthovanadate • Aqueous Mg-ion battery • Cathode materials • Energy storage

Introduction

With the rapid consumption of nonrenewable fossil fuels over the past decades, growing concerns on environmental issues and energy crisis [1, 2]. In order to achieve sustainable development, clean energy utilization, such as solar energy and wind energy and tidal energy, has become an urgent task of human society [3]. The exploitation of electric vehicles and smart grids is the general trend of world development. High-performance secondary batteries as the energy storage system play a key role in electric vehicles and smart grids [4]. At present, lithium-ion batteries (LIBs) as one of the most popular energy storage devices have been widely applied to portable electronics owing to its high energy density, long cycling stability, and mature market [5, 6]. However, the safety problems, limited lithium resource, and increased prices still hinder the further development of LIBs [7]. Therefore, there is an urgent need for alternative energy with abundant resources and low cost to tackle the above problems. Aqueous magnesium-ion batteries (AMIBs), an ideal substitute for LIBs, have a great potential for the large-scale energy storage applications because of its earth-abundant storage, high security, low price, low reduction potential of Mg metal (−2.4 V vs. SHE), and high theoretical volumetric capacity (3833 mAh cm^{-3} for Mg vs. 2046 mAh cm^{-3} for Li) [8]. Also, compared with organic electrolyte, aqueous electrolyte can greatly improve the ionic conductivity of salt and has the advantages of low cost, non-toxic, non-flammable, and low requirements for production environment [9–11]. Nevertheless, the divalent magnesium ions carry twice as much charge as the monovalent lithium ions while the ionic radius of magnesium (0.72 Å) is similar to lithium (0.76 Å), which causes strong polarization effect and sluggish diffusion kinetics of Mg^{2+} in host materials, resulting in unsatisfactory energy and power densities [12, 13]. Thus, one of the most crucial tasks in RMBs is searching for ideal cathode materials.

Vanadate is an important vanadium-based material. Compared with vanadium oxide, vanadate has more stable crystal structure, better cycle stability, and longer cycle life. This is because the metal cation has a high energy barrier, which is not easy to leave the lattice node position during charge and discharging process. It plays a role in supporting

X.-F. Ma · H.-Y. Li (✉) · D. Gao · W. Ren · J. Diao · B. Xie · G. Huang · J. Wang · F. Pan
College of Materials Science and Engineering, Chongqing University, Chongqing, 400044, China
e-mail: hongyi.li@cqu.edu.cn

H.-Y. Li · J. Diao · G. Huang · J. Wang · F. Pan
National Engineering Research Center for Magnesium Alloys, Chongqing University, Chongqing, 400044, China

S. Barela et al. (eds.), *Magnesium Technology 2023*, The Minerals, Metals & Materials Series,
https://doi.org/10.1007/978-3-031-22645-8_39

and stabilizing the crystal structure and is conducive to maintaining the cyclic stability of the electrode material [14, 15]. Simultaneously, vanadium has multiple chemical valence states (V^{3+}, V^{4+} and V^{5+}) and can be coordinated with oxygen to form various open V-O ligands. A variety of metal cations, such as alkali metal ions (Li^+, Na^+, K^+ and Rb^+), alkali earth metal ions (Ca^{2+}, Mg^{2+}, and Sr^{2+}), transition metal ions (Ag^+, Cu^{2+}, Zn^{2+}, Co^{2+}, etc.), and others (Al^{3+}, In^{3+}, Bi^{3+}, NH^{4+}, etc.) can be inserted into V-O polyhedron to directly composited with it, resulting vanadate display various electrochemical properties because of different components introduction [16–19]. Thus, vanadate has a broad application prospect in the field of energy storage.

In the present work, we report magnesium orthovanadate ($Mg_3V_2O_8$, MVO) prepared by a high temperature calcination method. MVO has a stable three-dimensional framework and large ion channels and is expected to become a high-performance energy storage electrode material. As expected, the corresponding results show that the MVO is an ideal cathode material. It displays an initial discharge capacity of 143 mAh g^{-1} and the average discharge voltage of ~2.34 V (vs. Mg^{2+}/Mg) using 4.5 M $Mg(NO_3)_2$ in deionized water as the electrolyte. Even at a high current density of 4 A g^{-1}, a reversible capacity of 61 mAh g^{-1} is obtained. In addition, the MVO cathode material exhibits an excellent cycling stability (10,000 cycles, 81% capacity retention) at high current density (3 A g^{-1}).

Experiment

Synthesis of Samples

All reagents were analytically pure and without further purification. $Mg_3V_2O_8$ was prepared by a high-temperature calcination method, which was reported by Li et al. before [20]. For a typical synthesis experiment, the raw materials V_2O_5 and MgO were first sieved to a uniform particle size of 45–75 μm. Then, MgO and V_2O_5 were mixed in a ceramic crucible (O.D. 9 cm, H 3.5 cm) in a molar ratio of 3 and calcined in a muffle furnace at 1223 K for 12 h. Finally, the white MVO powder was obtained and was used for the microstructural and the electrochemical characterizations.

Characterization

The crystallographic information of the prepared sample was measured by X-ray diffractometer (XRD, X'Pert PRO MPD) with Cu Kα radiation (λ = 0.1514178 nm) in 2θ range of 10–90° at 0.02° s^{-1} scan rate. The X-ray photoelectron spectroscopy (XPS) analysis was performed using a Thermo ESCALAB 250XI spectrometer with a monochromatized Al Kα X-ray. The morphology and microstructure of product was measured by field transmission electron microscopy (FESEM, TESCAN MIRA4) and transmission electron microscopy (TEM, Talos F200S).

Electrochemical Measurements

The electrochemical properties were measured at room temperature using CR2432 coin cells assembled in atmospheric air, which was fabricated by MVO electrode as cathode, activated carbon (AC) electrode as anode, saturated magnesium nitrate (4.5 M $Mg(NO_3)_2$) as aqueous electrolyte, and a Whatman Glass microfiber filter (Grade GF/A) was used as the separator. For the preparation of the electrodes, a slurry composed of 80 wt% MVO, 10 wt% acetylene black and 10 wt% polyvinylidene fluoride (PVDF) dispersed in 1-methyl-2-pyrrolidone (NMP). The obtained electrodes were prepared by coating the slurry onto carbon fiber papers (diameter 12 mm) with a mass loading of active material about 1 mg cm^{-2}, then drying in vacuum at 80 °C for 12 h. The cyclic voltammetry (CV) and electrochemical impedance spectroscopy (EIS) are collected using a electrochemical workstation (CHI 660E, Chenhua). The impedance data were recorded within the frequency range of 0.01–10,000 Hz. The galvanostatic charge–discharge tests were measured in the potential range of −1.1–1.4 V versus AC (1.3–3.8 V vs. Mg^{2+}/Mg) at different current densities using a multichannel battery testing machine (CT4008A, Neware). The assembled coin cells were aged for several hours before the charging/discharging process to ensure that the electrodes can be completely wetted.

Results and Discussion

The X-ray diffraction (XRD) patterns is displayed in Fig. 1a to investigate the crystallographic structure of the as-prepared sample. All the diffraction peaks are consistent with $Mg_3V_2O_8$ (JCPDS No 73-0207) with lattice parameters of a = 6.053 Å, b = 11.442 Å, c = 8.33 Å, $\alpha = \beta = \gamma$ = 90.0°, indicating the successful synthesis of MVO powder without any impurities. Meanwhile, the sharp diffraction peaks at 2θ equal to 19.70, 27.11, 29.67, 31.24, 35.21, 35.92, 43.62 and 63.14, which correspond to the (111), (112), (131), (040), (132), (023), (240), and (244) planes, indicating good crystallinity of the obtained MVO. Figure 1b demonstrates crystal structure of the orthorhombic MVO, where [VO_4] tetrahedra and edge-shared [MgO_6] octahedra are linked to oxygen atoms at the vertices. It forms a stable three-dimensional framework and a large ion channel, which is very conducive to the insertion/desertion of metal ions. The X-ray photoelectron spectrum (XPS) was

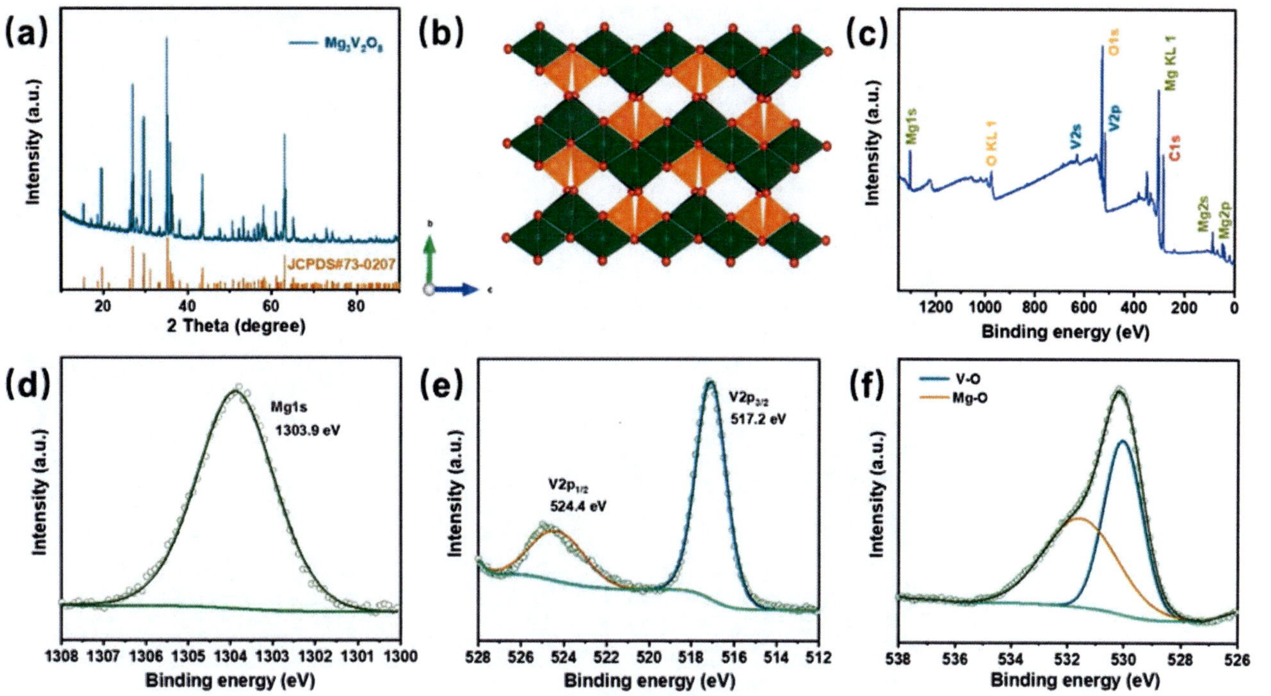

Fig. 1 XRD pattern of the MVO (**a**), crystal structure of MVO (**b**), XPS survey spectrum of MVO (**c**), XPS spectra of Mg 1s (**d**), V 2p (**e**), and O 1s (**f**) in MVO

used to further investigate the composition of the MVO. As is we can see in the survey spectrum (Fig. 1c), Mg, V, and O elements were detected. The Mg 1s, V 2p, and O 1s spectra was showed in Fig. 1d–f. The Mg 1s peak located at 1303.9 eV. In V 2p spectra, it shows two asymmetric peaks, which are related to V $2p_{1/2}$ (517.2 eV) and V $2p_{3/2}$ (524.4 eV) orbitals, belonging to V^{5+}. According to the O 1s core level, two distinct components at 530.1 eV and 533.2 eV correspond to V-O layers and Mg-O, respectively [21]. The morphology of the synthesized MVO is analyzed through SEM characterization, as shown in Fig. 2a–b, it is obvious that the products are irregular micron particles with size of 2–3 μm. According to energy dispersive X-ray spectroscopy (EDS) elemental mappings in Fig. 2c–f, the Mg, V and O atoms are homogeneously distributed in the MVO sample.

The magnesium storage electrochemical properties of MVO as cathode material were examined in the coin-type cells with 4.5 M $Mg(NO_3)_2$ as aqueous electrolyte, using the AC electrode as anode. Firstly, the typical cyclic voltammetry (CV) curves were measured at the scan rate of 0.1 mV s^{-1}, 0.2 mV s^{-1}, 0.3 mV s^{-1} and 0.4 mV s^{-1}, respectively, within the potential range from 1.1 to 3.8 V (vs. Mg^{2+}/Mg) (Fig. 3a). Three anodic peaks are observed around 1.8, 2.5 and 3.6 V, and three cathodic peaks are situated at 3.0, 2.3 and 1.5 V, which ascribed to the extraction and insertion potential of Mg^{2+}, respectively. The galvanostatic charge/discharge testing was used to further evaluated the electrochemical performance. As is shown in Fig. 3b, it exhibits the rate performance of the MVO at different current densities ranging from 0.05 to 4 A g^{-1}, delivering the reversible and stable capacities of 143, 108, 87, 82, 78, 76, 70, 65, and 61 mAh g^{-1}, respectively. When the current density returns to 0.05 A g^{-1}, the specific capacity retains 136 mAh g^{-1}, indicating outstanding structural stability of MVO electrode to withstand large-current tests, robust reaction kinetics and excellent rate capability. Figure 3c presents the corresponding galvanostatic charge–discharge curves of MVO at different current densities. As the current density increase, all these curves display similar voltage plateau behavior during charging and discharging process, a high average discharge voltage of ~2.34 V (vs. Mg^{2+}/Mg), relating to redox reactions responsible for insertion/extraction of Mg^{2+}. Meanwhile, the discharge specific capacity of cell decreased from 144 to 63 mAh g^{-1}, owing to the charge and discharge time is short at high current density, and the ion or electron transfer, oxidation–reduction reaction, etc., do not get enough time. In addition, high current density will lead to supersaturation or over consumption of protons in the electrolyte, and the internal resistance and ion resistance coefficient will increase. All these reasons will lead to the reduction of specific capacity at high current density. To further demonstrate the cycling stability of MVO for Mg^{2+} storage, long-term cycling performance was conduced at 3 A g^{-1}, as shown in Fig. 3d. After 10,000 cycles, an impressive 81%

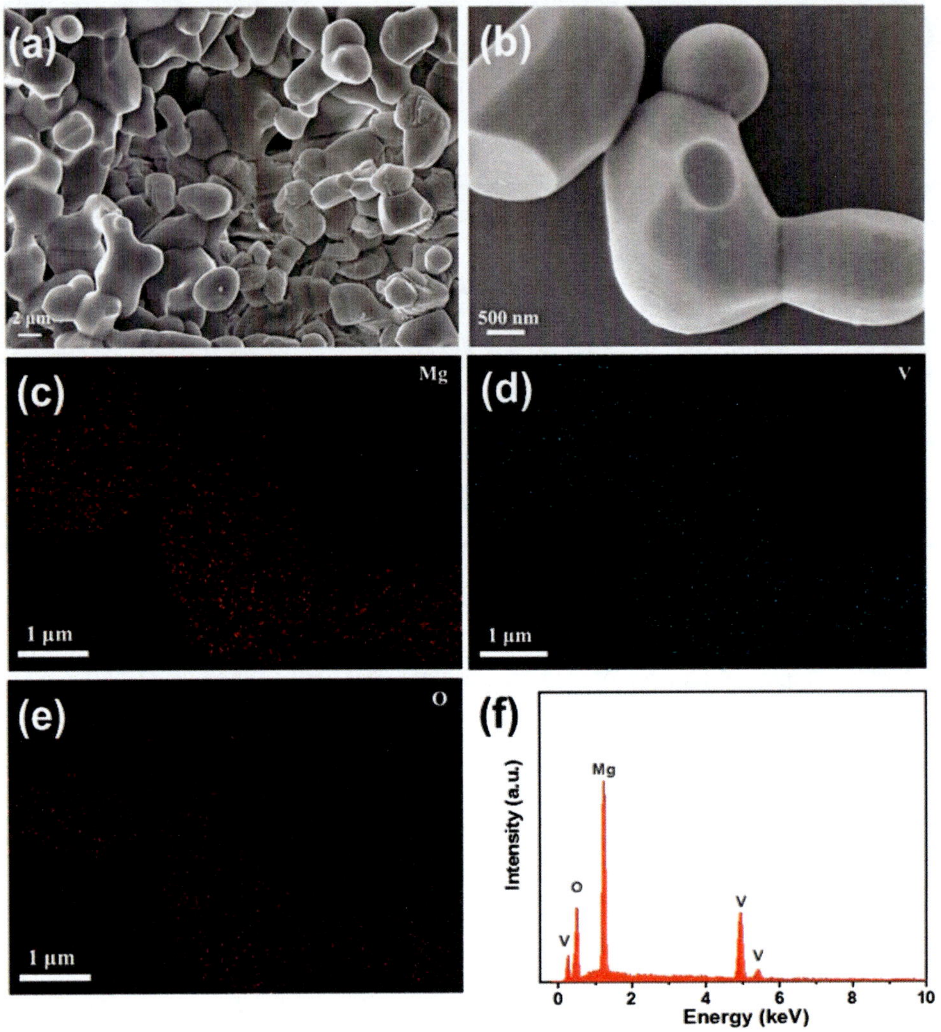

Fig. 2 SEM images of MVO (**a**–**b**), elemental mapping images of Mg (**c**), V (**d**), and O (**e**), energy dispersive X-ray (EDX) spectrum of the MVO (**f**)

retention of the initial capacity (68 mAh g^{-1}) indicates excellent structural stability, which may be attributed to the large three-dimensional ion channels in MVO. Concurrently, the high average Coulombic efficiency about 100% represents high reversibility of electrochemical Mg^{2+} storage and quantitative utilization of electrical charge. To study the ionic transportation kinetics, electrochemical impedance spectroscopy (EIS) experiments were employed. The intercept was corresponding to ohmic resistance (R_s) and the compressed semicircle in the high to medium frequency region was corresponding to charge transfer resistance (R_{ct}). An inclined line in the low-frequency range stands for the Warburg impedance (Z_w), which was related to the Mg^{2+} diffusion resistance in the active material. Among them, R_{ct} is the key factor that determines the charge and discharge rate performance. Zview software is used to establish the equivalent circuit model. As shown in Fig. 4a, the MVO electrode showed an ohmic resistance R_s of 4.38 Ω and charge transfer resistance R_{ct} of 3.72 Ω. The Mg^{2+} diffusion coefficient (D_{Mg}) could be calculated by Eqs. (1) and (2) to analyze the diffusion performance of Mg^{2+} in the electrode materials.

$$D_{Mg} = 0.5\left(RT/n^2F^2AC_{Mg}\sigma\right)^2 \quad (1)$$

$$Z' = R_s + R_{ct} + \sigma\omega^{-1/2} \quad (2)$$

where R is the gas constant, T is the absolute temperature, n is the number of the electrons per molecule attending the electronic transfer reaction, F is the Faraday constant, A is the surface area of the electrode, C_{Mg} is the concentration of magnesium ions, σ is the slope of the Z' versus $\omega^{1/2}$, which can be obtained from the line of $Z' \sim \omega^{1/2}$ (shown in Fig. 4b). The diffusion coefficients of Mg^{2+} was calculated by EIS method is about 9.15×10^{-13} cm s^{-2}. The small charge transfer impedance and the fast magnesium ion diffusion devoted to the preferable electrochemical rate performance.

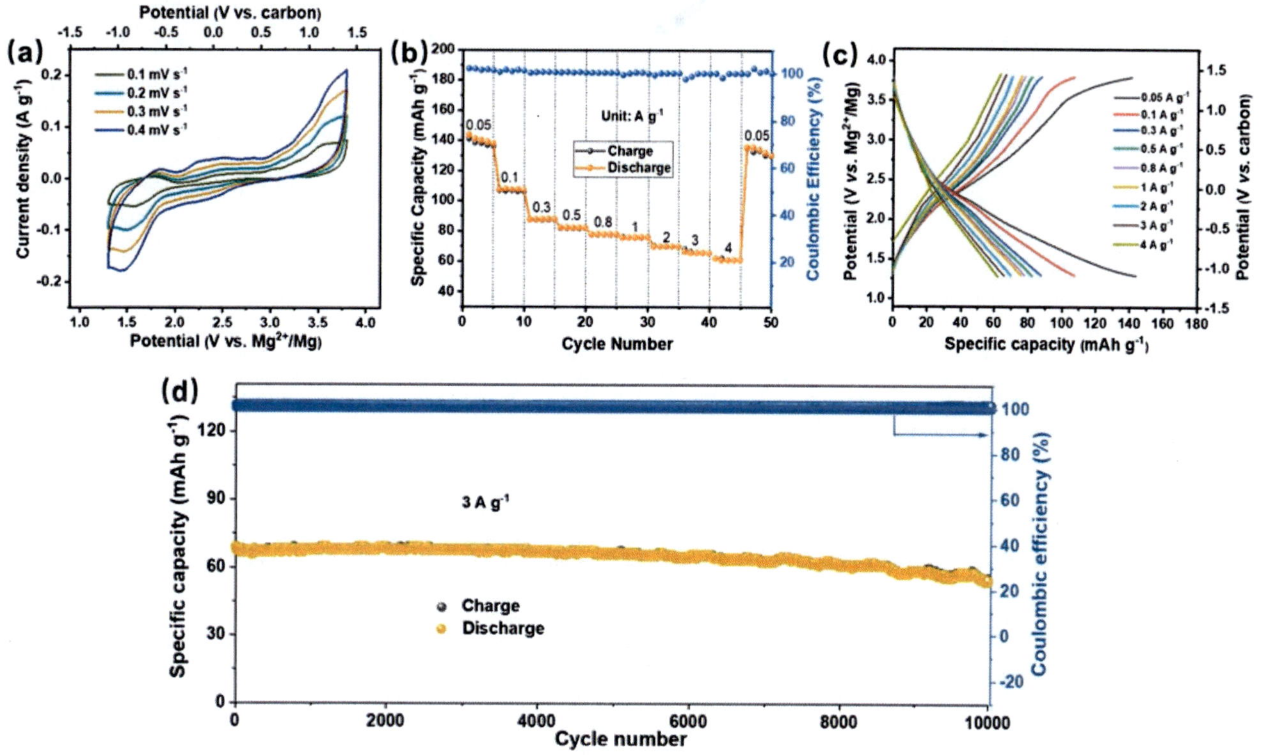

Fig. 3 CV curves of MVO at different sweep rates (**a**), rate performance (**b**), charge–discharge curves of MVO at different rates (**c**), cyclic performance of MVO at a current density of 3 A g^{-1} (**d**)

Fig. 4 Impedance spectra of the MVO electrode (**a**) and slope of the plots of Z' against $\omega^{-1/2}$ (**b**)

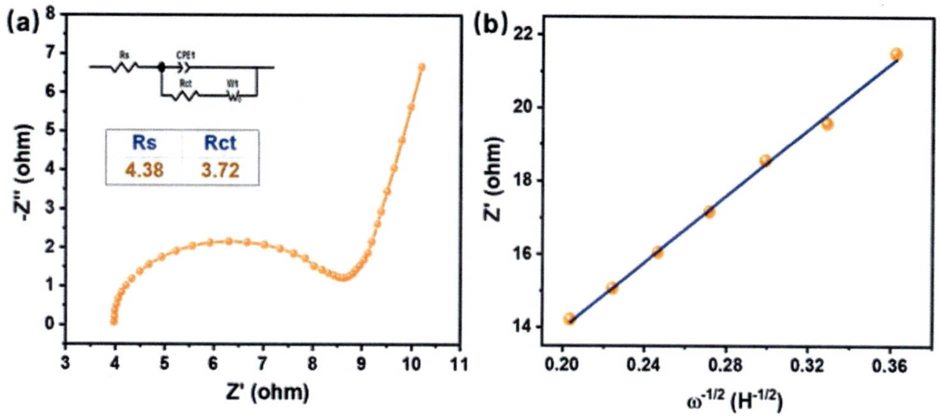

Conclusions

IN summary, the excellent electrochemical performance of MVO for magnesium-based battery are reported. The MVO as a cathode material exhibits initial specific capacity of 143 mAh g^{-1} at 0.05 A g^{-1} with a high average discharge voltage about 2.34 V (vs. Mg^{2+}/Mg) and shows remarkable long-term cyclic stability (81% capacity retention after 10,000 cycles at the current density of 3 A g^{-1}). It is worth noted that the MVO electrode delivers an impressive rate capacity, and the discharge capacities of MVO are 143, 108, 87, 82, 78, 76, 70, 65, and 61 mAh g^{-1} at 0.05, 0.1, 0.3, 0.5, 0.8, 1, 2, 3, and 4 A g^{-1}, respectively. When the current density returns to 0.05 A g^{-1}, the specific capacity is around 136 mAh g^{-1}, indicating outstanding structural stability of MVO electrode. This study demonstrates that MVO is a promising cathode material for magnesium ions storage, the excellent electrochemical performance can be attributed to its structure of stable three-dimensional framework and large ion channels.

Acknowledgements The authors would like to acknowledge the financial support by the National Natural Science Foundation of China (grant no. 52074050, 52222407), Chongqing Science and Technology Bureau (cstc2019jcyjjqX0006 and cstc2021ycjh-bgzxm0075).

References

1. H. Tang, C. Zuo, F. Xiong, C. Pei, S. Tan, P. Luo, W. Yang, Q. An, L. Mai, Flexible three-dimensional-networked iron vanadate nanosheet arrays/carbon cloths as high-performance cathodes for magnesium ion batteries, Sci. China Mater, 65 (2022): 2197–2206.
2. M. Karlsmo, R. Bouchal, P. Johansson, High-Performant All-Organic Aqueous Sodium-Ion Batteries Enabled by PTCDA Electrodes and a Hybrid Na/Mg Electrolyte, Angew Chem Int Ed Engl, 60 (2021): 24709–24715.
3. P. He, G. Zhang, X. Liao, M. Yan, X. Xu, Q. An, J. Liu, L. Mai, Sodium Ion Stabilized Vanadium Oxide Nanowire Cathode for High-Performance Zinc-Ion Batteries, Adv. Energy Mater, 8 (2018): e1702463.
4. H. Huang, D. Kundu, R. Yan, E. Tervoort, X. Chen, L. Pan, M. Oschatz, M. Antonietti, M. Niederberger, Fast Na-Ion Intercalation in Zinc Vanadate for High-Performance Na-Ion Hybrid Capacitor, Adv. Energy Mater, 8 (2018): e1802800.
5. Y. Liu, G. He, H. Jiang, I.P. Parkin, P.R. Shearing, D.J.L. Brett, Cathode Design for Aqueous Rechargeable Multivalent Ion Batteries: Challenges and Opportunities, Adv. Funct. Mater, 31 (2021): e2010445.
6. P. Canepa, G. Sai Gautam, D.C. Hannah, R. Malik, M. Liu, K.G. Gallagher, K.A. Persson, G. Ceder, Odyssey of Multivalent Cathode Materials: Open Questions and Future Challenges, Chem Rev, 117 (2017): 4287–4341.
7. H. Chen, J. Huang, S. Tian, L. Liu, T. Qin, L. Song, Y. Liu, Y. Zhang, X. Wu, S. Lei, S. Peng, Interlayer Modification of Pseudocapacitive Vanadium Oxide and Zn(H2 O)n (2+) Migration Regulation for Ultrahigh Rate and Durable Aqueous Zinc-Ion Batteries, Adv Sci, 8 (2021): e2004924.
8. L. Zhou, Q. Liu, Z. Zhang, K. Zhang, F. Xiong, S. Tan, Q. An, Y. M. Kang, Z. Zhou, L. Mai, Interlayer-Spacing-Regulated VOPO4 Nanosheets with Fast Kinetics for High-Capacity and Durable Rechargeable Magnesium Batteries, Adv. Mater, 30 (2018): e1801984.
9. D. Yang, Y. Zhou, H. Geng, C. Liu, B. Lu, X. Rui, Q. Yan, Pathways towards high energy aqueous rechargeable batteries, Coordin Chem Rev, 424 (2020): e213521.
10. F. Wang, X. Fan, T. Gao, W. Sun, Z. Ma, C. Yang, F. Han, K. Xu, C. Wang, High-Voltage Aqueous Magnesium Ion Batteries, ACS Cent Sci, 3 (2017): 1121–1128.
11. Y. Tang, X. Li, H. Lv, W. Wang, Q. Yang, C. Zhi, H. Li, High-Energy Aqueous Magnesium Hybrid Full Batteries Enabled by Carrier-Hosting Potential Compensation, Angew Chem Int Ed Engl, 60 (2021): 5443–5452.
12. Y. Liang, H.D. Yoo, Y. Li, J. Shuai, H.A. Calderon, F.C. Robles Hernandez, L.C. Grabow, Y. Yao, Interlayer-expanded molybdenum disulfide nanocomposites for electrochemical magnesium storage, Nano Lett, 15 (2015): 2194–2202.
13. X. Wang, X. Zhang, G. Zhao, H. Hong, Z. Tang, X. Xu, H. Li, C. Zhi, C. Han, Ether-Water Hybrid Electrolyte Contributing to Excellent Mg Ion Storage in Layered Sodium Vanadate, ACS Nano, 16 (2022): 6093−6102.
14. Q. An, Y. Li, H. Deog Yoo, S. Chen, Q. Ru, L. Mai, Y. Yao, Graphene decorated vanadium oxide nanowire aerogel for long-cycle-life magnesium battery cathodes, Nano Energy, 18 (2015): 265–272.
15. X. Wang, S. Zheng, S. Wang, Y. Zhang, H. Du, Self-anchoring dendritic ternary vanadate compound on graphene nanoflake as high-performance conversion-type anode for lithium ion batteries, Nano Energy, 22 (2016): 179–188.
16. H. Tang, Z. Peng, L. Wu, F. Xiong, C. Pei, Q. An, L. Mai, Vanadium-Based Cathode Materials for Rechargeable Multivalent Batteries: Challenges and Opportunities, Electrochem. Energy Rev, 1 (2018): 169–199.
17. M. Chen, Q. Liu, Z. Hu, Y. Zhang, G. Xing, Y. Tang, S.L. Chou, Designing Advanced Vanadium-Based Materials to Achieve Electrochemically Active Multielectron Reactions in Sodium/Potassium-Ion Batteries, Adv. Energy Mater, 10 (2020): e2002244.
18. S. Liu, L. Kang, J.M. Kim, Y.T. Chun, J. Zhang, S.C. Jun, Recent Advances in Vanadium-Based Aqueous Rechargeable Zinc-Ion Batteries, Adv. Energy Mater, 10 (2020): e2000477.
19. X. Xu, F. Xiong, J. Meng, X. Wang, C. Niu, Q. An, L. Mai, Vanadium-Based Nanomaterials: A Promising Family for Emerging Metal-Ion Batteries, Adv. Funct. Mater, 30 (2020): e1904398.
20. J. Cheng, C.-J. Wang, S. Shen, J. Diao, B. Xie, H.-Y. Li, Synthesis and Solubility Behavior of Magnesium Ortho-, Meta-, and Pyrovanadates, Jom, 74 (2021): 23–29.
21. Y. Zhu, G. Huang, J. Yin, Y. Lei, A.H. Emwas, X. Yu, O.F. Mohammed, H.N. Alshareef, Hydrated MgxV5O12 Cathode with Improved Mg2+Storage Performance, Adv. Energy Mater, 10 (2020): e2002128.

Author Index

S. Barela et al. (eds.), *Magnesium Technology 2023*, The Minerals, Metals & Materials Series,
https://doi.org/10.1007/978-3-031-22645-8

Subject Index

S. Barela et al. (eds.), *Magnesium Technology 2023*, The Minerals, Metals & Materials Series,
https://doi.org/10.1007/978-3-031-22645-8